AF789504

Topics in Applied Physics
Volume 91

Topics in Applied Physics is part of the SpringerLink service. For all customers with standing orders for Topics in Applied Physics we offer the full text in electronic form via SpringerLink free of charge. Please contact your librarian who can receive a password for free access to the full articles by registration at:

springerlink.com → Orders

If you do not have a standing order you can nevertheless browse through the table of contents of the volumes and the abstracts of each article at:

springerlink.com → Browse Publications

Springer-Verlag Berlin Heidelberg GmbH

Topics in Applied Physics

Topics in Applied Physics is a well-established series of review books, each of which presents a comprehensive survey of a selected topic within the broad area of applied physics. Edited and written by leading research scientists in the field concerned, each volume contains review contributions covering the various aspects of the topic. Together these provide an overview of the state of the art in the respective field, extending from an introduction to the subject right up to the frontiers of contemporary research.

Topics in Applied Physics is addressed to all scientists at universities and in industry who wish to obtain an overview and to keep abreast of advances in applied physics. The series also provides easy but comprehensive access to the fields for newcomers starting research.

Contributions are specially commissioned. The Managing Editors are open to any suggestions for topics coming from the community of applied physicists no matter what the field and encourage prospective editors to approach them with ideas.

Managing Editors

Dr. Claus E. Ascheron

Springer-Verlag Heidelberg
Topics in Applied Physics
Tiergartenstr. 17
69121 Heidelberg
Germany
Email: ascheron@springer.de

Dr. Hans J. Koelsch

Springer-Verlag New York, Inc.
Topics in Applied Physics
175 Fifth Avenue
New York, NY 10010-7858
USA
Email: hkoelsch@springer-ny.com

Assistant Editor

Dr. Werner Skolaut

Springer-Verlag Heidelberg
Topics in Applied Physics
Tiergartenstr. 17
69121 Heidelberg
Germany
Email: skolaut@springer.de

Takeshi Kobayashi Hisao Hayakawa
Masayoshi Tonouchi (Eds.)

Vortex Electronis and SQUIDs

With 259 Figures

Springer

Professor Takeshi Kobayashi
Graduate School of Engineering Science
Osaka University
1-3 Machikaneyama, Toyonaka
Osaka 560-8531, Japan
kobayasi@sup.ee.es.osaka-u.ac.jp

Professor Masayoshi Tonouchi
Research Center
for Superconductor Photonics
Osaka University
2-1 Yamadaoka, Suita
Osaka 565-0871, Japan
tonouchi@rcsuper.osaka-u.ac.jp

Professor Hisao Hayakawa
Graduate School of Engineering
Nagoya University
Furo-cho, Chikusa-ku
Nagoya-City 464-8601, Japan
hhayakaw@nuee.nagoya-u.ac.jp

Library of Congress Cataloging-in-Publication Data

Vortex electronics and SQUIDs / [edited by] T. Kobayashi, H. Hayakawa, M. Tonouchi.--
1st ed.
p. cm. -- (Topics in applied physics, ISSN 0303-4216 ; v. 91)
Includes bibliographical references and index.
ISBN 3-540-40231-4 (alk. paper)
1. High temperature superconductors. 2. Vortex-motion. 3. Quantum theory. I.
Kobayashi, T. (Takeshi), 1944- II. Hayakawa, H. (Hisao), 1940- III. Tonouchi, M.
(Masayoshi), 1959- IV. Series.

QC611.98.H54V66 2003
537.6'23--dc21 2003050682

Physics and Astronomy Classification Scheme (PACS): 85.25.Dq; 74.25.Qt

ISSN print edition: 0303-4216
ISSN electronic edition: 1437-0859
ISBN 978-3-662-30775-5 ISBN 978-3-540-44876-1 (eBook)
DOI 10.1007/978-3-540-44876-1

springeronline.com

Originally published by Springer-Verlag Berlin Heidelberg New York in 2003.

Softcover reprint of the hardcover 1st edition 2003

Typesetting: DA-TEX · Gerd Blumenstein · www.da-tex.de
Production: LE-TEX Jelonek, Schmidt & Vöckler GbR, Leipzig
Cover design: *design & production* GmbH, Heidelberg

Printed on acid-free paper 57/3141/YL 5 4 3 2 1 0

Preface

With increase in demand for ultrahigh sensitivity of signal sensing, ultrahigh speed of data processing, ultralow power dissipation of computer components and so on, quantum effect electronics has become more and more promising. In this context vortex electronics has begun to offer an important role in twenty-first century electronics. Vortex itself is a tiny magnetic flux quantum and exists in the coherent electron wave system like superconductors. The basics and electronics applications of vortex in high-T_c superconductors are the main issues of this book. Most authors of the book were members of a research project in the priority area named "Vortex Electronics" supported by the Ministry of Education, Science, Sports and Culture of Japan from 1998 to 2000. The major contents of this book are outcomes of the project. For the publication of the book we are grateful to the Japan Society for the Promotion of Science (JSPS) for financial support.

Osaka, October 2003

T. Kobayashi
H. Hayakawa
M. Tonouchi

Contents

Introduction

T. Kobayashi

Graduate School of Engineering Science, Osaka University
1-3 Machikaneyama, Toyonaka, Osaka 560-8531, Japan
kobayashi@sup.ee.es.osaka-u.ac.jp

In the history of modern physics and engineering, the electron quantum mechanical effect and its utilization in electronics have attracted much attention. With increase in demand for ultrahigh sensitivity of signal sensing, ultrahigh speed of data processing, ultralow power dissipation of computer components and so on, quantum effect electronics has become more and more promising. Since the middle of the twentieth century most of vacuum electronics have been replaced by the solid state electronics, where the electron dynamics are well described by the band-diagram description. Solid state electronics itself and band-diagram description were originally in the category of quantum mechanics. However, the treatment of carriers based on the band-diagram inside semiconductors used the classical model where the electron wave concept has been missing. One should perceive it to be also true even for semiconductor superlattices. In the late twentieth century, along with the extremely miniaturized scale of semiconductor devices, mesoscopic physics and mesoscopic electronics have boomed. In this research area, one attempts to manipulate directly the coherent electron wave instead of the electron particle. In ordinary semiconductors, however, only the nanoscale dimension allowed us to handle electron wave interference, interaction and so on to some extent. Besides, success was obtained only at cryogenic temperatures. Collisions between electrons or between electron and lattice (or impurity) was recognized again as a forcible enemy of electron wave engineering. It was a time when both researchers and engineers in the field were urged forward to search for better physical and engineering stages where no electron collisions take place thereby maintaining electron wave coherency over very long distances in the appropriate temperature range.

In 1950, London indicated theoretically the possible existence of vortices in the ideal coherent electron wave system free from collisions (superfluid state). By taking type-II superconductor properties into account, Abrikosov derived more realistic features of vortices and their dynamics in 1957. Along with their predictions, some clever experiments suggested the existence of vortices in the electron wave system as a magnetic flux quantum. The superconductor was the only stage on which vortices were entirely developed and behaved in accordance with the theoretical prediction (in a pure physical field,

T. Kobayashi et al. (Eds.): Vortex Electronics and SQUIDs,
Topics Appl. Phys. **91**, 1–4 (2003)

vortex is also found in superfluid liquid hellium). Afterward, in the boom of mesoscopic electronics, a quasi-vortex appeared in an ultrasmall metal ring cooled to cryogenic temperatures. After the discovery of high temperature superconductors in 1986, the superconductor itself has become very familiar to researchers of electronics and, simultaneously, the vortex has been widely known as a magnetic flux quantum which comprises the minimum magnetic flux unit.

Indeed, the flux amount of a vortex is as low as 1.5×10^{-15} Wb. The radius of a vortex is dependent on a material parameter "the penetration depth" that differs from material to material. By roughly estimation, it is around 0.1 µm in ordinary metal superconductors and 0.5–1 µm in high temperature oxide superconductors. In any case, it can be definitely said that the vortex has a tiny body whose radius is limited within the submicron scale. Under the rules of electromagnetic theory, there is a strong correlation between the vortex and electron wave phase via coupling of the current-vector potential. Besides, the phase sensing can be now done very sensitively through interference technique. One can thus detect the magnetic field using vortex technology with ultrahigh resolution. In a similar manner, when one uses the vortex as a bit signal, an ultrasmall flux value such as 1.5×10^{-15} Wb is really sufficient to judge "yes or no" of the circuit state. Based on this basic knowledge, the new device concepts of Superconducting Quantum Interference Device (SQUID) and Single Flux Quantum digital circuit (SFQ) were born. The former will be a powerful tool for diagnosis of the human body and the environment and is now expanding its commercial market. The latter is in the laboratory research and development stage. In any case, the vortex will serve as the key technology of twenty-first century microelectronics.

As is well known, vortex stability is a measure of the critical current in the supercurrent. Originally, the superconductor wire was expected to carry unlimited high current without an accompanying voltage drop. This is surely true when low current flows inside the wire. However, when there is a much higher current, a voltage suddenly appears along the wire. According to close examination, the superconductor matrix still keeps its superconductivity even when voltage appears. The mechanism of voltage appearance is now fully understood by the vortex motion model. Namely, in the still state where the current is below the critical value, the vortex is entirely pinned. On the other hand, on increase in the current beyond the critical value, the vortex begins to move by Lorentz force inside the superconductor wire which, in turn, induces the voltage along the wire. In other words, the critical current means the current value at which the Lorentz force balances the pinning force working on the vortex. Tight pinning of vortex is also the important issue for establishment of low-noise vortex devices (SQUID) as well as for ultrahigh-field magnets. So far, much effort has been expended on how the pinning force can be made high. One can see the sophisticated cross-section of the commercially used NbTi wire which consists of multiple NbTi thin filaments embedded in a Cu pinning body. Switching the object from ordinary superconductors to high-T_c oxide superconductors, basic knowledge concerning the pinning design is still less for the oxide system. The complexity of the oxide crystal structure including the two-dimensional conducting nanolayer is a reason for this. A more serious and complicated problem arises from the anisotropy of the wave function called the *d*-wave peculiar to the oxide system. Compared to the isotropic *s*-wave system in the metal superconductors, many things remained unsolved in the *d*-wave system. Features of vortices in the high-T_c oxide superconductors were one of them. Particularly, the vortex behavior in the vicinity of the *d*-wave superconductor junction was in the question.

As to the high-T_c vortex electronics application, the SQUID sensor chip and SQUID system are commercially available. The superiority of high-T_c SQUIDs is expressed by "easy handling". Although performance is sacrificed to some extent, 77 K operation of the SQUID is fully outstanding in the magnetosensor. It can really cover sensing of human magnetocardiography without requiring special techniques other than SQUIDs. To date, one can note an expanding application of high-T_c SQUIDs from the biomedical to semiconductor production fields. Moreover, there occurred a new stream of study, realizing the mobile SQUID, toward more practical use of SQUID equipment. Since the mobile SQUID needs no magnetic shield against environmental magnetic noise, restrictions imposed on SQUID utility could be largely relaxed with the mobile version. Owing to this rich harvest, a burst of applications of high-T_c mobile SQUIDs including industrial use is expected.

This publication was edited with the aim to convey to all readers the latest scientific and technological knowledge on vortices in *d*-wave high-T_c oxide superconductors, vortex observation, vortex manipulation, SQUID fundamentals and applications, and processing technologies mainly for oxide thin film growth. In the following, the contents of chapters from 2 to 7 are summarized.

In the chapter by *Kuriki* et al., the fundamentals of vortices in high-T_c superconductors are described. Due to 2-dimensionality and anisotropy of the electronic state, peculiar to high-T_c superconductors, vortex features are distinct in various aspects from conventional superconductors, though common in their basics. Based on a huge volume of data, an appropriate equilibrium phase diagram is given that differs greatly differed from the conventional one. In addition, vortex dynamics are demonstrated for bulk material, thin films and grain boundaries.

The following chapter by *Ohshima* et al. observation of vortex in high-T_c superconductors is dealt with. As mentioned above, the vortex is a tiny quantized magnetic flux. How can we visualize it? In the text, four advanced methods are introduced: the high resolution Bitter method, the scanning SQUID microscope method, the magneto-optical imaging method and the terahertz-radiation-imaging method. Using these methods, even the dynamical behavior of vortices in the vicinity of the grain boundary and junction is demonstrated.

In the chapter by *Tonouchi* et al., several topics related to new aspects of vortices observed in high-T_c superconductors are reviewed. A recent new finding is the space charge inside the vortex core. In ordinary superconductors, the vortices have long been believed to be electrically neutral, and it is really true. However, in high-T_c superconductors, the situation seems to substantially differ from the ordinary one. Namely, the vortex in high-T_c materials is a coaxial ring-shaped electric dipole. In this chapter, details about the charged vortex are given. In addition to this, vortex dynamics in the intrinsic Josephson junction and vortex manipulation by ultrafast laser shot are described.

Enpuku et al. devote their chapter to introducing fundamentals of high-T_c SQUID as an electronic application of vortex. In principle, a SQUID's resolution limit reaches the flux quantum or less. Actually, however, residual vortex noise spoils the net performance of SQUIDs even when the environmental noise is completely suppressed. The chapter denotes many pages to showing why vortex noise is generated and how to minimize it for bicrystal Josephson junctions in SQUIDs and superconductor film platelets. In addition, the development of a high-T_c SQUID microscope is described.

Following the above, in the chapter real applications of high-T_c SQUIDs are demonstrated by *Itozaki* et al. They are (1) SQUID on the commercial market, (2) challenge to shield-free Open-SQUID, (3) application to biological immunoassay, (4) application to measurement of precursory phenomena of earthquake and volcano eruption, (5) laser-SQUID microscope for LSI diagnosis, and (6) small scale digital applications. All these topics are ahead of their time.

In the last chapter by *Kobayashi* et al., the discussion concerns the high-T_c superconductor material process. Pulsed laser ablation (PLD), metalorganic chemical vapor deposition (MOCVD), sputtering and molecular beam epitaxy (MBE) methods are the topics dealt herein. All of them treat subjects from atomic layer epitaxy to multiple heteroepitaxy.

Vortices in High-T_c Superconductors

S. Kuriki[1], S. Hirano[1*], A. Maeda[2], and T. Kiss[3]

[1] Research Institute for Electronic Science, Hokkaido University
N12-W6, Kita-ku, Sapporo 060-0812, Japan
sk@es.hokudai.ac.jp
[2] Department of Basic Science, The University of Tokyo
3-8-1 Komaba, Meguro-ku, Tokyo 153-8902, Japan
maeda@maildbs.c.u-tokyo.ac.jp
[3] Graduate School of Information Science and Electrical Engineering,
Kyushu University
Higashi-ku, Hakozaki 6-10-1, Fukuoka 812-8581, Japan
kiss@sc.kyushu-u.ac.jp

Abstract. In this chapter we describe the behavior of vortices in high-T_c superconductors. Because of two-dimensional-like and spatially anisotropic electronic structures, the vortices in these materials exhibit characteristic equilibrium and dynamic states, which have common basic parts, but are distinct in various aspects from conventional superconductors.

1 Introduction

We first delineate the physical pictures of vortices in type-II superconductors. There are three important energy scales that determine the vortex behavior, including the interaction between vortices, which favor the ordered state such as vortex lattice, pinning by disorders, which hinders the ordering and favors the glassy state, and thermal energy, which also hinders vortices from ordering. In conventional superconductors the first term is much stronger than the others so that ordered vortex lattice dominates in the phase diagram. Contrary to this, these energy levels and the Josephson coupling between adjacent CuO_2 layers are of the same order of magnitude in high-T_c superconductors, which results in a largely altered equilibrium phase diagram.

The anisotropic crystal structure and the two-dimensional electronic structure of high-T_c superconductors produces peculiar features of vortices such as the topography of "pancakes" and strings combination. Further, the *d*-wave character of the two-dimensional superconducting wavefunction causes significant changes in the electronic structure of the vortex core. Here there are bound states of quasiparticles in the vortex core of conventional s-wave superconductors. However, the existence of the bound states is one the basic

* Present address: Superconducting Research Laboratory, International Superconducting Technology Center, 10-13 Shinonome, Koto-ku, Tokyo 135-0062, Japan, shirano@istech.or.jp

T. Kobayashi et al. (Eds.): Vortex Electronics and SQUIDs,
Topics Appl. Phys. **91**, 5–51 (2003)

questions of high-T_c superconductors, which is based on the anisotropic gap structure of the vortex core given by the d-wave characteristics. Including this, several new topics on the electronic structure of the vortex core will be discussed.

The vortex lattice moves elastically under a driving force in the conventional picture of vortex dynamics. When a pinning force exists, flux creep of flux bundles takes place. The resistivity generated by this flux creep is ohmic in the low driving force limit. This phenomenon was readily observed in high-T_c superconductors, as thermally assisted flux flow. Further, new aspects of vortex dynamics have been elucidated using various imaging and noise measurement techniques. Those observations suggest that the dynamic state of the vortices changes in such a way from pinned state, plastic flow, stochastic motion of the vortex bundle to coherent motion. These features agree well in some aspects with theoretically proposed dynamic phase diagrams of vortices in high-T_c superconductors.

Next we describe the dissipation mechanism in the mixed state of high-T_c superconductors based on a study of the vortex dynamics. Previous studies suggested that the dissipation should be reexamined taking into account the influence of pinning and thermal fluctuation. A theoretical study has been made of the effects of thermal agitation on the vortex pinning that has a random distribution of the strength. It has been found that the percolation of depinned fluxoids becomes appreciable at the transition temperature, at which the critical current density becomes zero due to the activated thermal motions of the fluxoids.

A voltage is induced by the local plastic flow along a narrow percolation path. The transport E (electric field) versus J (current density) characteristics are governed by this percolation process. Here, the vortex glass-liquid transition is identical to the thermal depinning determined by the percolation of depinned clusters. The iso therm scaling of the E–J characteristics obtained from the thermal depinning is the same as that of a thermodynamic phase transition, but it is directly related to the pinning nature. It is confirmed that the critical exponents agree with the pin parameters.

In the final section we describe the behaviors of vortices in a grain boundary junction of high-T_c superconductor films. Measurements of the magnetic flux that is generated by the vortices trapped/moving in the wide grain boundary have been made, based on direct flux detection, in a monopole approximation, using a superconducting thin-film pickup coil and a low-T_c SQUID. The dynamic behavior of single-vortex motion along the long grain boundary junction is observed as a random switching noise in the time trace above a threshold magnetic field. Long-distance motion is inferred over a distance of 20–500 μm. It is also found that the long vortex motion can be suppressed to lengths within 1 μm by a slot structure along the long grain boundary, which forms a parallel array of SQUIDs that can hold multiple fluxoids stably.

2 Basic Physics of Vortices

2.1 Type-II Superconductivity and Vortex

The concept of vortex in superconductors was first predicted by Abrikosov [1]. It is a characteristic feature of so-called type-II superconductors, where the superconductivity survives up to the upper critical field, B_{c2}, even after perfect diamagnetism (Meissner-Ochsenfeld effect) is broken at the lower critical field, B_{c1} [2].

Between B_{c1} and B_{c2} the magnetic field penetrates into the superconductor as quantized vortices, each with a magnetic flux of

$$\Phi_0 = \frac{hc}{2e} \,. \tag{1}$$

Φ_0 is called a flux quantum. Supercurrent circulates around the vortex. These are shown schematically in Fig. 1.

By combining the London equation,

$$\boldsymbol{j} = -\frac{1}{\lambda^2}\boldsymbol{A} \,, \tag{2}$$

($\boldsymbol{A}$ is the vector potential, $\boldsymbol{j}$ is the current density and λ is the penetration depth) and the Maxwell equation, the local magnetic field distribution, $h(r)$, of a single vortex was found to be

$$h(r) = \frac{\Phi_0}{2\pi\lambda^2} K_0\left(\frac{r}{\lambda}\right) \,, \tag{3}$$

where K_0 is a zero-order modified Hankel function and r is the distance from center of the vortex. On the other hand, in the center of the vortex, which is called a vortex core, the superconducting order parameter, Φ, becomes small. Originally, the core was thought to be a normal metal [3]. The core size is of the order of the GL coherence length [4], ξ_{GL} (below we will drop the suffix GL unless there is a possibility of misunderstanding), whereas the length scale of the magnetic field distribution is the penetration depth, λ. When $\lambda \gg \xi$, the surface energy of the superconductor-normal metal interface is negative. This is the reason why the magnetic field penetrates into the superconductor as

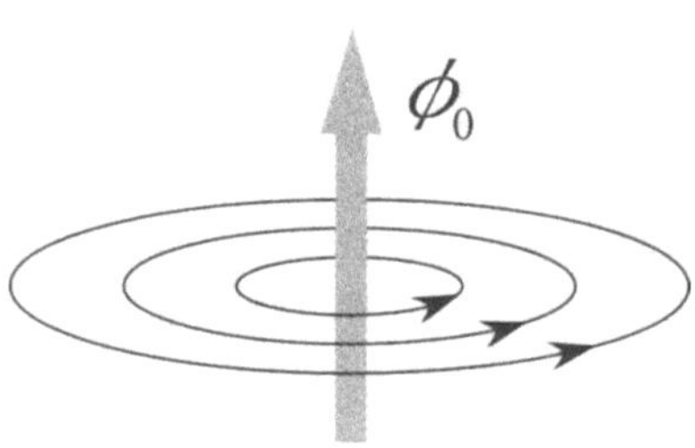

Fig. 1. Schematic feature of a vortex in superconductor

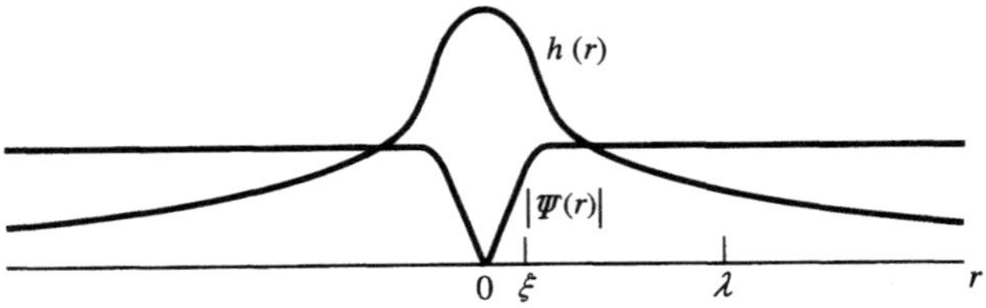

Fig. 2. Magnetic field- and order-parameter profiles of a vortex

the quantized vortices. A simple analysis using the so-called GL theory gives that the superconductor is type-II for a temperature independent parameter

$$\kappa \equiv \frac{\lambda}{\xi} \tag{4}$$

(GL parameter) larger than $1/\sqrt{2}$. The profile of magnetic field and the superconducting order parameter are shown in Fig. 2.

2.2 Vortex Lattice, Bragg Glass

Vortices interact with each other through electromagnetic interaction, F_{int},

$$F_{\text{int}} = \frac{\Phi^2}{8\pi^2\lambda^2} K_0\left(\frac{r_{12}}{\lambda}\right), \tag{5}$$

where r_{12} is the distance between the two vortices. Since F_{12} is repulsive, vortices tend to form a lattice. A simple argument tells us that vortices form a triangular lattice, which was confirmed by the magnetic decoration technique, etc. [5,6][1].

In real crystals there is always a finite amount of disorder. Such disorder introduces two effects. First, the disorder pins each vortex, since the vortex core favors the location of disorder where the superconducting order parameter deteriorates. Second, the competition between pinning and the vortex-vortex interaction destroys the long-range order of the vortex lattice (VL).

Larkin and *Ovchinikov* [7] argued that the VL was divided into domains with a finite correlation length, L. Thus observation of the triangular lattice in real crystals suggests that L is larger than (or of the same order as) the sample dimension. However, it was found that the presence of the host lattice made L less than the interatomic distance, a [8]. Recent renormalization group analysis by *Giamarch* and *Le Doussal* [9] clarified the puzzle. They showed that the vortices exhibited the *quasi*-long range order even in the presence of disorder. This quasi-long range ordered state is called a *Bragg glass*, which

[1] Abrikosov originally concluded that vortices formed a square lattice. Later this was found to be wrong. The difference of the free energy between a triangular lattice and square lattice is rather small

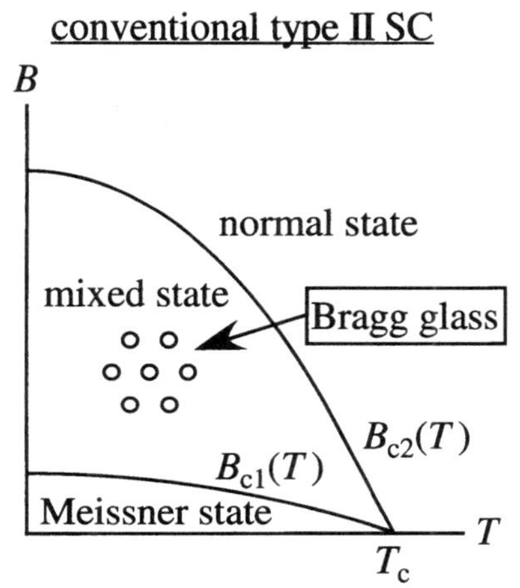

Fig. 3. $B-T$ phase diagram of conventional superconductors

is characterized by a power-law decay of the correlation function. Indeed, a recent experiment claimed that such a power law decay was observed in $Ba_{1-x}K_xBiO_3$ [10].

The equilibrium phase diagram for conventional superconductors (CSCs) is shown in Fig. 3. This will be discussed again in the next section for high-T_c cuprates.

2.3 Motion of a Vortex: Flux Flow and Flux Creep

Vortices move under the presence of a driving current, $\boldsymbol{j}$, since the Lorentz force, $\boldsymbol{f}_L$, acts on vortices as

$$\boldsymbol{f}_L = \boldsymbol{j} \times \frac{\Phi_0}{c} . \tag{6}$$

When they move, energy is dissipated at the core (*flux flow*). *Bardeen* and *Stephen* [3] were the first to calculate the energy dissipation by the flux flow. They assumed that the core can be regarded as a normal metal, and derived a formula for the flux-flow resistivity, ρ_f, which was found to be ohmic and given by

$$\rho_f = \rho_n \frac{B}{B_{c2}} , \tag{7}$$

where ρ_n is the resistivity in the normal state. More generally, energy dissipation by the flux flow is closely related to the electronic structure in the vortex core. This will be discussed again later.

If there is finite pinning, thermal energy play an essential role. Thermal energy can make vortices escape from the pinning potential well, which generates fluctuating voltages (*flux creep*). Application of the driving current introduces an unbalance between left going current and right going current and leads to a net creep velocity in one direction,

$$v \propto \sinh \frac{fVL}{k_B T}, \tag{8}$$

where f, V and L are the force density, the domain volume and the domain length, respectively. This causes a finite voltage [11]. Thus, in the strict sense of the word, the vortex state is no longer superconducting (with zero resistance). Indeed, (8) is ohmic in the limit of $f \to 0$. However, effectively, we can define a finite critical currents that overcomes the pinning, since (8) changes very sharply at some current value.

On the other hand, for high-T_c superconductors well defined ohmic resistivity was observed for a wide range of driving current because of the large thermal fluctuations and the small coherence volume, V. This is called thermally assisted flux flow (TAFF) [12].

2.4 Microscopic Electronic Structure of Vortex Core

At the vortex core the superconducting pair potential is weak. This means that there are bound states in the vortex core, analogous to the potential well problem in textbooks on quantum mechanics. An exact analysis was first made by *Carrori* et al. [13]. They found a series of bound states with a minimum energy of

$$E_{\min} = \frac{\Delta_0^2}{E_F} \equiv \hbar\omega_0 , \tag{9}$$

where Δ_0 is half of the mean-field superconducting gap and E_F is the Fermi energy (Fig. 4a). For CSCs, $E_{\min}$ is of the order of $1 \sim 10\,\mu\text{eV}$, which is hardly resolved by any experimental technique available. Indeed, an STM observation in $NbSe_2$ [14] showed an envelope curve of these states peaked at zero energy (Fig. 4b).

Disorder introduces a finite life time τ for these bound states. Usually, $\omega_0\tau \ll 1$, which makes the original *Bardeen–Stephen* assumption correct.

In the opposite limit (very clean limit) where $\omega_0\tau \gg 1$ in addition to the force of (6), it is believed that the so-called Magnus force acts on vortices, $\boldsymbol{f}_M$, which acts in the same direction as the vortex flow. Taking this Magnus effect into account, the vortex motion is characterized by the Hall angle θ, which is known to be given by [15]

$$\tan\theta = \omega_0\tau . \tag{10}$$

The Hall angle, θ, can be measured directly by the Hall effect. Alternatively, an effective viscosity η_{eff} defined as

$$\eta_{\text{eff}} = \eta + \alpha_H^2/\eta , \tag{11}$$

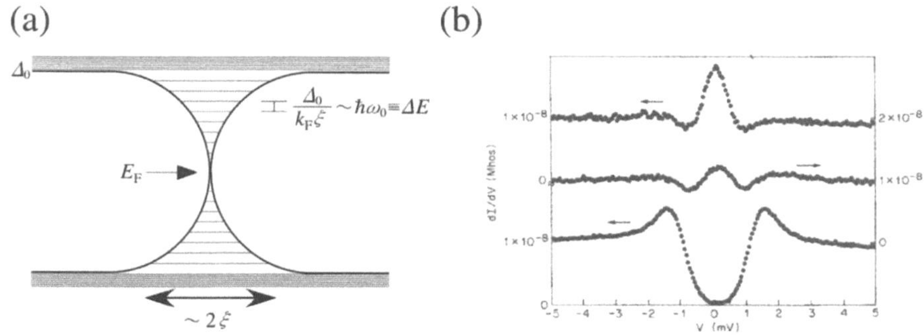

Fig. 4. (**a**) Quantized levels in the vortex core. (**b**) STS map of $NbSe_2$ (taken from [14])

where η and α_H are the viscosities in the perpendicular and parallel directions, respectively, also gives θ through the relationship

$$\frac{\eta_{\text{eff}}}{\pi n \hbar} = \omega_0 \tau \,, \tag{12}$$

where n is the carrier density. This will be discussed in more detail in Sect. 4.5.

Recently many exotic superconductors with highly anisotropic superconducting order parameters have become known. For such superconductors, the physics discussed above may be changed in many aspects. One example is the energy levels of the quasiparticles in the vortex core. For anisotropic superconductors, there are nodes in the superconducting gap. This introduces largely extended quasiparticle density of states in the direction of nodes [16,17]. We will discuss d-wave case in the next section for high-T_c cuprates. If the pairing of electrons are spin-triplet, as has been found in UPt_3 [18] and Sr_2RuO_4 [19], rich structure is introduced in the vortex core. These are the topics of current interest, and some more years will be needed to clarify the details.

3 Vortices in High-T_c Superconductors

3.1 Characteristic Aspects of the Mixed State of High-T_c Superconductors

High-T_c cuprate superconductors have provided many treasures for physicists. In addition to an extraordinary amount of studies for understanding the mechanism of high-T_c superconductivity, the materials brought us a revolution in our understanding of the vortex state. A detailed review has been made by *Blatter* et al. [20].

As has already been described, in general, there are three important energy scales to discuss the physics of vortex matter. They are (a) interaction between vortices, which favors the ordered state such as VL, (b) pinning by

(a) (b)

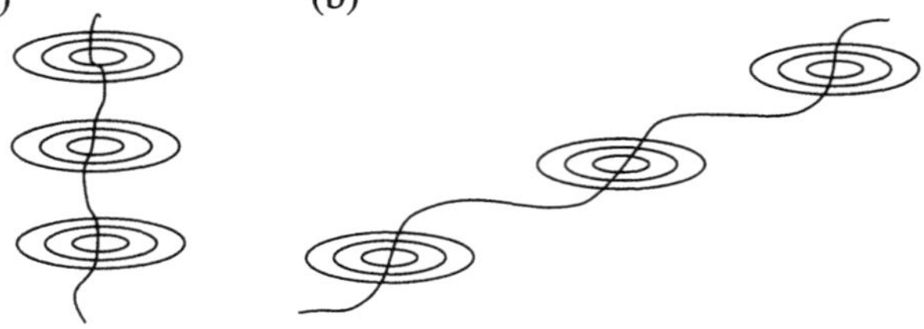

(c)

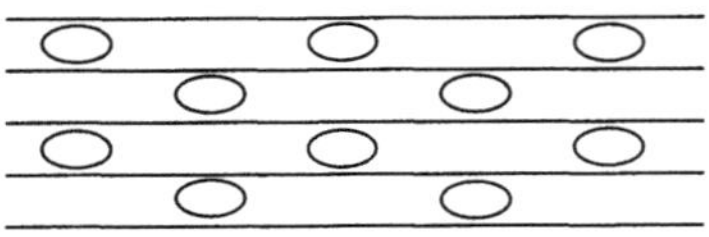

Fig. 5. (**a**) Pancake vortices and strings. (**b**) Vortices of high-T_c cuprates for inclined magnetic field. (**c**) Josephson vortices

disorder, which hinders ordering and favors the glassy state, (c) thermal energy, which also hinders the vortices ordering. For CSCs, (a) is much stronger than the other two, and the ordered VL dominates in the resultant phase diagram, as in Fig. 3.

In high-T_c cuprates, the following characteristics change the physics of the vortex matter to a large extent. First, the crystal structure and the electronic structure are quasi-two dimensional, which makes a vortex the combination of "pancakes" and strings (Fig. 5) [21]. The strong two dimensionality also introduces new features for vortices that are generated in tilted and parallel magnetic fields. Second, all the energies of (a)–(c) and the Josephson energy between adjacent layers are of the same order of magnitude, which alters the equilibrium phase diagram greatly. Third, the *d*-wave character of the condensate wave function [22] causes significant changes in the electronic structure of the vortex core.

Below we will discuss each of these in more detail.

3.2 Vortices in High-T_c Superconductors

In high-T_c cuprate, due to the strong two dimensionality, the circulating current around the quantized vortex is confined in each CuO_2 plane. These confined vortices are called pancake vortices (Fig. 5a and 5b) [21]. Pancake vortices in adjacent layers interact with each other through Josephson coupling.

Competition between magnetic coupling and Josephson coupling leads to dimensional crossover at a field [23]

$$B_{cr} \sim \frac{\Phi_0}{\gamma^2 s^2}, \tag{13}$$

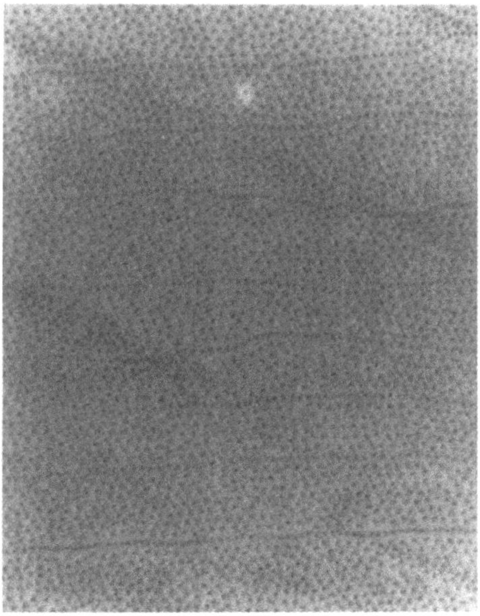

Fig. 6. Crossing lattice (from [27])

where $\gamma \equiv \lambda_c/\lambda_{ab} = (m_c/m_{ab})^{1/2}$ is the anisotropy parameter, λ_c, λ_{ab}, m_c, and m_{ab} are c-axis and in-plane penetration depths, and c-axis and in-plane effective masses, respectively, and s is the spacing between adjacent CuO_2 layers. The second equality assumes the validity of the effective mass model. For instance, in $Bi_2Sr_2CaCu_2O_y$ (BSCCO), which is one of the most anisotropic high-T_c cuprates, it is believed that $\gamma \simeq 500 \sim 1000$. Thus vortices in this material are considered to be two dimensional pancakes.

When the magnetic field is applied parallel to the CuO_2 plane we can have coreless Josephson vortices (Fig. 5c) [24,25]. The coreless feature leads to ultra high speed motion in the presence of a driving current. Josephson vortices will be discussed in Chap. 5 in more detail.

By changing the angle of the magnetic field from the CuO_2 plane some new features show up. First, vortices continue to be in parallel to the plane when the angle is very small. This is intrinsic pinning [25]. On further increasing the angle, a strongly anisotropic material shows a new feature; the crossing lattice (Fig. 6) [26]. There a dilute lattice of Josephson vortices coexists with a dense lattice of pancake vortices. The existence of the crossing lattice is confirmed by the magneto-optic technique [27].

3.3 Equilibrium Phase Diagram

For high-T_c cuprate the high-temperature scale and large anisotropy alter the equilibrium phase diagram of the mixed state greatly. Figure 7 shows a summary of our current understanding of the equilibrium phase diagram

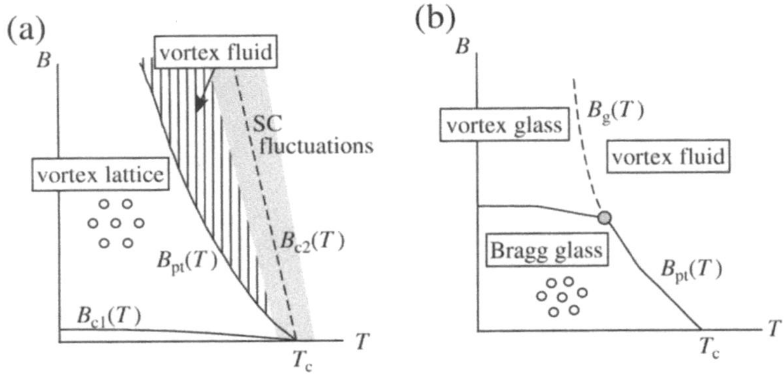

Fig. 7. Equilibrium phase diagram of high-T_c cuprate for (**a**) a disorder-free sample and (**b**) a sample with a finite amount of disorder, when a magnetic field is applied perpendicular to CuO_2 plane

of the vortex state of high-T_c cuprate for the perpendicular magnetic field, constructed both theoretically and experimentally.

Figure 7a is the theoretically predicted $B-T$ phase diagram in the absence of disorder [28]. There is only one true thermodynamic phase transition between VL and the vortex liquid, which is known to be of first order. There is no qualitative distinction between the vortex liquid and the normal state. Thus, the upper critical field, B_{c2}, is no longer a true phase transition field, and only becomes a barometer of the development of the fluctuations of superconductivity. Experimentally, resistive broadening has been observed around the mean-field B_{c2} (Fig. 8) [29]. Because of the resistive broadening it became impossible to determine B_{c2} from the resistivity data.

In real crystals we always have a finite amount of disorder and the equilibrium phase diagram becomes quite different. In Fig. 7b the consensus formed through experimental and theoretical studies on high-T_c cuprate since their discovery is summarized. In the presence of a finite amount of disorder the onset of irreversibility introduces much complexity in the interpretation of the experimental data, and it took long time to form a consensus. It is the same that B_{c2} is not a phase transition field even for finite disorder. We also have the first order transition, where definite thermodynamic anomalies were observed [30,31]. Experimentally, thermodynamic anomalies faded out at some critical point. Around the critical point, irreversibility sets in because of the pinning of vortices. Below this critical temperature an almost horizontal boundary appeared. Therefore, the so-called peak effect took place [32]. Recent experiments [33,34] and numerical simulations [35] suggested this peak effect boundary also corresponds to a first order phase transition. Thus, the concave upward phase boundary in the disorder free case changes into a new concave downward phase boundary, where also a first order transition exists. Together with the theoretical considerations introduced above, it is natural

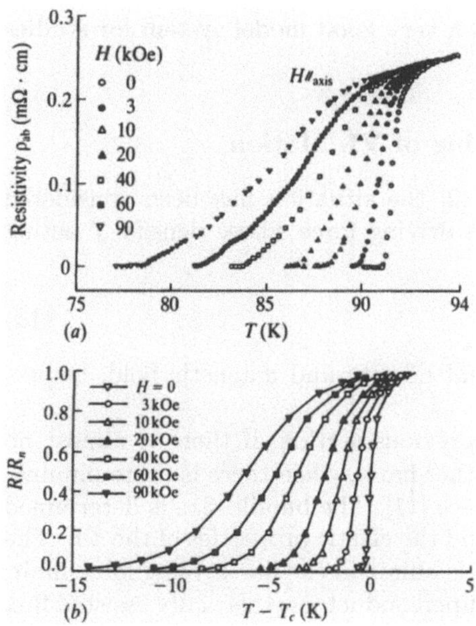

Fig. 8. Resistive broadening of high-T_c superconductor (from [2] after [29])

to assume that the hatched area in Fig. 7b corresponds to the moving Bragg glass. For direct evidence, power-law correlation decay should be confirmed in experiments such as neutron diffraction, etc. We can consider that the high-temperature first order transition separates the Bragg glass and the vortex liquid (vortex melting), whereas the low-temperature almost horizontal first order transition separates the Bragg glass and the vortex glass. In this region the dominant driving factor for the phase transition is disorder.

If we do not have any other phase transition, the vortex glass phase and the vortex liquid phase are qualitatively the same. Whether or not the other phase transition exists should be clarified in succeeding work.

3.4 Dynamic Phase Diagram of Driven Vortices

The physics of current-driven vortices has attracted much recent attention [36]. This subject concerns the physics of a driven system under random pinning. Much work has been focused on the nature of the moving state and also on the dynamic phase diagram, namely the phase diagram in the $B - T - F$ (B is the magnetic field, T is the temperature and F is the driving force) space. The physics of driven vortices has many common aspects to the dynamics of charge and spin-density waves (CDW and SDW) in quasi-one dimensional materials [37], to those of Wigner crystals in correlated materials [38,39,40,41,42,43,44,45,46], to magnetic bubble arrays [47] and also to solid friction [48]. This means that the subject probes a quite general, important issue in condensed-matter physics. Also, from an application point of view, the physics of driven vortices is a very good model system for studies of friction between interfaces.

3.4.1 Conventional Understanding of VL Motion

In a textbook on superconductivity [2] the situation has been considered where a VL moves elastically under a driving force whose density $\boldsymbol{f}$ acting on vortices is expressed as

$$\boldsymbol{f} = \boldsymbol{j} \times \boldsymbol{B}\,, \tag{14}$$

where $\boldsymbol{j}$ and $\boldsymbol{B}$ are the driving current density and magnetic field, respectively.

As was already described in the previous section, if there is almost no pinning, flux flow occurs [3]. On the other hand, when there is finite pinning, flux creep of a vortex bundle takes place [11]. The bundle size is determined by the competition between pinning and the elastic properties of the VL. The resistivity generated by the flux creep is ohmic in the low-driving force limit, which was easily observed in high-T_c superconductors (thermally assisted flux flow: TAFF [12]). Collective pinning theory [7] predicted non-ohmic resistivity, which explained some experiments in high-T_c superconductors.

3.4.2 Recent Development of Theoretical Understanding

In both cases of flux flow and flux creep, vortices were considered to move in an elastic bundle. Recently, however, more complex forms of vortex motion have been considered. Several theoretical studies [54,55,56,57] and numerical simulations [58,59,60,61,62] have discussed in detail the possible dynamical phases of driven vortices in superconductors. Many of them agree with each other regarding the following aspects. If the driving force increases, depinned vortices start moving. When the driving force is small, vortices move in a plastic manner (plastic flow (PF) [58]) where there are channels in which vortices move with a finite velocity, whereas in other channels vortices remain pinned. Thus, between moving channels and static channels there are dislocations in the VL. With further increasing driving current, vortices tend to re-order. This is a common feature of the theoretical studies listed above. However, regarding the exact details, there has been no consensus on what the dynamical phase of the vortices should be.

Koshelev and *Vinokur* [54] proposed a dynamic melting under a finite driving force. Through this dynamic melting, a stationary VL changes into a moving VL via the PF. In addition, the critical field (or the critical temperature) corresponding to the first order transition of the VL in the equilibrium state does not depend on the driving force. These ideas are summarized schematically in Fig. 9.

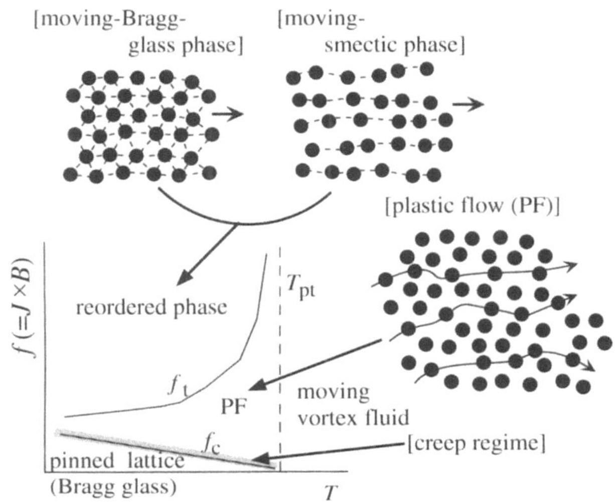

Fig. 9. Schematical dynamic phase diagram of driven vortices expected by theories

Giamarchi and *Le Doussal* [56] proposed another ordered phase, the moving-Bragg-glass. This phase is characterized by the presence of translational order whose spatial correlation decays also in a power law both parallel and perpendicular to the translational motion of the vortices, as in the static case. According to this theory, the transition between the Bragg glass and the moving-Bragg-glass is not a phase transition, but rather a gradual crossover. They also argued that the Bragg glass changes into the moving Bragg glass directly, without a region of PF, if the disorder is weak. On the other hand, if the disorder is strong, the equilibrium phase is the vortex glass. After going through the PF, it changes into a moving transverse glass, which does not possess translational order.

Phenomenologically, the moving-Bragg-glass phase can be characterized by the presence of a so-called washboard oscillation, which is a periodic velocity modulation caused by the interaction between an almost uniform translation of a periodic object and pinning centers (Fig. 10). Thus, the frequency of the washboard oscillation, f_W is simply given by

$$f_W = v/a\,, \tag{15}$$

where v is the velocity and a is the spatial periodicity of the object. This phenomenon is well known in the CDW and the SDW systems [37].

On the other hand, *Balents* et al. [55] proposed a smectic phase which lacks translational order in the longitudinal direction of the flow even in the case of weak disorder. This was based on the experimental fact that washboard noise had not been observed in the vortices of superconductors before, except in a system with artificially introduced strong periodic pinning centers [63].

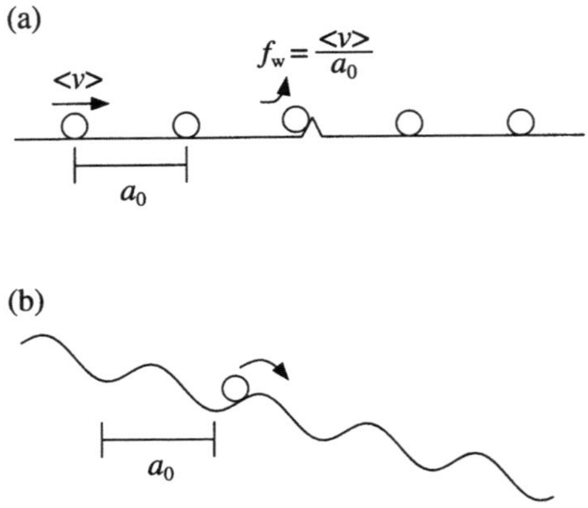

Fig. 10. Schematical picture of washboard motion. (**a**) Real space picture (**b**) a converted picture

For randomly pinned systems, two so-called interference experiments, one for a CSC Al [64] and the other for a high-T_c cuprate YBCO [65] showed the existence of the washboard frequency. In these experiments, however, the presence of the large-amplitude ac driving force altered the dynamical state of the vortices and made it difficult to explore the true dynamical state to be compared with the theories described above.

Quite recently, washboard oscillation was observed in the noise spectrum generated by vortex motion under random pinning in superconductors, $NbSe_2$ [66] and BSCCO [67] Fig. 10. This supports the notion of the development of translational order along the flow direction and should force reconsideration of current theories.

3.4.3 Brief Look at the Experiments on the Dynamics of Vortices

To elucidate the dynamics of vortices in terms of a dynamic phase diagram, several experimental efforts have been made [66,67,68,69,70,71,72,73,74,75] [76,77,78,79,80]. In particular, in a conventional layered superconductor, $NbSe_2$, various imaging techniques such as small angle neutron scattering (SANS) [78], magnetic decoration [80] and STM [66] have established the existence of various states of driven vortices.

Other than these techniques, observation of low frequency noise has been a popular tool to study the vortex motion both in conventional and high-T_c superconductors [67,68,69,70,71,72,73,81,82,83,84,85,86], and such noise has been considered to have some relationship with various low frequency phenomena [87,88].

To get more local information, *the local-density fluctuation* of vortices has been discussed. *Yeh* et al. [86] already studied density fluctuations using a SQUID technique both in a CSC and in a high-T_c superconductor. However, they did not concentrate on the dependence on the location nor on the spatial correlations. *Tsuboi* et al. studied the local-density noise (LDN) of

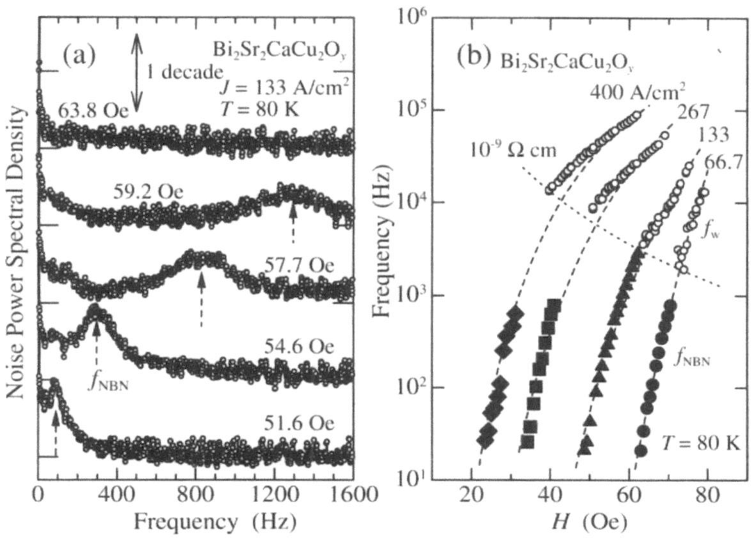

Fig. 11. Washboard noise observed in $Bi_2Sr_2CaCu_2O_y$. (**a**) Noise spectrum. (**b**) The peak frequency vs magnetic field together with the expected washboard frequency estimated from the resistivity data (taken from [67])

vortices generated in a small area in the sample, measured by a micro-Hall probe array [68,69,70,71]. With this technique, together with conventional conduction noise (CN) (*velocity fluctuation*) measurements, we can investigate the dependence of the density noise on the location in the sample. It is also possible to discuss the spatial correlations of the noise generated at different places both parallel and perpendicular to the flow direction. With this technique it was suggested that the dynamic state of the vortices of BSCCO changed as follows: (1) pinned state, (2) plastic flow, represented by large broadband-LDN, (3) stochastic motion of a vortex bundle, represented by large broadband-CN, which is the precursor to coherent motion, and (4) coherent motion, represented by the washboard noise. These features agreed well with most of the theoretically proposed dynamical phase diagrams of vortices in superconductors.

3.5 Electronic States of Vortex Cores of High-T_c Superconductor

As was described in Sect. 3.4, in CSCs with s-wave gap symmetry there are bound states of quasiparticles (QPs) localized in the vortex core. The level separation between the bound states, ΔE, is

$$\Delta E \equiv \hbar\omega_0 \sim \Delta_0/k_F\xi\,, \qquad (16)$$

where Δ_0 is the bulk gap energy and k_F and ξ are the Fermi wavenumber and the GL coherence length, respectively [89]. Here ω_0 is regarded as the minimal gap frequency inside the core.

As was already discussed, in CSC, T_c is below 10 K and $\xi \sim 100$ Å, which leads to $\Delta E \leq 0.1$ K. This could not be resolved by available experimental techniques. Since $\delta E \gg \Delta E$ in CSC, the vortices in CSC can be regarded as tubes with a normal metal of radius ξ. In such a case, the vortex motion was well described by the *Bardeen–Stephen* (B–S) model [3].

In HTSC, such a simple description of the vortex core is incorrect for several reasons. First, the energy gap has d-wave symmetry in HTSC [22]. Since the amplitude of the gap at the node is zero, QPs are not localized in the vortex core but extended along the node direction [16]. A theoretical calculation suggested that there were no truly localized states in the vortex core in pure $d_{x^2-y^2}$ superconductors [17]. On the other hand, recent STM results in HTSC showed that the peak at zero energy vanished and that the density of states at a finite energy below the gap Δ_0 was enhanced near the center of the vortex core [90,91,92]. Therefore, the presence or the absence of localized states in the vortex core of HTSC is still one of the central basic questions.

The second different feature of the vortex core in HTSC is the semi-quantum nature of the core. Since $\Delta_0 \sim 20$ meV for HTSC, ξ is about 20 Å. This gives $k_F\xi \sim 20$, which is quite different from the case of CSC (for which $k_F\xi \sim 200$, typically). Indeed, if the peak observed in the STM experiment mentioned above corresponds to the quantum levels in the core, it gives $k_F\xi \sim 2$. The physics of such a quantum core has been undeveloped up to now. In such a highly anomalous situation, to the best of our knowledge there have been no rigorous calculations of the dynamic properties of vortices. So the appearance of some new effect may be expected.

Such anomalous aspects of the electronic structure around the core in HTSC should affect the dynamic response of the vortices. To address the problem, let us repeat the discussion made above (Sect. 2.4). If the vortex motion is described phenomenologically, the dissipation can be discussed in terms of the frictional force,

$$\boldsymbol{F}_v = \eta \boldsymbol{v}_L + \alpha_H \boldsymbol{v}_L \times \hat{\boldsymbol{z}}\,, \tag{17}$$

which is proportional to the velocity of vortices, $\boldsymbol{v}_L$. Here η and α_H are the viscous drag and the Hall coefficients, respectively, and $\hat{\boldsymbol{z}}$ is the unit vector parallel to an applied magnetic field. The dissipation mechanism is, in turn, closely related to the characteristic scattering time of QPs in the vortex core, τ. The vortex viscosity η, and Hall constant α_H were calculated [20,93] for arbitrary values of $\omega_0\tau$, and it was found that

$$\eta = \pi\hbar n \frac{\omega_0\tau}{1+(\omega_0\tau)^2}\,, \tag{18}$$

$$\alpha_H = \pi\hbar n \frac{(\omega_0\tau)^2}{1+(\omega_0\tau)^2}\,, \tag{19}$$

where n is the QP concentration. The ratio of the level spacing ΔE to the level width δE,

$$\frac{\Delta \mathrm{E}}{\delta E} = \omega_0 \tau \,, \tag{20}$$

is used as an index of the cleanness in the vortex core. In the dirty limit ($\omega_0 \tau \ll 1$), the Hall effect (19) is negligible and (18) is expressed as $\eta = \pi \hbar n \omega_0 \tau = B_{c2} \Phi_0 / \rho_n$, where $\rho_n = m^*/ne^2\tau$, and B_{c2} is the upper critical field. This is the B-S expression for the viscosity, η [3]. In the opposite limit, $\omega_0 \tau \gg 1$ (the superclean limit), the Hall effect is dominant and η is expressed as $\eta \simeq \pi \hbar n / \omega_0 \tau$. According to [15] the diagonal component of the vortex resistivity under constant transport current, $\boldsymbol{J}$, was expressed as

$$\rho_v = \frac{B\Phi_0}{\eta + \alpha_H^2/\eta} = \frac{B\Phi_0}{\eta_{\text{eff}}} \,, \tag{21}$$

where the effective viscosity η_{eff} was defined as

$$\eta_{\text{eff}} = \eta + \alpha_{\text{H}}^2/\eta \,. \tag{22}$$

This means that the experimentally obtained viscous coefficient under the condition $\boldsymbol{J} =$ const. was η_{eff}. (18), (19), and (22) yield

$$\omega_0 \tau = \frac{\eta_{\text{eff}}}{\pi \hbar n} \,. \tag{23}$$

Therefore we can determine the QP lifetime in the vortex core from estimates of the viscosity of the vortex motion in the mixed state of HTSC.

There have been many experimental efforts to determine the vortex dynamics parameters of $YBa_2Cu_3O_y$ [94,95,96,97,98,99,100,101]. A review of earlier results is provided in [15]. Among them, one report [96] caused a debate. They measured R_s as a function of magnetic field up to 7 T at low temperatures and deduced a very large viscosity, which was two orders of magnitude larger than in CSCs. This huge value was interpreted as evidence that the core of YBCO was in the superclean regime, $\omega_0 \tau \gg 1$. However, their estimation of the viscosity was based on the assumption that the pinning frequency ω_p is much smaller than the measurement frequency ω. Under that assumption, they estimated the viscosity only from the data on the surface resistance, R_s (the real part of the surface impedance, Z_s). In CSC, the assumption $\omega \gg \omega_p$ was reasonable in the microwave region since the characteristic pinning frequency $\omega_p \sim 100\,\text{MHz}$ [97]. In HTSC, however, the condition $\omega \gg \omega_p$ might not be satisfied. Indeed, experimentally estimated values of the pinning frequency ω_p at low temperatures in previous reports [94,98,99] were of the order of $\sim 10\,\text{GHz}$. In such a case estimation of η from R_s alone leads to an incorrect value for η and measurement of Z_s as a complex value is essential.

To discuss the problem exactly, *Tsuchiya* et al. [101] measured the complex surface impedance as a function of magnetic field, frequency and temperature up to higher magnetic fields. Through comparison with a theoretical calculation by *Coffey* and *Clem* [104], they estimated the pinning frequency, vortex viscosity and pinning constant. Estimated values of η at 10 K are $\sim$ 4–5×10^{-7} Ns/m^2. These values correspond to $\omega_0 \tau \sim 0.3$–0.5, which suggests that the vortex core of YBCO is not in the deeply superclean regime but in the moderately clean regime.

In $Bi_2Sr_2CaCu_2O_y$ it was reported that a rapid increase of surface reactance, X_s, took place at the first order transition (FOT) of the VL [102]. Since X_s is proportional to the real part of the effective penetration depth, it was proposed that this increase may be ascribed to a decrease in the superfluid density rather than to the loss of pinning. This suggested that the change in the structure of the VL affected the electronic structure of the vortex strongly. Unfortunately, in YBCO, in the data at microwave frequencies from the high temperature region above 60 K we cannot find any obvious anomaly around the FOT in both R_s and X_s. This is in contrast to the results observed in BSCCO. Thus, the anomaly may depend on the magnitude of anisotropy and frequency.

Quite recently, the existence of a collective mode in the microwave frequency range was predicted in unCSCs with mixed symmetry order parameters [103]. Such a mode has not yet been observed in any experiments. This is another interesting problem.

Another recent interest in the electronic structure of vortex cores is the problem of the other related ordered structures, such as antiferromagnetic correlations [105,106,107,108] and checker board like order [109].

Some of these will be discussed later in this book (Sect. 5).

4 Vortex Dynamics in HTSC Thin Film

One of the noticeable features of HTSC compared with metallic superconductors is that the effects of thermal agitation on the vortex dynamics are quite severe, especially at higher temperatures. Namely, the strength of the thermal fluctuations is comparable to that of pinning and vortex-vortex interaction [110]. This fact leads to the appearance of the melting temperature, T_m, of the Abrikosov flux line lattice (FLL) far below the critical temperature T_c [111]. In a disordered sample, the melting temperature deviates from T_m due to the influence of flux pinning. Assuming random point pins, Fisher and coworkers [112,113,114] predicted that the Abrikosov lattice will change into a glassy state (vortex glass) due to flux pinning, and the vortex glass will melt to become a vortex liquid at a transition temperature, T_{gl}. Based on the scaling hypothesis in a second order phase transition, they predicted that the temperature dependent electric field (E) vs. current density (J) curves in the vortex glass phase and the liquid phase can be scaled to the respective master curves with the aid of the static critical exponent, ν, and the dynamic critical exponent, z. Their prediction was first confirmed by *Koch* et al. in dc transport measurements on $Y_1Ba_2Cu_3O_{7-\delta}$ (YBCO) thin film. Then similar scaling properties were also observed in other kinds of HTSC thin films such as the Bi-system [115,116], and the Tl-system [117].

The vortex glass-liquid (GL) transition is essentially determined by flux pinning phenomena where the individual pinning force can be regarded as a kind of stochastic variable. Nevertheless, the influence of pinning strength, S_{pin}, on T_{gl} is not clear in these phase transition theories. *Dorsay* et al. derived an expression for T_{gl} as a function of S_{pin}, and predicted that $T_{gl}(S_{pin} \rightarrow 0) = 0$ [118]. On the other hand, *Blatter* et al. pointed out that $T_{gl}(S_{pin} \rightarrow 0)$ should be equal to T_m, though they did not derive concrete expression for T_{gl} [110]. From these theories it may also be difficult to explain the observed tendencies that the scaling holds over extremely wide ranges of temperature, T, and magnetic field, *Misat* et al. pointed out that the scale of correlation time reached as long as 60 decades apparently in an a-axis oriented YBCO film [119]. Lack of universality of the critical exponents has also been reported [120,121,122]. Furthermore, in an experiment on a-axis oriented YBCO film, *Misat* et al. pointed out that the transition temperature and critical exponents show anisotropy depending on the crystallographic inplane orientation parallel to the b-axis or c-axis even though they probed the same vortices in the same thermodynamic condition [119]. These results indicate that the GL transition and the dissipation mechanism in HTSC should be reexamined by taking into account the influence of pinning as well as thermal fluctuations.

In this section, vortex dynamics in HTSC thin film have been examined on the basis of theoretical studies on vortex pinning in random media under the influence of thermal fluctuations. It is pointed out that the percolation process of thermally depinned vortices essentially affects the dissipation mechanism in the mixed state of HTSC.

4.1 Thermally Depinning Transition

One of the important effects of thermal fluctuations on vortex pinning is the effective decrease of pinning strength due to thermal motions of fluxoids inside the essential pinning potentials. The critical current density per flux bundle in the presence of thermal agitation becomes smaller than the intrinsic value [123]. When the effective critical current density in the presence of thermal agitation becomes zero due to the active thermal motions of the fluxoids, the flux bundle concerned can be regarded as thermally depinned. This means that the flux bundle concerned becomes a candidate for making the flux flow. In order to see the detailed character of the thermal depinning phenomena, we carried out a Monte Carlo simulation for a two-dimensional fluxoid lattice, where the fluxoids are assumed for simplicity to be continuous and straight along their directions [124,125]. This assumption holds widely for thin films less than a few micro meter since the magnetic correlation length along the fluxoid is known to be the order of 10 μm even in the most anisotropic HTSC cuprate, i.e., Bi-2212 [126].

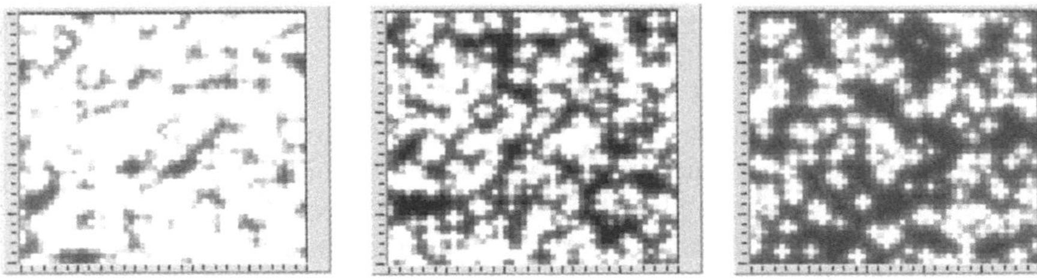

(a) (b) (c)

Fig. 12. Percolative growth of unpinned clusters under the influence of thermal agitation (**a**) $T < T_\mathrm{p}$, (**b**) $T \approx T_\mathrm{p}$, and (**c**) $T > T_\mathrm{p}$ in a random pin medium obtained by a Monte Carlo simulation, where T_p is the percolation temperature. The white regions indicate pinned clusters whereas the black regions indicate unpinned clusters

Figure 12 shows the percolation process of unpinned clusters under the influence of thermal agitation obtained by the Monte Carlo simulation. The white regions indicate the pinned clusters while the black regions indicate the thermally depinned clusters. As the temperature is increased the depinned clusters grow gradually, and then these clusters percolate at the transition temperature, T_p, where the percolation threshold becomes zero. The correlation length, ℓ, among the unpinned clusters diverges inversely proportional to the power of $|1 - T/T_\mathrm{p}|$ near the percolation temperature T_p:

$$\ell = \ell_0 \, |1 - T/T_\mathrm{p}|^{-\nu} \,, \tag{24}$$

where ν is the critical index and ℓ_0 is a numerical constant. At the percolation temperature macroscopic plastic flow occurs along the percolation path even in the limit where $J \rightarrow 0$. Therefore, the minimum value of the critical current density in the flux bundle row, which is denoted by J_cm, becomes zero. While $T < T_\mathrm{p}$ the value of J_cm is positive since the locally depinned clusters are still pinned by the surrounding pinned clusters. When $T > T_\mathrm{p}$, the sign of J_cm apparently becomes negative. This indicates that the thermally depinned clusters have already percolated even in the limit of $J \rightarrow 0$, and therefore linear resistivity will be induced due to the localized flux flow along the percolation path in the limit of $J \rightarrow 0$.

From the Monte Carlo simulation, we have obtained a universal property of the pinning phenomena in the random medium. We obtained the depinning probability, Q, for the flux bundle row as a function of current density where the macroscopic flux flow occures along the percolation path. The results of Q obtained at various condition of temperatures are shown in Fig. 13a. The depinning probability at each temperature is scaled as shown in Fig. 13b if we normalize the bias current density by the nondimensional quantity $(J - J_\mathrm{cm})/J_0$, where J_0 is the scale parameter corresponding to the half width of

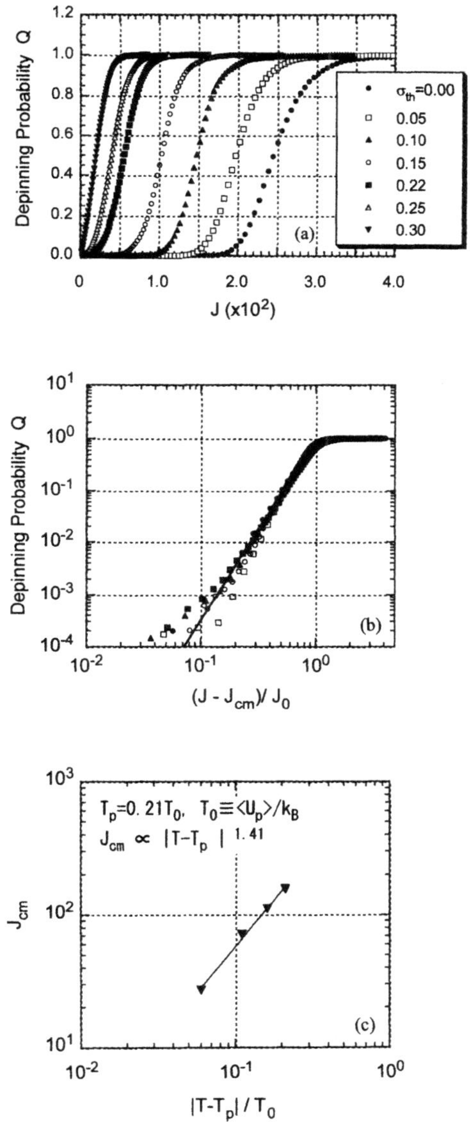

Fig. 13. (**a**) Depinning probability under the influence of thermal fluctuation $\sigma_{\rm th} \equiv kT/\langle U_{\rm p}\rangle$ obtained by the Monte Carlo study shown in Fig. 12, where $\langle U_{\rm p}\rangle$ is the average value of the pinning potential and k is the Boltzman constant. (**b**) Scaling behavior of the depinning probability. $J_{\rm cm}$ is the percolation threshold while J_0 is scale parameter indicationg the half width of the $J_{\rm c}$ distribution. The *full curve* indicates (25) with $m = 3.5$. (**c**) Temperature dependence of $J_{\rm cm}$ in the vicinity of the percolation temperature $T_{\rm p}$

the J_c distribution. The universal curve shown in Fig. 13b by the full curve can be described by the Weibull function

$$Q(J) = 1 - \exp\left[-\left(\frac{J - J_{cm}}{J_0}\right)^m\right] S\left(\frac{J - J_{cm}}{J_0}\right) , \tag{25}$$

where $S(x)$ is the step function defined as $S(x) = 1$ for $x \geq 0$ and $S(x) = 0$ for $x < 0$, and the parameter m represents the shape of the statistic J_c distribution. As shown in Fig. 13c, simulation results indicate that the percolation threshold J_{cm} approaches zero in proportion to a power of $|1 - T/T_p|$ as T approaches T_p. It can be expressed with the aid of the critical index ν as

$$J_{cm} = J_t \, |1 - T/T_p|^{(D-1)\nu} , \tag{26}$$

where J_t is a numerical parameter and D is the dimensionality of FLL. In the present case D is 3 since the correlation length along the fluxoid is longer than the film thickness. These characteristics reflect the divergence of the correlation length among unpinned clusters [124] in a similar way to a spin glass as pointed out by *Fisher* [112]. Moreover, the relationship (26) is essentially the same as the scaling of the macroscopic pinning force density in the vicinity of the transition point. A detailed discussion about the scaling will be presented in Sect. 4.3.

Using the coherent potential approximation theory, *Matsushita* and *Kiss* showed theoretically that the depinning transition is second-order [127]. Since the GL transition was originally predicted to be the second order, this agreement is reasonable. It will be also shown in Sect. 4.3 that the isotherm scaling of the E–J characteristics can be described by thermal depinning. These facts show that the GL transition predicted by *Fisher* et al. as a kind of second-order phase transition can better be interpreted as the percolation transition of thermally depinning clusters, where the correlation length among the depinned vortices diverges. Namely, T_p is identical to T_{gl}. It is to be noted that the existence of the GL transition was proposed based on a model in which the effect of vortex pinning is not shown explicitly. Further, the model has been confirmed experimentally only by the behavior of the dc and ac E–J characteristics in film samples or tapes, for which the phase diagrams of the vortices are not yet known very clearly. On the other hand, thermal depinning phenomena are expected to occur whatever the configuration of pinned vortices is. Therefore, the observed transition of the E–J characteristics in film samples is possibly attributed to the thermal depinning transition. However, note that the theoretical origin of the glass state is different. The present model indicates that the pinning is more fundamental to determine the nature of the vortices.

4.2 Electric Field vs Current Density Characteristics

We have pointed out in Sect. 4.1 that the flux flow in HTSC is governed by the percolation process of unpinned clusters due to the random distribution

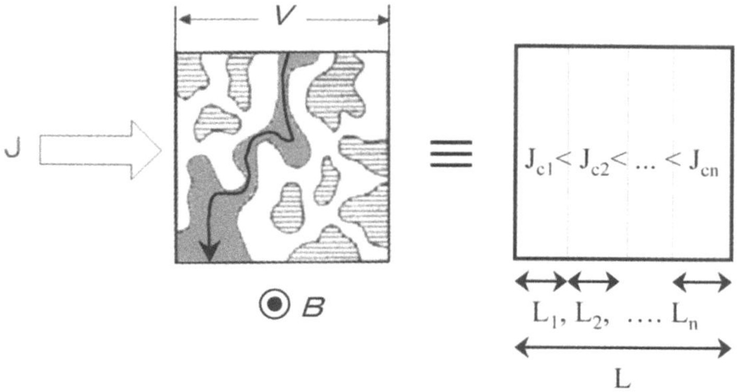

Fig. 14. Schematic diagram of the percolative flux flow and the equivalent series domain model. The length of each domain, L_i, is proportional to the probability density of J_c value in each domain, denoted by J_{ci}

of pinning strength and thermal fluctuation [124]. Voltage, V, is induced by the local plastic flow along a narrow percolation path even though most other clusters are still pinned. The macroscopic E–J characteristics of the system can be described by a series network which consists of domains having different values of J_c, where the value of J_c in each domain is denoted by J_{ci} as schematically shown in Fig. 14. The length of each domain, L_i, is proportional to the probability density function, P, of J_c in each domain (flux bundle rows), that is, $P(J_{ci}) = L_i/L$ where L is the total length between the voltage terminals.

Note that the flux flow region is localized in space and therefore the local electric field is much larger than the nominal value, V/L. If we compare the dynamic resistance, dE/dJ, in the vicinity of the percolation threshold to the saturated value in the uniform flux flow state obtained by applying high enough current density [128], the percolative flow region in the glass state is estimated to be of the order of 10^{-3} to 10^{-4} of the total length of the sample when the electric field is in the range around 10^{-4} V/m. Namely, the local electric field at the plastic flow region reaches as high as 10^{-1} to 1 V/m even if the apparent electric field is only around 10^{-4} V/m. Therefore, it would be a reasonable assumption that the elementary E–J characteristic in each domain, denoted by $E_e(J, J_c)$, can be expressed simply by the flux flow as long as we consider the typical transport E–J data observed by the four probe method.

$$E_e(J, J_c) = \begin{cases} \rho_{ff}(J - J_c), & \text{for} \quad J > J_c \\ 0, & \text{for} \quad J \le J_c \end{cases} \tag{27}$$

where ρ_{ff} is the flux flow resistivity in the uniform flow. Then the macroscopic E–J characteristics can be described based on the statistical distribution of J_c,

$$E(J) \equiv \frac{V(J)}{L} = \frac{1}{L}\sum_{i=1}^{n} L_i E_e(J, J_{ci}) = \int_{J_{cm}}^{\infty} P(J_c)E_e(J, J_c)dJ_c \,. \tag{28}$$

The statistical probability density function $P(J_c)$ in the random system can be described by the derivation of the Weibull function [124].

$$P(J_c) \equiv \frac{dQ}{dJ_c} = \begin{cases} \frac{m}{J_0}\left(\frac{J_c - J_{cm}}{J_0}\right)^{m-1} \exp\left[-\left(\frac{J_c - J_{cm}}{J_0}\right)^m\right], & J_c \geq J_{cm} \\ 0, & J_c < J_{cm} \end{cases} \tag{29}$$

In the vicinity of the percolation threshold $P(J_c)$ can be approximated by a simple power function. Therefore, the E–J characteristics can be written in analytical form as follows [124].

$$E(J) = \frac{\rho_{\text{ff}}}{m+1} J \left(\frac{J}{J_0}\right)^m \left(1 - \frac{J_{cm}}{J}\right)^{m+1}, \quad \text{for glass} \tag{30}$$

$$E(J) = \frac{\rho_{\text{ff}}}{m+1} |J_{cm}| \left(\frac{|J_{cm}|}{J_0}\right)^m \left[\left(1 + \frac{J}{|J_{cm}|}\right)^{m+1} - 1\right], \quad \text{for liquid} \tag{31}$$

In Fig. 15, we show an example of a comparison between measured E–J characteristics and the analytical expression shown by (30) and (31), where ρ_{ff} and m have constant values; 10 μΩ cm and 5.4, respectively.

4.3 Scaling Laws of Pinning and Iso-therm Scaling of *E*–*J*

In the case of practical low T_c superconducting materials, the J_c distribution is very sharp because the influence of thermal agitation is negligible compared to the strength of vortex pinning. It corresponds to the limit of $J_0 \rightarrow 0$ and $J_c \cong J_{cm}$ in the present model. Moreover, J_c is thought of as a physical quantity which can be attributed to the upper critical field, B_{c2}. Then the B, T-dependence of J_c can be described by the scaling of the macroscopic pinning force density $F_p \equiv J_c B$ as a function of the reduced magnetic field B/B_{c2} [129,130]

$$F_p \equiv J_c B = A B_{c2}(T)^{\varsigma} \left(\frac{B}{B_{c2}(T)}\right)^{\gamma} \left[1 - \left(\frac{B}{B_{c2}(T)}\right)\right]^{\delta}, \tag{32}$$

where A, ζ, γ, and δ are the pin parameters characterizing the pinning mboxmechanism.

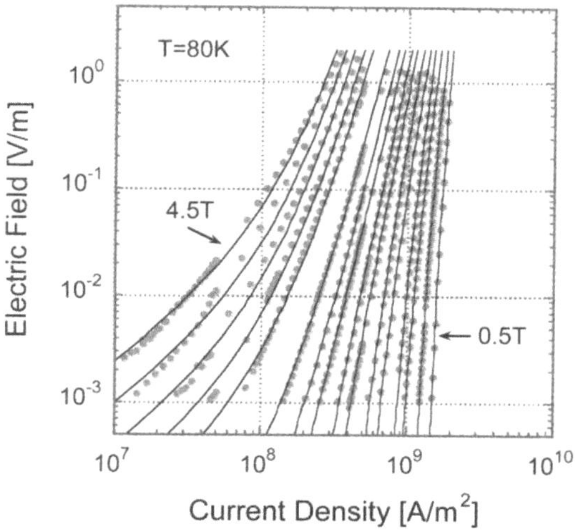

Fig. 15. Comparison between the percolation model (*solid lines*) and measured results (*dots*) in a *c*-axis oriented YBCO thin film

If we assume that the basic mechanism of the vortex pinning itself is essentially the same in HTSC, similar scaling will be valid for the pinning force density in HTSC [131,132]. However, note that, it will be necessary to take into account the significant influence of thermal agitation and the stochastic distribution of the elementary pinning strength. The J_c distribution can be characterized by two parameters; one is the minimum value of J_c denoted by J_{cm}, and the other is $J_k \equiv (J_0 + J_{cm})$ which is a measure of the average value of J_c. These parameters are expected to obey similar kinds of the scaling laws given by

$$\begin{aligned} F_{pm} \equiv J_{cm}B &= AB_{gl}(T)^{\varsigma}\left(\tfrac{B}{B_{gl}}\right)^{\gamma}\left(1-\tfrac{B}{B_{gl}}\right)^{\delta}, \qquad \text{for} \quad B \le B_{gl} \\ &\cong -AB_{gl}(T)^{\varsigma}\left(1-\tfrac{B}{B_{gl}}\right)^{\delta}, \qquad \text{for} \quad B > B_{gl} \end{aligned} \tag{33}$$

$$F_{pk} \equiv J_k B = AB_k(T)^{\varsigma}\left(\frac{B}{B_k}\right)^{\gamma}\left(1-\frac{B}{B_k}\right)^{\delta}, \qquad \text{for} \quad B \le B_k \tag{34}$$

where B_{gl} and B_k are the transition fields determined by the criteria $J_{cm} = 0$ and $J_k = 0$, respectively. The parameter sets may vary for F_{pm} and F_{pk} depending on the pinning properties.

It is important to consider the differences between B_{c2}, B_{gl} and B_k. In the case of practical low T_c superconductors, the depinning line is almost the same as that of B_{c2} to a good approximation because the thermal fluctuation is negligible compared to the pinning strength. As a result, J_c is a quantity that can be ascribed to B_{c2}. In the case of HTSC, on the other hand, the

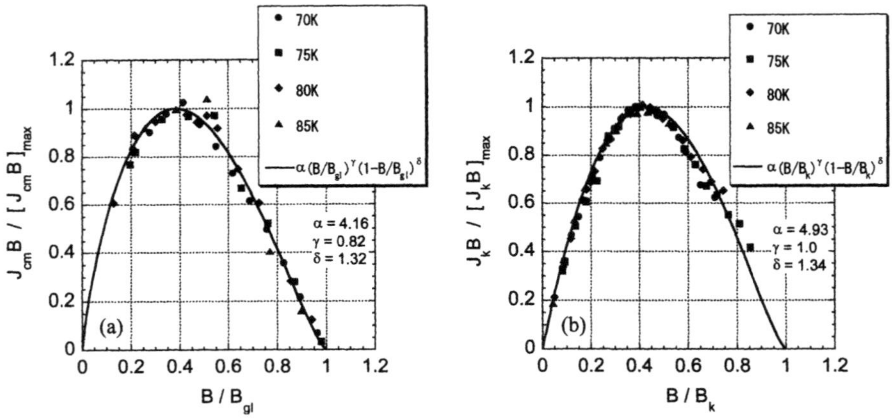

Fig. 16. Scaling of the pinning force density corresponding to (**a**) the minimum value of J_c denoted by J_{cm} and (**b**) the typical value of J_c denoted by J_k. The *solid lines* in (**a**) and (**b**) are (33) and (34), respectively

depinning line decreases significantly due to thermal fluctuations, and then the pinning strength distributes widely as can be seen in the results of Monte Carlo simulation mentioned in Sect. 4.1 Therefore, B_{c2} is replaced by B_{gl} and B_k.

As shown in Fig. 16, the scaling holds well in the measured range. The temperature dependence of the maximum values $F_{pm,max}$ and $F_{pk,max}$ are proportional to $B_{gl}(T)^{1.7}$ and $B_k(T)^{1.7}$, respectively. The temperature dependence of $B_{gl}(T)$ and $B_k(T)$ was proportional to $[1-(T/T_c)^2]^{1.5}$ in the measured temperature region from 70 K to 90 K. The extracted values at $T = 0$ are $B_{gl}(0) = 30$ T, and $B_k(0) = 75$ T. The values of B_{gl} and B_k might increase more significantly at low temperature less than about 30 K because of the *d*-wave symmetry of the order parameters [133]. Using the scaling, the critical surface in the YBCO film can be described by the critical current densities J_{cm} and J_k as shown in Fig. 17.

The analytical expressions (30) and (31) along with the scaling allows us to describe the E–J characteristics at arbitrary B and T. A comparison between the calculation results for the iso-J_c lines and measurements is shown in Fig. 18. As can be seen, the calculation results agree well with the measured results over a wide range of B and T.

The GL transition seems to be supported by the isotherm scaling of the observed E–J curves. According to the present percolation model, however, the same scaling foucan be derived [124]. The point in the present scenario is that the scaling indices will be influenced strongly by the pinning properties whereas they are supposed to follow a universal relationship in the phase transition theory [112]. If we rewrite (33) as a function of T in the vicinity of

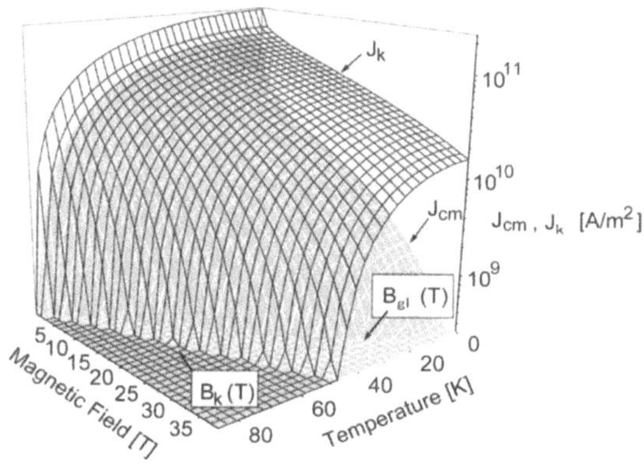

Fig. 17. Critical surface in YBCO thin film. The boundary between the superconductive state and the dissipative state cannot be separated clearly by a single surface because of the J_c distribution. However, it can be characterized by the minimum value and the typical value of J_c, i.e., J_{cm} and J_k

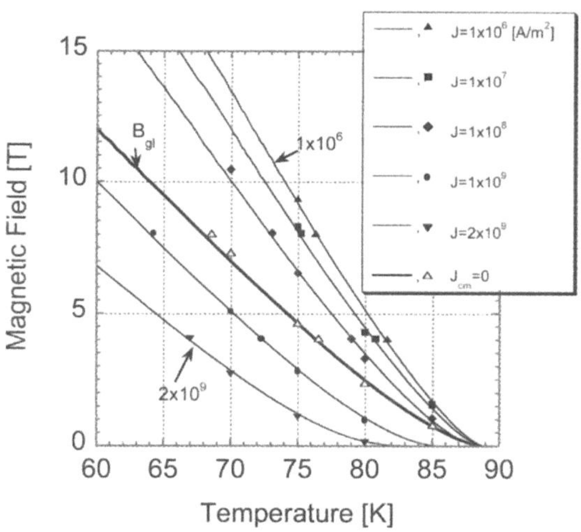

Fig. 18. Iso-J_c lines in the YBCO thin film shown in Figs. 15 to 17. The electric field criterion was 10^{-3} V/m. The *solid lines* are analytical expression whereas the *markers* are measured results

B_{gl}, we can see that ν is directly related to the minimum value of the pinning strength.

$$J_{cm} = J_t \left| 1 - \frac{T}{T_{gl}(B)} \right|^{\delta} , \tag{35}$$

$$J_t \approx A B_{gl} (T_{gl})^{\varsigma - 1} \left(\left| \frac{\partial T_{gl}}{\partial B} \right| \frac{B_{gl}}{T_{gl}} \right)^{-\delta} . \tag{36}$$

Therefore,

$$v = \frac{1}{2}\delta\,. \tag{37}$$

T_{gl} is the transition temperature where $B = B_{gl}$. From the power index of the E–J curve at the transition point, the dynamic index z can also be related to m according to

$$z = 2m + 1\,. \tag{38}$$

From the data on the magnetic field dependent E–J characteristics shown in Fig. 15 we can determine the parameters $\delta = 1.32$ and $m = 5.4$ using the percolation model given by (30) and (31). These results suggest that the critical indices will be $\nu = 0.66$ and $z = 11.8$. The iso thermal scaling of the E–J curves obtained in the same sample shown in Figs. 15 to 18 was shown in Fig. 19. The temperature dependent E–J curves were measured from 53.5 K to 74.2 K at a constant magnetic field of 8T. It can be seen that the critical indices agreed well with the prediction of the pin parameters. The relationship (37) has also been confirmed by the geometric effect on the scaling as will be described in Sect. 4.4. Notice that in the present analysis the value of z is extremely large to be explained within the framework of the phase transition theory [112]. These results strongly suggest that the scaling is influenced by the sharp distribution of the pinning strength in the present sample.

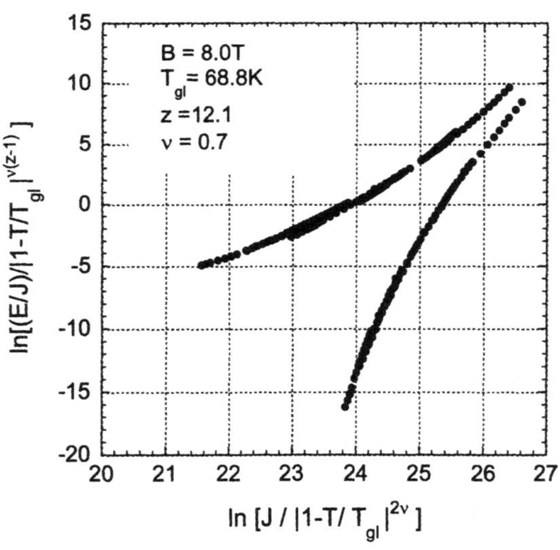

Fig. 19. Iso thermal scaling of the E–J characteristics obtained in the same sample shown in Figs. 15 to 18. The critical exponents z and ν agree well with the pin parameters along with the relationships (37) and (38)

4.4 Geometric Effect

When the sample width, W, limits the divergence of the correlation length, ℓ, among unpinned clusters, the transition temperature depends on the sample width according to

$$T_{gl}(W) = T_{gl,\infty}\left[1 - \left(\frac{\ell_0}{W}\right)^{\frac{1}{\nu}}\right], \tag{39}$$

where $T_{gl,\infty}$ is the transition temperature when the sample size is infinity. Therefore, we can directly confirm the divergence of the correlation length ℓ from the width dependence of the transition temperature. In the vicinity of the transition point, the same formulation with (24) and (39) holds for the magnetic field with the same critical index ν [125].

Using different width bridges formed in the same YBCO film deposited on a 1 cm by 1 cm $SrTiO_3$ (100) single crystal substrate, we examined the geometric effect. The ratio between the width, W, and the bridge length was about 2.5 for all cases, while the different W were 10 µm, 120 µm, and 1200 µm, respectively.

The value of T_c in the narrowest bridge was slightly higher than that in the widest bridge, but the difference between the bridges was less than 0.5 K. The average value of T_c was 87.8 K. In the low magnetic field, J_c in the bridges was almost the same or the narrowest bridge had a lightly larger value of J_c, whereas in the high magnetic field the J_c value in the narrowest bridge reduced more significantly than the wider bridges. We extracted J_{cm} from the extended E–J characteristics using the percolation model, and then compared the value of the transition field B_{gl}, where $J_{cm} = 0$. As the bridge width becomes narrower, B_{gl} reduced at all measured temperatures from 70 K to 85 K. Figure 20 shows the scaling of the width dependent B_{gl}. The W-dependence of B_{gl} is collapsed on the same line independent of temperatures as predicted by (39). From the measured results, the critical index ν is estimated to be 0.91. The magnetic field dependence of J_{cm} in the vicinity of B_{gl} is also shown in Fig. 21. J_{cm} approached zero according to a power of $1 - B/B_{gl}$ as shown by (33). Comparing the results of Figs. 20 and 21, we can see that the power index $\delta = 1.8$ agrees well with 2ν as predicted by (37). These results support the percolation model.

4.5 Summary

The dissipation mechanism in the mixed state of HTSC film has been discussed based on a study of vortex dynamics under the influence of vortex pinning and thermal fluctuation. It has been shown that the transport E–J characteristics are governed by a percolation process of thermally depinned vortex clusters and that the vortex glass-liquid transition is identical to the thermal depinning determined by the percolation of depinned clusters. The

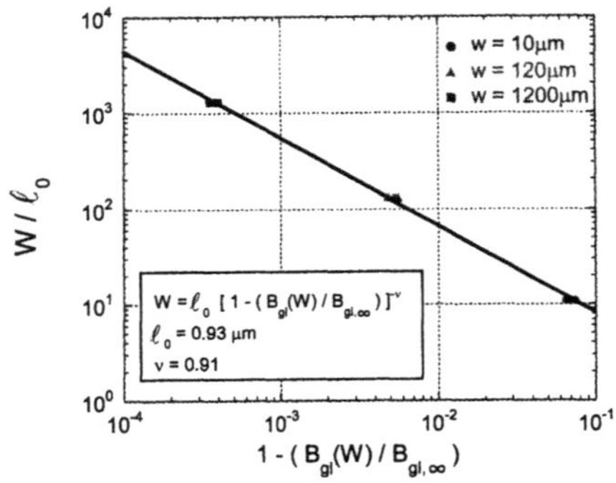

Fig. 20. Width dependence of the transition field $B_{\rm gl}$ in a YBCO thin film. The width dependence obtained in 10 μm-, 120 μm-, and 1200 μm-wide bridges at various temperature in the range of 70 K to 85 K is scaled as shown by (39)

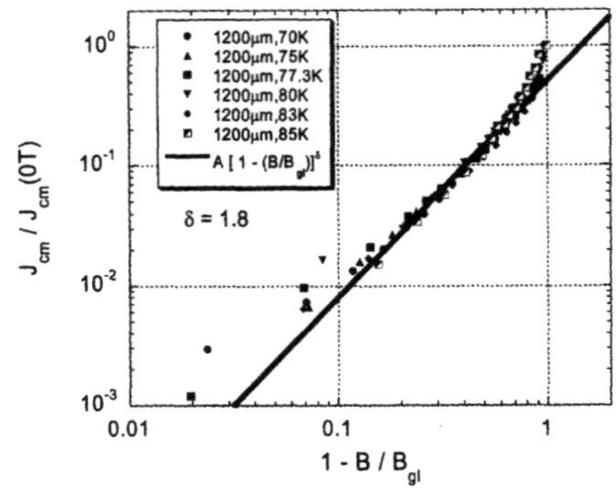

Fig. 21. Power law behavior of $J_{\rm cm}$ in the vicinity of $B_{\rm gl}$ in the widest bridge shown in Fig. 20. Note that the power index δ agrees with 2ν

iso therm scaling of the E–J characteristics obtained from the thermal depinning is the same to that of a thermodynamic phase transition. However, it is directly related to the nature of the pinning. It has been confirmed that the critical exponents agree with the pin parameters. An analytical expression for the E–J characteristics has been derived based on the statistical probability of the thermal depinning. The measured data agreed well with the calculation results over a wide range of temperature and magnetic field. The results of the width dependence of the transport E–J characteristics also support the percolation model.

5 Vortices in High-$T_{\rm c}$ Grain Boundary Junction

5.1 Direct Flux Detection Method

Among new techniques that have been developed to directly observe moving vortices in a type-II superconductor in the mixed state, *Yeh* and *Kao* [134] were the first to measure the fluctuations of magnetic flux associated with vortex motion by directly detecting flux changes using a superconducting pick-up coil placed on the surface of a low-$T_{\rm c}$ superconducting foil sample and magnetically coupled to the SQUID magnetometer. This detection scheme, called the *direct flux detection method* [134], offers a characterization technique to analyze the dynamical behavior of vortices moving in superconducting thin films.

Ferrari et al. [135] studied magnetic flux fluctuations in high-$T_{\rm c}$ superconducting thin films using the SQUID washer as a detection coil. They achieved a high magnetic coupling to the SQUID by placing the film close

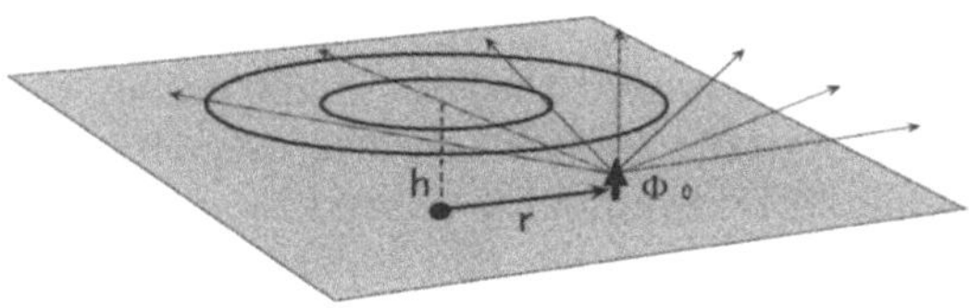

(a)

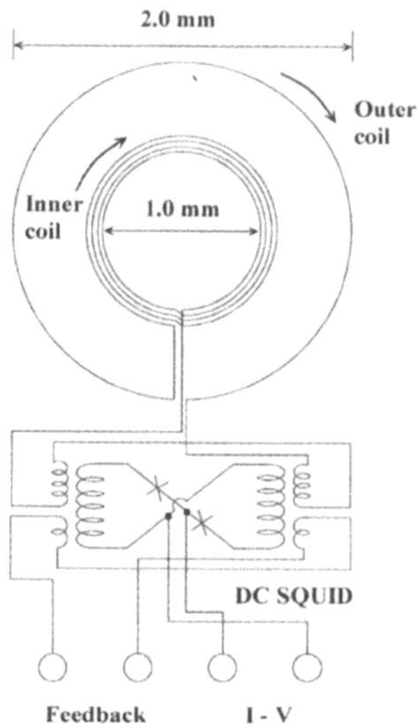

(b)

Fig. 22. (**a**) Schematic diagram of the basic arrangement of the direct flux detection method. (**b**) Equivalent circuit of the integrated dc SQUID/gradiometer

to the SQUID washer. Here we describe recent developments in the direct flux detection method using a thin film planar gradiometer and a dc SQUID integrated on a single chip [136]. The direct flux detection method has been applied to study vortex motion in high-T_c grain boundary junctions (GBJs), as discussed in the following sections.

The basic arrangement of the direct flux detection method is shown schematically in Fig. 22. Above the broad surface of a superconducting thin film sample a superconducting thin-film pickup coil is closely placed, as illustrated in Fig. 22a. This pick-up coil constitutes a flux transformer that magnetically couples to a dc SQUID shown in Fig. 22b. The dc SQUID detects flux fluctuations associated with vortex motion in the sample. The key factor in achieving a high signal-to-noise ratio with the direct flux detection method is to configure the sample and the superconducting pick-up coil so that the monopole approximation holds. Above the surface of an infinitely large superconductor at a distance $r \gg \lambda$, where λ is the penetration depth, the magnetic field due to an isolated vortex is approximated by the field produced by a point monopole (monopole approximation) [137], as

$$B(r) = \frac{\mu_0}{4\pi} \frac{g}{r^2} \hat{r} , \tag{40}$$

where r is the distance from the origin at which the vortex is located, μ_0 is the permeability of free space and g is the magnetic charge of a monopole

$$g = \frac{2\Phi_0}{\mu_0} . \tag{41}$$

To reduce the ambient magnetic field, the thin-film pickup coil constitutes a concentric planar gradiometer [138] with one-turn outer and four-turn inner coils connected in series and wound in the opposite sense to each other (Fig. 22b). The dc SQUID also has the gradiometric configuration of double square washers [139].

A numerical simulation has been made for the magnetic flux coupled to the gradiometer [136] assuming a single vortex in an infinite superconducting film. The distance between the gradiometer and the film is 200 μm. A similar simulation is also made in the opposite limit where the magnetic field due to a single vortex is approximated by the field produced by a magnetic dipole made of two magnetic charges $\pm g$ at either end separated by a distance (20 nm) of the thickness of the superconducting thin film. Figure 23a and b show the calculated magnetic flux for the monopole and dipole approximations, respectively, as a function of the vortex position, r, from the center of the gradiometer. It is found that the magnetic flux coupled to the gradiometer is much larger, by a factor of 10^4, for the monopole than the dipole. Therefore, realization of the monopole approximation is essential in the direct flux detection method in order to obtain sizable signals.

Abrikosov vortices trapped in type-II superconducting thin films are assumed in the above simulations. The size of the vortex is of the order of the London penetration depth or effective penetration depth of $\lambda_{\text{eff}}(= \lambda^2/d)$ in thin film geometry, where d is the thickness of the film. The monopole approximation may also be valid for Josephson vortices existing in Josephson junctions or weak links since the size of the gradiometer is much larger than the size of the Josephson vortices of λ_{J} (Josephson penetration depth). *Clem* has shown [140] that the monopole approximation holds for a pancake vortex in high-T_{c} superconducting thin films.

5.2 Magnetic Flux From a Single Vortex in Slotted YBCO

The direct flux detection method has been applied [141] to a single-layer YBCO film (200 nm in thickness) that has wide grain boundaries. Figure 24

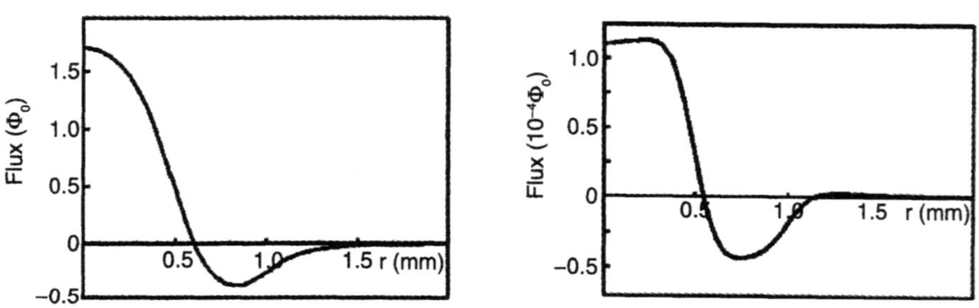

Fig. 23. Magnetic flux of a single vortex coupled to the gradiometer as a function of distance from the center of the gradiometer (**a**) in the monopole approximation and (**b**) in the dipole approximation

shows the geometry of the YBCO film and different regions for which the measurements were performed. Here, the washer-shaped film mimics a pickup coil of a directly coupled dc SQUID magnetometer [142]. The washer coil has two 3 mm wide grain boundary junctions (GBJ) on both sides along the bicrystal line of a $SrTiO_3$ substrate. The whole area of the coil, except for the vicinity of one of the two GBJs, is slotted to form a mesh structure with 5 mm wide YBCO films and slots with an area of 45 μm × 5 μm. The measured regions include three different areas of region I where there is no GBJ beneath the gradiometer, and regions II and III where the gradiometer covers unslotted and slotted GBJs at its center, respectively (right side of Fig. 24).

From the 5 μm linewidth of the superconducting film, trapping of vortices when the thin-film sample is cooled in a magnetic field would be suppressed below a threshold of 65 μT according to the formula $\pi\Phi_0/4w^2$ [143]. On the other hand, a cooling field of 9 μT is sufficient to incorporate a single flux quantum in the SQUID. One of the GBJs has slots of the same size as the mesh structure across the bicrystal line. The total number of slots along the GBJ is about 300. A parallel array of SQUIDs is formed along the slotted

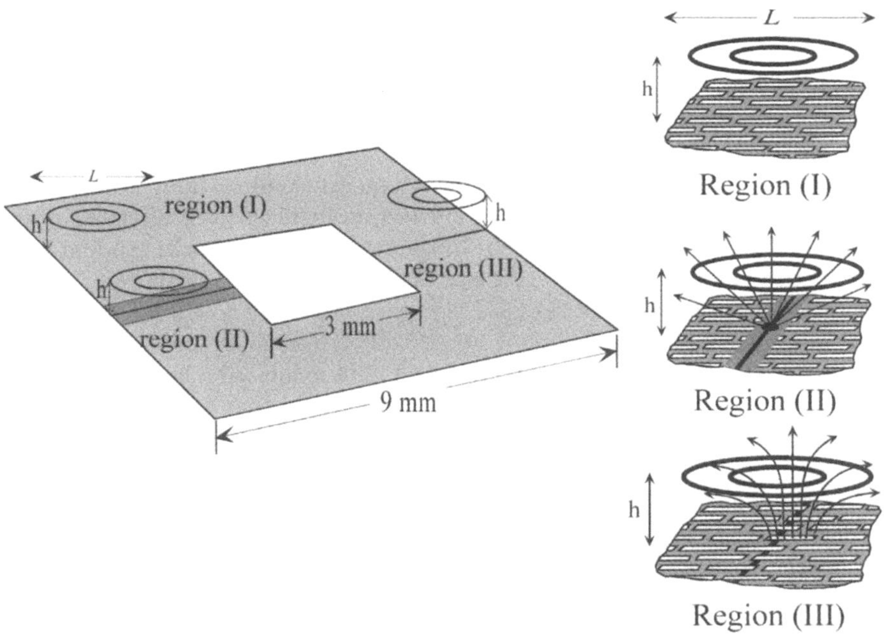

Fig. 24. YBCO thin film sample. The *straight lines* indicate the grain boundary junction (GBJs). *Circles* show the positions of the gradiometer in regions I, II, and III, which contain no GBJ, unslotted GBJ, and slotted GBJ, respectively, as indicated in enlarged figures. The gradiometer was placed at a liftoff of h. L is the length of a sensing area

GBJ, where the inductance parameter, $\beta_L (= 2\pi L I_c/\Phi_0)$ of each SQUID is estimated to be about 30. It is expected that a single flux quantum and also multi-flux quanta can stay stably in these SQUID arrays [144].

The supercurrent due to a flux quantum circulating around the rectangular slot in the YBCO film or grain boundary has a different distribution from that of a flux vortex within the superconducting film. In calculating the magnetic flux, which is coupled to the gradiometer, generated from a flux captured in a slot, a circular hole of radius, a, that has the same inductance as the rectangular slot is assumed for simplicity. Using the current distribution around the circular hole of [145]

$$K(r') = \frac{\Phi_0}{\mu_0 \pi r' \sqrt{r'^2 - a^2}}, \tag{42}$$

the total magnetic flux coupled to the gradiometer is obtained as $1.7\Phi_0$ at $r = 0$, which is almost equal to the case of the single vortex in the superconducting film. Therefore, the same profile of the coupled magnetic flux as that in Fig. 23a can be assumed for the flux quantum in the slot.

5.3 Long-distance Vortex Motion in Wide Grain Boundary Junction

Figure 25a shows the flux noise power spectrum, $S_\Phi(f)$, of regions I and II of the sample when it was cooled in a static magnetic field of $+10\,\mu$T and the magnetic field was changed by different amounts of ΔB_{ex} up to $-20\,\mu$T [141]. The $S_\Phi(f)$ of region I (indicated by A) on the bulk of film is about the same level as the system noise and showed no increase even after the maximum ΔB_{ex} of -20μT was applied. In the real-time trace, only random noise of peak-to-peak amplitude $0.1\Phi_0$ was observed. The flux noise $S_\Phi(f)$ of region II (B) over the unslotted GBJ is slightly higher at low frequencies than that of region I after field cooling at $+10\,\mu$T. It remained unchanged for the applied ΔB_{ex} below a threshold of $+1.4\,\mu$T, but exhibited a large Lorentzian-like power spectrum (C) at the threshold. With further increases in ΔB_{ex}, $S_\Phi(f)$ shows a rapid increase and a $1/f^2$-type spectrum (D). The value of S_Φ at 1 Hz was 10^4 larger than that after initial cooling. When the applied field was kept constant above the threshold, the flux noise decreased slowly by an order of magnitude in a few hours but did not decay further.

In the real-time trace of the flux for ΔB_{ex} at and above the threshold random switching noises were observed, as shown in Fig. 25b. Only a limited number of discrete heights, i.e., six at the threshold ΔB_{ex} (C) and nine above the threshold (D), appearing repeatedly in a range of 0.1–$1.3\Phi_0$. Among the six discrete switching heights at the threshold, the most frequently appearing height was $0.7\Phi_0$. For negative field change, similar behavior was observed where the threshold field was $-1.5\,\mu$T.

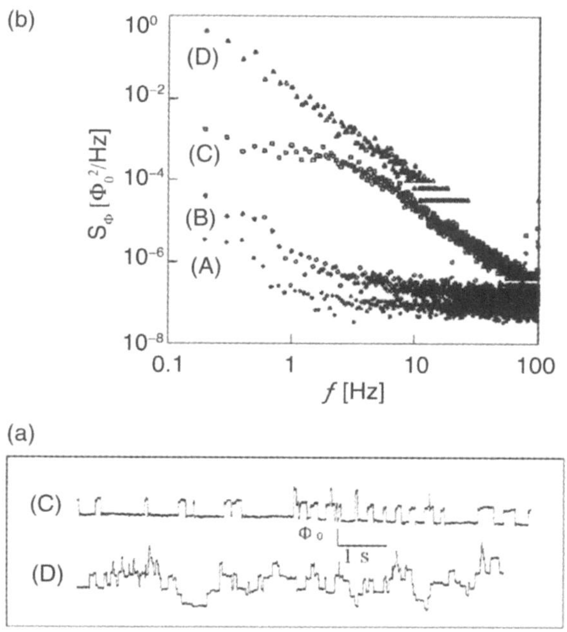

Fig. 25. (**a**) Flux noise power spectra of (A) the region I and (B)–(D) region II. Measurements were made after initial cooling at $+10\,\mu T$ in the case of (A) and (B), at the threshold $\Delta B_{ex} = +1.4\,\mu T$ in the case of (C), and above the threshold $\Delta B_{ex} = +1.5\,\mu T$ in the case of (D). (**b**) Real-time traces of the flux of region II for (C) $\Delta B_{ex} = +1.4\,\mu T$ and (D) $\Delta B_{ex} = +1.5\,\mu T$

Comparison between the flux noise power $S_\Phi(f)$ of region I and that of region II when ΔB_{ex} was applied (Fig. 25a) indicates that the major contribution to the low frequency flux noise arises from the GBJ. The threshold ΔB_{ex} of $+1.4\,\mu T$ and $-1.5\,\mu T$ for the appearance of the switching noise of region II may correspond to the fields at which random entry/exit of vortices is initiated at the edge of the GBJ. The random switching noises in the real-time trace at and above the threshold suggest uncorrelated hopping of vortices in one direction among several trapping sites in the GBJ. At the threshold, the observed six switching heights may correspond to the hopping processes among several trapping sites at six different distances. The maximum number of moving vortices was estimated to be three. Above the threshold, a larger number of vortices may be involved, where waveforms of up to nine switching heights are observed (Fig. 25b D).

The distances of vortex hopping in the unslotted GBJ are estimated from the switching heights by directly comparing them with the simulation of the flux coupled to the gradiometer (Fig. 23). At the threshold the observed switching heights are 0.1–$1.3\Phi_0$, which give hopping distances of 20 to 500 μm. Such long-distance hopping may result in significant contribution to the low frequency noise of directly coupled dc SQUID magnetometers when they include wide weak links in the pick-up loop.

The power spectral density, $S_{\Phi}^{\mathrm{GM}}(f)$, of discrete switching of the flux detected by the gradiometer is assumed to have a form [146]

$$S_{\Phi}^{\mathrm{GM}}(f) \approx n \sum_{i=1}^{m} \left(\frac{\delta \Phi_i^{\mathrm{GM}}}{\delta r_i} \right)^2 S_{\mathrm{r}}^i(f) , \tag{43}$$

where n is the average number of moving vortices, m is the number of hopping processes of vortices, $S_{\mathrm{r}}^i(f)$ is the power spectral density of the i-th hopping process and δr_i is the corresponding hopping distance. It is difficult to identify individual hopping processes from the observed real-time traces, but the dominant switching had roughly a height of Φ_0 at the threshold. Approximating the random switching to a single process with a height of $1.0\Phi_0$, the most frequent hopping distance was estimated to be 200 mm and S_{r} at 1 Hz was calculated to be about $2 \times 10^{-11}\,\mathrm{m^2/Hz}$ from the flux noise S_{Φ} at the threshold [(C) in Fig. 25a].

Nucleation and penetration of vortices into the GBJ occurs when the magnitude of the applied magnetic field at the edge reaches the lower critical field, B_{c1}, [147]. The B_{c1} of the GBJ is given by $B_{\mathrm{c1}} = (4\mu_0 J_{\mathrm{c}} \lambda_{\mathrm{J}})/\pi$, where J_{c} and λ_{J} are the critical current and Josephson penetration depth of the GBJ, respectively, and estimated to be about 300 μT from the values of $J_{\mathrm{c}} \sim 10^4\,\mathrm{A/cm^2}$ and $\lambda_{\mathrm{J}} \sim 2\,\mu\mathrm{m}$. However, the applied magnetic field is strongly enlarged due to the demagnetization effect in the thin-film geometry. The enlarged field at the film edge is given by [148]

$$B_{\mathrm{edge}} \approx \sqrt{\frac{w}{d}} B_{\mathrm{a}} , \tag{44}$$

where w and d are the width and the thickness of the film, respectively, and B_{a} is the net applied field. It is then found that B_{c1} is reached at a net applied field of only 1.4 μT in the geometry of the washer coil of Fig. 24. This threshold value is much lower than the static field (10 μT) applied to the washer coil during cooling. However, the slotted structure in the bulk of the thin film substantially reduces the demagnetizing effect for the cooling field. Therefore, it is expected that the field at the edge of the GBJ is roughly given by the sum of the initial cooling field and the enlarged edge field that is produced by the subsequent field change of ΔB_{ex} [149]. The sum of these fields may be responsible for the entry of vortices in the BGJ.

5.4 Suppression of Long-distance Vortex Motion in Grain Boundary Junctions

Figure 26a and b shows the power spectrum and real-time trace of flux noise, respectively, observed from the region III with a slotted GBJ that had been

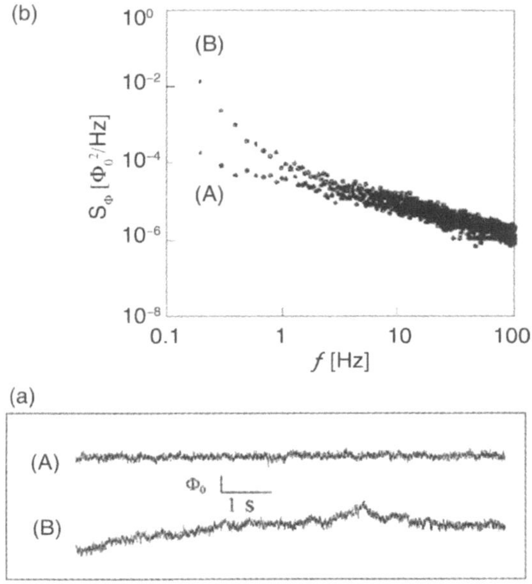

Fig. 26. (**a**) Power spectra of the magnetic flux of region III after initial cooling at $+10\,\mu$T (A), and at $\Delta B_{\text{ex}} = -20\,\mu$T (B). (**b**) Real-time traces of the flux of region III after initial cooling (A) and for $\Delta B_{\text{ex}} = -20\,\mu$T (B)

cooled in a static magnetic field of $+10\,\mu$T. Since the magnetic flux coupled to the gradiometer of a flux quantum in the slot is roughly the same as that of a vortex in the superconducting film, the detected flux noise can be directly compared between the regions II without and III with slots. Then, it is found that the slotted GBJ has stronger (30 times in power) flux noise (A in Fig. 26a) than the unslotted GBJ after initial field cooling. This excess flux noise may be caused by thermally activated flux motion of trapped flux quanta in the slots, which will be discussed later. When a large field change ($\Delta B_{\text{ex}} = -20\,\mu$T) was applied after $+10\,\mu$T field cooling, the flux noise power (B in (a)) showed only a moderate increase below a few Hz. The power spectrum is approximately a $1/f^\alpha$ type, where α is roughly equal to 0.9. Real-time traces (Fig. 26b) at $\Delta B_{\text{ex}} = 0\,\mu$T (A) and $-20\,\mu$T (B) have irregular waveforms without any switching noise.

These results suggest that long-distance vortex motion in the GBJ was suppressed by the slots, i.e., array of SQUIDs. Since it is difficult to estimate the vortex displacement directly from the observed real-time traces, the power spectrum is used in the following calculation. Letting the flux change due to a displacement, δr, of a single vortex along the GBJ be $\delta\Phi^{\text{GM}} = (\delta\Phi^{\text{GM}}/\delta r)\delta r$, the power spectral density of the flux fluctuation can be expressed as [141]

$$\int S_\Phi^{\text{GM}}(f)df = \langle(\delta\Phi^{\text{GM}})^2\rangle \approx N \cdot \langle(\delta r)^2\rangle \cdot \int_L \left(\frac{\delta\Phi^{\text{GM}}(r)}{\delta r}\right)^2 dr\,, \tag{45}$$

where N is the average density of vortices in the GBJ, L is the length ($\sim 3\,\mathrm{mm}$) along the GBJ in the sensing area of the gradiometer and the bracket $\langle\rangle$ indicates the time average. The density N is estimated to be $7\times 10^4\,\mathrm{m}^{-1}$ for $B_{\mathrm{ex}} = 10\,\mu\mathrm{T}$, which corresponds to a total number of vortices in the GBJ $Nw = 210$. The value of the integral in (45) is estimated to be $\sim 1.5\times 10^4 \Phi_0^2/\mathrm{m}$ for $h = 200\,\mu\mathrm{m}$ from the numerical simulation in Fig. 23. The value of $\langle(\delta\Phi^{\mathrm{GM}})^2\rangle$ is given by the integral of measured $S_{\Phi}^{\mathrm{GM}}(f)$ in Fig. 26a over a measurement frequency range between 0.1 and 100 Hz. Finally, the root-mean-squared distance $\delta r_{\mathrm{rms}} \equiv \langle(\delta r)^2\rangle^{1/2}$ of the vortex movement is obtained as about 800 nm. Compared with the hopping distance of 20 to 500 μm and the most frequent hopping distance of 200 μm in the unslotted GBJ, this value of δr_{rms} confirms that the long-distance flux motion is substantially suppressed by the slots. Further, S_{r} at $f = 1\,\mathrm{Hz}$ is estimated to be about $1\times 10^{-13}\,\mathrm{m}^2/\mathrm{Hz}$, which is two orders of magnitude smaller than that of the unslotted GBJ.

5.5 $1/f$ Behavior of the Flux Noise of Slotted Grain Boundary Junctions

Figure 27a shows the spectrum of flux noise power of region III measured at different temperatures after the sample was cooled in a field of 10 μT [150]. The flux noise power at 61 K (A) is comparable to the detection system noise but becomes larger at higher temperatures [(B)–(E)] where the spectrum scales roughly as $1/f$ between 70 and 77 K. The real-time trace of the flux noise showed only random noise, such as that in Fig. 26b, without any switching noise at all temperatures. The temperature dependence of the flux noise power S_Φ at 1 Hz is summarized in Fig. 27b. It is found that $S_\Phi(1\,\mathrm{Hz})$ is proportional to e^T in the temperature range of 70–77 K. Above 77.8 K, the flux noise became irreproducible with a shift by almost two orders of magnitude (shaded region in Fig. 27b).

The fact that no switching noise appeared in the real-time trace indicates that there exists no long-distance vortex motion in the slotted GBJ at all temperatures. The e^T dependence of the flux noise at 70–77 K suggests that the flux noise is generated by thermally activated vortex motion among the slots along wide GBJ. Here the cooling field (10 μT) gives an average vortex-to-vortex distance of about 14 μm, suggesting that roughly 70% of 300 slots are occupied and 30% empty. According to the thermal activation model of dc SQUIDs [151], a single vortex in a slot can make a transition over the potential energy barrier to its neighboring slot when it is empty. Then, $1/f$ behavior of the flux noise may be well explained by the Dutta, Dimon, and Horn (DDH) model [152], which assumes that the $1/f$ noise is caused by a superposition of thermally activated random processes over an activation energy E. The characteristic time τ of the activation process at a temperature T is determined by

$$\tau = \tau_0 \exp(E/k_{\mathrm{B}}T)\,, \tag{46}$$

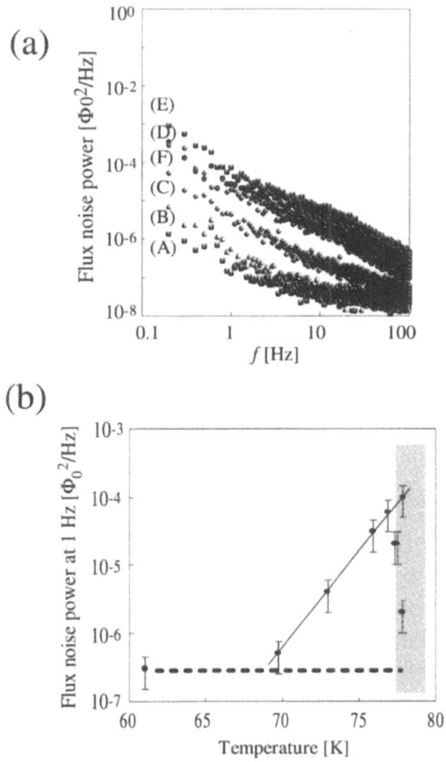

Fig. 27. (**a**) Flux noise power spectra of the slotted GBJ at temperatures (A) 61.0 K, (B) 69.7 K, (C) 73.0 K, (D) 75.9 K, (E) 76.9 K, and (F) 77.5 K. (**b**) Flux noise power at 1 Hz versus temperature. The *straight line* indicates the e^T dependence. The *dotted line* indicates the system noise level at 1 Hz and the *shaded area* shows the region where the flux noise was irreproducible

where τ_0 is an inverse attempt frequency. It is assumed that there is a distribution $D(E)$ of the activation energies, slowly varying over the energy range of $k_B T$.

Applying these models to the vortices in the slotted GBJ, the spectral density of the flux noise power is given by [153]

$$S_\Phi(f,T) \propto \int_0^\infty dED \left\{ \frac{4\tau_0 \exp(E/k_B T)}{1 + (2\pi f \tau_0)^2 \exp(2E/k_B T)} \right\} , \tag{47}$$

where E is the barrier height of the potential energy of a dc SQUID at zero bias current and τ_0 is the inverse of an attempt frequency of the order of the Josephson frequency, $f_J \approx I_c R_N / \Phi_0$, of the 5 µm GBJ. The term in the curly bracket { } is a sharply peaked function of E, centered at E_f given by

$$E_f = k_B T \ln \left(\frac{1}{2\pi f \tau_0} \right) . \tag{48}$$

If we assume that $D(E)$ is a slowly varying function compared with the energy width of the order of k_BT of the peaked function, we obtain a $1/f$ dependence of

$$S_\Phi(f,T) \propto \frac{k_BT}{f} D(E_f) \,. \tag{49}$$

The bandwidth (0.1–100 Hz) of the measurement of flux noise determines the range of potential energy barriers, E_f, under observation to be 130–180 meV. These values are smaller than the nominal Josephson coupling energy (200–300 meV) estimated for a 5 µm wide GBJ. Thus such slots that have low activation energies may contribute to the observed $1/f$ noise. Since a broad distribution function is assumed as $D(E)$ over about 300 slots in the GBJs, only a limited number of low-energy barrier slots may be responsible for the observed $1/f$ noise.

5.6 Flux Noise of Directly Coupled SQUID Magnetometers

In this section we estimate the magnitude of flux noise of a directly coupled dc SQUID magnetometer, which has a wide washer coil and GBJ (Fig. 24), using the observed values of the noise power of one-dimensional vortex motion. The vortex motion along the GBJ in the pickup coil of width w may give rise to the power spectral density of flux noise of [146]

$$S_\Phi(f) \approx \left(\frac{\alpha L_s}{L_p}\right)^2 (Nw) S_r(f) \frac{\Phi_0^2}{w^2} \,, \tag{50}$$

where α is the coupling constant and L_s and L_p are the inductance of the SQUID and the pickup coil, respectively. In (50) a crude approximation of $\delta\Phi \sim \Phi_0 \delta r/w$ is implicitly assumed [154]. Considering the case of unslotted GBJ under an applied field of threshold (Fig. 25a, 1.4 µT), the average number of moving vortices is of the order of one ($Nw = 1$) and $S_r(1\,\mathrm{Hz}) \sim 2 \times 10^{-11}\,\mathrm{m^2/Hz}$. Then we obtain $\Phi_n (= S_\Phi^{1/2}) \sim 3 \times 10^{-5} \Phi_0/\mathrm{Hz}^{1/2}$, where the values of $w = 3\,\mathrm{mm}$, $\alpha = 0.9$, $L_s = 100$ pH, $L_p = 5$ nH were used [146]. This Φ_n corresponds to a magnetic field noise B_n of about $150\,\mathrm{fT/Hz^{1/2}}$ at 1 Hz, assuming an effective area of about $0.4\,\mathrm{mm^2}$. Therefore, the long-distance motion of vortices in the GBJ may cause a slight increase in the low frequency field noise when a field change of threshold is applied. However, a field change (1.5 µT) slightly above the threshold increases the low frequency noise more than an order of magnitude (see Fig. 25a).

It is implied from (50) that the flux noise of the magnetometer increases as w is reduced. When w is comparable to or smaller than the hopping distance of vortices in the GBJ, it is likely that the vortices can pass across its width, driven by the Lorentz force of screening current, causing a strong low frequency noise. "Flux jump" observed in multiple-loop pickup coil magnetometers [155] may be due to such long-distance movement of single vortices across the flux dam, i.e., GBJ, inserted in the pickup loop.

The flux noise Φ_n of the same magnetometer but with slotted GBJ in the pickup coil can be estimated in a similar manner. We consider the case that the magnetometer is cooled in a magnetic field of 10 μT. Substituting $Nw = 210$ and $S_r(1\,\mathrm{Hz}) \sim 1 \times 10^{-13}\,\mathrm{m^2/Hz}$, for the slotted GBJ (50) gives $\Phi_n \sim 3 \times 10^{-5}\Phi_0/\mathrm{Hz}^{1/2}$ at 1 Hz. This value is equal to the flux noise of the magnetometer with unslotted GBJ at the threshold field-change of 1.4 μT. Above the threshold, however, the slots suppress the increase of low frequency flux noise by more than an order of magnitude from the noise of the magnetometer without slots (cf. Figs. 25a D and 26a B). Such suppression of the low frequency flux noise may be required in unshielded operation of the magnetometer, where tilting or displacement of the magnetometer in a spatially non-uniform field environment can readily reach the threshold field-change. The slotted structure of GBJ in the magnetometer would also be helpful to reduce the flux noise originating in the long-distance vortex motion that might otherwise occur in a wide GBJ in a noisy environmental field.

References

1. A. A. Abrikosov: Zh. Eksperim. Teor. Fiz. **32**, 1442 (1957) [Sov. Phys. JETP **5**, 1174 (1957)]
2. For physics of superconductivity, see, for example, M. Tinkham: *Introduction to Superconductivity* (McGraw-Hill, New York 1975)
3. J. Bardeen, M. J. Stephen: Phys. Rev. **140**, A1197 (1965)
4. V. L. Ginzburg, L. D. Landau: Zh. Eksperim. Teor. Fiz. **20**, 1064 (1950)
5. H. Trauble, U. Essmann: Phys. Stat. Sol. **25**, 395 (1968)
6. R. P. Huebener: *Magnetic Flux Structures in Superconductors*, Springer Ser. Solid State Sci. **6**, (Springer, Berlin, Heidelberg 1978)
7. A. I. Larkin, Y. N. Ovchinikov: J. Low. Temp. Phys. **34**, 409 (1979)
8. T. Nattermann: Phys. Rev. Lett. **64**, 2452 (1990)
9. T. Giamarchi, P. Le Doussal: Phys. Rev. B **55**, 6577 (1997)
10. T. Klein, I. Joumard, S. Blanchard, J. Marcus, R. Cubitt, T. Giamarchi, P. Le Doussal: Nature **413**, 404 (2001)
11. P. W. Anderson: Phys. Rev. Lett. **9**, 309 (1962)
    P. W. Anderson, Y. B. Kim: Rev. Mod. Phys. **36**, 39 (1964)
12. P. H. Kess, J. Aarts, J. van den Berg, C. J. van der Beek, J. A. Mydosh: Supercond. Sci. Technol. **1**, 242 (1989)
13. C. Caroli, P. G. de Gennes, J. Matricon: Phys. Lett. **9**, 307 (1964)
14. H. F. Hess, R. B. Robinson, R. C. Dynes, J. M. Valles, J. V. Waszczak: Phys. Rev. Lett. **62**, 214 (1989)
15. For a review, M. Golosovsky, M. Tsindlekht, D. Davidov: Supercond. Sci. Technol. **9**, 1 (1996)
16. G. E. Volovik: Pis'ma Zh. Eksp. Teor. Fiz. **58**, 457 (1993) [JETP Lett. **58**, 469 (1993)]
    N. Schopohl, K. Maki: Phys. Rev. B **52**, 490 (1995)
    M. Ichioka, Y. Hayashi, N. Enomoto, K. Machida: Phys. Rev. B **53**, 15316 (1996)
17. Y. Wang, A. H. MacDonald: Phys. Rev. B **52**, R3876 (1995)

18. For example, D. R. Tilley, J. Tilley: *Superfluidity and Superconductivity* (Hilger, New York 1990) Chap. 10.4
19. For a review, Y. Maeno: J. Supercond. **12**, 535 (1999)
20. G. Blatter, M. V. Feigel'man, V. B. Geshkenbein, A. I. Larkin, V. M. Vinokur: Rev. Mod. Phys. **66**, 1125 (1994)
21. A. I. Buzdin, D. Feinberg: J. Physique **51**, 1971 (1990)
S. N. Artemenko, A. N. Kruglov: Phys. Rev. Lett. A **143**, 485 (1990)
J. R. Clem: Phys. Rev. B **43**, 7837 (1991)
22. D. J. van Harlingen: Rev. Mod. Phys. **67**, 515 (1995)
23. L. I. Glazman, A. E. Koshelev: Phys. Rev. B **43**, 2835 (1991)
24. J. R. Clem, M. W. Coffey: Phys. Rev. B **42**, 6209 (1990)
25. M. Tachiki, S. Takahashi: Solid State Commun. **72**, 1083 (1989)
26. L. N. Bulaevskii, M. Ledvij, V. G. Kogan: Phys. Rev. B **46**, 362 (1992)
M. Benkraouda, M. Ledvij: Phys. Rev. B **51**, 6123 (1995)
A. E. Koshelev: Phys. Rev. Lett. **83**, 187 (1999)
27. C. A. Bolle, P. L. Gammel, D. G. Grier, C. A. Murray, D. J. Bishop, D. B. Mitzi, A. Kapitulnik: Phys. Rev. Lett. **66**, 112 (1991)
28. X. Hu, S. Miyashita, M. Tachiki: Phys. Rev. Lett. **79**, 3498 (1997); Phys. Rev. B **58**, 3438 (1998)
Y. Nonomura, X. Hu, M. Tachiki: Phys. Rev. B **59**, R11657 (1999)
29. Y. Iye, T. Tamegai, H. Takeya, H. Takei: Jpn. J. Appl. Phys. **26**, L150 (1987)
K. Kitazawa, S. Kambe, M. Naito, I. Tanaka, H. Kojima: Jpn. J. Appl. Phys. **28**, L555 (1989)
30. H. Pastoriza, M. F. Goffman, A. Arribere, F. ed la Cruz: Phys. Rev. Lett. **72**, 2951 (1994)
E. Zeldov, D. Majaer, M. Konczykowski, V. B. Geshkenbein, V. M. Vinokur, H. Shtrikman: Nature **375**, 373 (1995)
T. Hanaguri, T. Tsuboi, A. Maeda, T. Nishizaki, N. Kobayashi, Y. Kotaka, J. Shimoyama, K. Kishio: Physica C **263**, 434 (1996)
U. Welp, J. A. Fendrich, W. K. Kwok, G. W. Crabtree, B. W. Veal: Phys. Rev. Lett. **76**, 4809 (1996)
31. A. Schilling, R. A. Fisher, N. E. Phillips, U. Welp, D. Dasgupta, W. K. Kwok, G. W. Crabtree: Nature **382**, 791 (1996)
32. For a recent review, S. Battacharya, T. Giamarchi: pre-print, references cited therein
33. T. Shibauchi, T. Nakao, M. Sato, T. Kisu, N. Kameda, N. Okuda, S. Ooi, T. Tamegai: Phys. Rev. Lett. **83**, 1010 (1999)
M. B. Gaifullin, Y. Matsuda, N. Chikumoto, J. Shimoyama, K. Kishio: Phys. Rev. Lett. **84**, 2945–2948 (2000)
34. N. Avraham, B. Khaykovich, Y. Myasoedov, M. Rappaport, H. Shtrikman, D. E. Feldman, T. Tamegai, P. H. Kes, M. Li, M. Konczykowski, K. van der Beek, E. Zeldov: Nature **411**, 451 (2001)
35. R. Sugano, T. Onogi: Physica C **341–348**, 1113 (2000); Physica B **284–288**, 803 (2000)
Y. Nonomura, X. Hu: Phys. Rev. Lett. **86**, 5140 (2001)
36. See for instance, G. W. Crabtree, D. R. Nelson: Phys. Today **50**, 38 (1997)
A. Maeda, Y. Togawa, T. Hanaguri, H. Kitano: *Studies of High Temperature Superconductors* **41**, ed. by A. Narlikar (Nova Science, New York 2002) Chap. 10, pp. 247–274

37. G. Grüner: Rev. Mod. Phys. **60**, 1129 (1988); *Density Waves in Solids* (Addison-Wesley, Reading 1994)
38. E. Y. Andrei, G. Deville, D. C. Glattlie, F. I. B. Williams, E. Paris, B. Etienne: Phys. Rev. Lett. **60**, 2765 (1988)
39. V. J. Goldman, M. Santos, M. Shayengan, J. E. Cunningham: Phys. Rev. Lett. **65**, 2189 (1990)
40. Y. P. Li, T. Sajoto, L. W. Engel, D. C. Tsui, M. Shayegan: Phys. Rev. Lett. **67**, 1630 (1991)
41. Y. P. Li, D. C. Tsui, L. W. Engel, M. Santos, T. Sajoto, M. Shayegen: Solid State Commun. **96**, 379 (1995); ibid. **99**, 255 (1996)
42. C. C. Li, L. W. Engel, D. Shahar, D. C. Tsui, M. Shayegen: Phys. Rev. Lett. **79**, 1353 (1997)
43. Y. P. Li, D. C. Tsui, T. Sajoto, L. W. Engel, M. Santos, M. Shayegen: Solid State Commun. **95**, 619 (1995)
44. L. W. Engel, C. C. Li, D. Shahar, D. C. Tsui, M. Shayegen: Solid State Commun. **104**, 167 (1997)
45. H. Fukuyama, P. A. Lee: Phys. Rev. B **18**, 6245 (1978)
46. R. Chitra, T. Giamarchi, P. Le Doussal: Phys. Rev. Lett. **80**, 3827 (1998)
47. R. Seshadri, R. M. Westervelt: Phys. Rev. Lett. **70**, 234 (1993)
48. H. Matsukawa, H. Fukuyama: Phys. Rev. B **49**, 17286 (1994)
49. H. Pastoriza, M. F. Goffman, A. Arribere, F. ed la Cruz: Phys. Rev. Lett. **72**, 2951 (1994)
    E. Zeldov, D. Majaer, M. Konczykowski, V. B. Geshkenbein, V. M. Vinokur, H. Shtrikman: Nature **375**, 373 (1995)
    T. Hanaguri, T. Tsuboi, A. Maeda, T. Nishizaki, N. Kobayashi, Y. Kotaka, J. Shimoyama, K. Kishio: Physica C **263**, 434 (1996)
50. U. Welp, J. A. Fendrich, W. K. Kwok, G. W. Crabtree, B. W. Veal: Phys. Rev. Lett. **76**, 4809 (1996)
    A. Schilling, R. A. Fisher, N. E. Phillips, U. Welp, D. Dasgupta, W. K. Kwok, G. W. Crabtree: Nature **382**, 791 (1996)
51. X. Hu, S. Miyashita, M. Tachiki: Phys. Rev. Lett. **79**, 3498 (1997); ibid. Phys. Rev. B **58**, 3438 (1998)
    Y. Nonomura, X. Hu, M. Tachiki: Phys. Rev. B **59**, R11657 (1999)
52. R. Sugano, T. Onogi: Physica C **341–348**, 1113 (2000); Physica B **284–288**, 803 (2000)
    Y. Nonomura, X. Hu: Phys. Rev. Lett. **86**, 5140 (2001)
53. T. Nishizaki, T. Naito, N. Kobayashi: Phys. Rev. B **58**, 11169 (1998); Phys. Rev. B **61**, 3649 (2000)
    J. Low Temp. Phys. **117**, 1375 (1999); Physica C **341–348**, 957 (2000)
54. A. E. Koshelev, V. M. Vinokur: Phys. Rev. Lett. **73**, 3580 (1994)
55. L. Balents, M. P. A. Fisher: Phys. Rev. Lett. **75**, 4270 (1995)
    L. Balents, M. C. Marchetti, L. Radzihovsky: Phys. Rev. B **57**, 7705 (1998)
56. T. Giamarchi, P. Le Doussal: Phys. Rev. Lett. **76**, 3409 (1996); Phys. Rev. B **55**, 6577 (1997)
    P. Le Doussal, T. Giamarchi: Phys. Rev. B **57**, 11356 (1998)
57. L. Balents, M. C. Marchetti, L. Radzihovsky: Phys. Rev. Lett. **78**, 751 (1997)
    T. Giamarchi, P. Le Doussal: Phys. Rev. Lett. **78**, 752 (1997)
58. H. Jensen, A. Brass, A. J. Berlinsky: Phys. Rev. Lett. **60**, 1676 (1988)
59. K. Moon, R. T. Scalettar, G. T. Zimányi: Phys. Rev. Lett. **77**, 2778 (1996)

60. D. Domínguez, N. Grønbech-Jensen, A. R. Bishop: Phys. Rev. Lett. **78**, 2644 (1997)
61. S. Ryu, M. Hellerqvist, S. Doniach, A. Kapitulnik, D. Stroud: Phys. Rev. Lett. **77**, 5114 (1996)
62. C. J. Olson, C. Reichardt, F. Nori: Phys. Rev. Lett. **81**, 3757 (1998)
63. P. Martinolli, O. Daldini, C. Leemann, B. van den Brandt: Phys. Rev. Lett. **36**, 382 (1976)
64. A. T. Fiory: Phys. Rev. Lett. **27**, 501 (1971)
65. J. M. Harris, N. P. Ong, R. Gagnon, L. Taillefer: Phys. Rev. Lett. **74**, 3684 (1995)
66. A. M. Troyanovski, J. Aarts, P. H. Kes: Nature **399**, 665 (1999)
67. Y. Togawa, R. Abiru, K. Iwaya, H. Kitano, A. Maeda: Phys. Rev. Lett. **85**, 3716 (2000)
68. T. Tsuboi, T. Hanaguri, A. Maeda: Phys. Rev. Lett. **80**, 4550 (1998)
69. T. Tsuboi, T. Hanaguri, A. Maeda, R. Abiru, K. Iwaya, H. Kitano: Physica B **284–288**, 843 (2000)
70. A. Maeda, T. Tsuboi, T. Hanaguri, Y. Togawa, R. Abiru, Y. Tsuchiya, K. Iwaya: J. Low Temp. Phys. **117**, 1329 (1999)
71. A. Maeda, T. Tsuboi, T. Hanaguri, Y. Togawa, R. Abiru, Y. Tsuchiya, K. Iwaya: Phys. Rev. B **65**, 54506 (2002)
72. Y. Togawa, R. Abiru, K. Iwaya, H. Kitano, A. Maeda: Physica C **341–348**, 1215-1216 (2000)
73. Y. Togawa, H. Kitano, A. Maeda: J. low Temp. Phys. **131**, 907 (2003)
74. S. Battacharya, M. J. Higgins: Phys. Rev. Lett. **70**, 2617 (1993); Phys. Rev. B **52**, 64 (1995)
75. W. Henderson, E. Y. Andrei, M. J. Higgins, S. Battacharya: Phys. Rev. Lett. **77**, 2077 (1996); ibid. **81**, 2352 (1998)
Z. L. Xiao, E. Y. Andrei, M. J. Higgins: Phys. Rev. Lett. **83**, 1644 (1999); ibid. **85**, 3265 (2000); ibid. **86**, 2431 (2001)
76. J. A. Fendrich, U. Welp, W. K. Kwok, A. E. Koshelev, G. W. Crabtree, B. W. Veal: Phys. Rev. Lett. **77**, 2073 (1996)
77. T. Tsuboi, T. Hanaguri, A. Maeda: Phys. Rev. B **55**, R8709 (1997)
78. U. Yaron, P. L. Gammel, D. A. Huse, R. N. Kleiman, C. S. Oglesby, E. Bucher, B. Batlogg, D. J. Bishop, K. Mortensen, K. N. Clausen: Nature **376**, 753 (1995)
79. T. Matsuda, K. Harada, H. Kasai, O. Kamimura, A. Tonomura: Science **271**, 1393 (1996)
80. F. Pardo, F. de la Cruz, P. L. Gammel, E. Bucher, D. J. Bishop: Nature **396**, 348 (1998)
81. For a review of the noise study in conventional superconductors, see for example, J. R. Clem: Phys. Rep. **75**, 1 (1981)
82. A. C. Marley, M. J. Higgins, S. Battacharya: Phys. Rev. Lett. **74**, 3029 (1995)
83. R. D. Merithew, M. W. Rabin, M. B. Weissman, M. J. Higgins, S. Battacharya: Phys. Rev. Lett. **77**, 3197 (1996)
84. G. D'Anna, P. L. Gammel, H. Safar, G. B. Alers, D. J. Bishop, J. Giapintzakis, D. M. Ginsberg: Phys. Rev. Lett. **75**, 3521 (1995)
85. H. Safar, P. L. Gammel, D. A. Huse, G. B. Alers, D. J. Bishop, W. C. Lee, J. Giapintzakis, D. M. Ginsberg: Phys. Rev. B **52**, 6211 (1995)
86. W. J. Yeh, L. K. Yu, M. Yang, L. W. Song, Y. H. Kao: Physica C **195**, 367 (1992); Phys. Rev. B **44**, 360 (1991)

87. Y. Paltiel, E. Zeldov, Y.N. Myasoedov, H. Shtrikman, S. Battacharya, M.J. Higgins, Z.L. Xiao, E.Y. Andrei, P.L. Gammel, D.J. Bishop: Nature **403**, 398 (2000); Phys. Rev. Lett. **85**, 3712 (2000)
88. S.N. Gordeev, P.A.J. de Groot, M. Oussena, A.V. Volkozub, S. Pinfold, R. Langan, R. Gangon, L. Taillefer: Nature **385**, 324 (2000); Phys. Rev. B **58**, 527 (1998)
89. C. Caroli, P. G. de Gennes, J. Matricon: Phys. Lett. **9**, 307 (1964)
90. I. Maggio-Aprile, Ch. Renner, A. Erb, E. Walker, O. Fischer: Phys. Rev. Lett. **75**, 2754 (1995)
91. Ch. Renner, B. Ravaz, K. Kadowaki, I. Maggio-Aprile, O. Fischer: Phys. Rev. Lett. **80**, 3606 (1998)
92. S.H. Pan, E.W. Hudson, A.K. Gupta, K.W. Ng, H. Eisaki, S. Uchida, J.C. Davis: Phys. Rev. Lett. **85**, 1536 (2000)
93. N.B. Kopnin, V.E. Kravtsov: Pis'ma Zh. Eksp. Teor. Fiz. **23**, 631 (1976) [Sov. Phys. JETP Lett. **23**, 578 (1976)]
94. M.S. Pambianchi, D.H. Wu, L. Ganapathi, S.M. Anlage: IEEE Trans. Appl. Supercond. **3**, 2774 (1993)
95. D.C. Morgan, K. Zhang, D.A. Bonn, R. Liang, W.N. Hardy, C. Kallin, A.J. Berlinsky: Physica C **235-240**, 2015 (1994)
96. Y. Matsuda, N.P. Ong, Y.F. Yan, J.M. Harris, J.B. Peterson: Phys. Rev. B **49**, 4380 (1994)
97. J.I. Gittleman, B. Rosenblum: Phys. Rev. Lett. **16**, 734 (1966)
98. S. Revenaz, D.E. Oates, D. Labb'e-Lavigne, G. Dresselhaus, M.S. Dresselhaus: Phys. Rev. B **50**, 1178 (1994)
99. M. Golosovsky, M. Tsindlekht, H. Chayet, D. Davidov: Phys. Rev. B **50**, 470 (1994)
100. B. Parks, S. Spielman, J. Orenstein, D.T. Nemeth, F. Ludwig, J. Clarke, P. Marchant, D.J. Lew: Phys. Rev. Lett. **74**, 3265 (1995)
101. Y. Tsuchiya, K. Iwaya, K. Kinoshita, T. Hanaguri, H. Kitano, A. Maeda, K. Shibata, T. Nishizaki, N. Kobayashi: Phys. Rev. B **63**, 184517 (2001)
102. T. Hanaguri, T. Tsuboi, Y. Tsuchiya, K. Sasaki, A. Maeda: Phys. Rev. Lett. **82**, 1273 (1999)
103. A.V. Balatsky, P. Kumar, J.R. Schrieffer: Phys. Rev. Lett. **84**, 4445 (2000)
104. M.W. Coffey, J.R. Clem: Phys. Rev. Lett. **67**, 386 (1991)
105. A. Himeda, M. Ogata, Y. Tanaka, S. Kashiwaya: J. Phys. Soc. Jpn. **66**, 3367 (1997)
M. Ogata: Int. J. Mod. Phys. B **13**, 3560 (1999)
J.H. Han, D.H. Lee: Phys. Rev. Lett. **85**, 1100 (2000)
106. B. Lake, G. Aeppli, K.N. Clausen, D.F. McMorrow, K. Lefmann, N.E. Hussey, N. Mangkorntong, M. Nohara, H. Takagi, T.E. Mason, A. Schroder: Science **291**, 1759 (2001)
107. K. Kakuyanagi, K. Kumagai, Y. Matsuda: Phys. Rev. B **65**, 060503 (2002)
108. V.F. Mitrovic, E.E. Sigmund, M. Eschrig, H.N. Bachman, W.P. Halperin, A.P. Reyes, P. Kuhns, W.G. Moulton: Nature **413**, 501 (2001)
109. J.H. Hoffman, E.W. Hudson, K.M. Lang, V. Madhaven, S.H. Pan, H. Eisaki, S. Uchida, J.C. Davis: Science **295**, 466 (2002)
110. G. Blatter, M.V. Feigelman, V.B. Geshkenbein, A.I. Larkin, V.M. Vinokur: Rev. Mod. Phys. **66**, 1125 (1994)
111. P.L. Gammel, L.F. Schneemeyer, J.V. Waszczak, D.J. Bishop: Phys. Rev. Lett. **61**, 1666 (1988)

112. M. P. A. Fisher: Phys. Rev. Lett. **62**, 1415 (1989)
113. D. S. Fisher, M. P. A. Fisher, D. A. Huse: Phys. Rev. B **43**, 130 (1991)
114. A. T. Dorsay, M. P. A. Fisher: Phys. Rev. B **68**, 694 (1992)
115. Q. Li, H. J. Weismann, M. Suenaga, L. Motowidlo, P. Haldar: IEEE Trans. Appl. Superconductivity **5**, 1713 (1995)
116. H. Yamasaki, K. Endo, Y. Mawatari, S. Kosaka, M. Umeda, S. Yoshida, K. Kajimura: IEEE Trans. Appl. Superconductivity **5**, 1888 (1005)
117. T. Nabatame, O. B. Hyun, H. Suhara, S. Koike, N. Nakamura, I. Hirabayashi: Advances in Superconductivity **VII**, 517 (Springer, Tokyo 1995)
118. A. T. Dorsay, M. Huang, M. P. A. Fisher: Phys. Rev. B **45**, 523 (1992)
119. S. Misat, P. J. King, D. Fuchs, J. C. Villegrier, R. P. Campion, P. S. Czerwinka: Physica C **331**, 241 (2000)
120. Y. Ando, H. Kubota, S. Tanaka: PRL **69** (19), 2851 (1992)
121. T. Kiss, T. Matsushita, F. Irie: SuST **12**, 1079 (1999)
122. P. S. Czerwinka, R. P. Campion, K. F. Horbelt, P. J. King, S. Misat, S. M. Morley, H.-U. Habermeier, B. Leibold: Physica C **324**, 96 (1999)
123. K. Yamafuji, T. Fujiyoshi, K. Toko, T. Matsuno, K. Kishio, T. Matsushita: Physica C **212**, 424 (1993)
124. K. Yamafuji, T. Kiss: Physica C **258**, 197 (1996)
125. K. Yamafuji, T. Kiss: Physica C **290**, 9 (1997)
126. M. Kiuchi, H. Yamamoto, T. Matsushita: Physica C **269**, 242 (1996)
127. T. Matsushita T. Kiss: Physica C **315**, 12 (1999)
128. M. N. Kunchur: Mod. Phys. Lett. B **9**, 399 (1995)
129. A. M. Campbell, J. E. Evetts: Adv. Phys. **21**, 199 (1972)
130. T. Matushita, J. W. Ekin: *Composite Superconductors* (Marcel Dekker, New York 1994)
131. S. Awaji, K. Watanabe, N. Kobayashi, H. Yamane, T. Hirai: Proc. 8th IWCC, 183 (1997)
132. J. D. Hettinger, A. G. Swanson, W. J. Skocpol, J. S. Brooks, J. M. Graybeal, P. M. Mankiewich, R. E. Howard, B. L. Stroughn, E. G. Burkhardt: Phys. Rev. Lett. **62**, 2044 (1989)
133. S. Awaji, K. Watanabe: Jpn. J. Appl. Phys. **40**, L1022 (2001)
134. W. J. Yehl, Y. H. Kao: Phys. Rev. B **44**, 360 (1991)
W. J. Yeh, Y. H. Kao: Phys. Rev. Lett. **53**, 1590 (1984)
135. M. J. Ferrari, M. Johnson, F. C. Wellstood, J. Clarke, P. A. Rosenthal, R. H. Hammond, M. R. Beasley: Appl. Phys. Lett. **53**, 695 (1988)
M. Johnson, M. J. Ferrari, F. C. Wellstood, J. Clarke, M. R. Beasley, A. Inam, X. D. Wu, L. Nazar, T. Venkatesan: Phys. Rev. B **42**, 10792 (1990)
M. J. Ferrari, M. Johnson, F. C. Wellstood, J. J. Kingston, T. J. Shaw, J. Clarke: J. Low Temp. Phys. **94**, 15 (1994)
136. S. Hirano, Y. Hirata, S. Kuriki, M. Matsuda, T. Morooka, S. Nakayam: J. Appl. Phys. **85**, 7819 (1999)
137. J. R. Clem: Phys. Rev. B **1**, 2140 (1970)
138. S. Kuriki, Y. Isobe, Y. Mizutani: J. Appl. Phys. **61**, 781 (1987)
139. T. Morooka, S. Nakayama, A. Odawara, N. Shimizu, K. Chinone, T. Ataka, N. Kasai: Jpn. J. Appl. Phys. **35**, L486 (1996)
140. J. R. Clem: Phys. Rev. B **43**, 7837 (1991)
141. S. Hirano, H. Oyama, S. Kuriki, T. Morooka, S. Nakayama: Appl. Phys. Lett. **78**, 1715 (2001)

142. H. Oyama, S. Hirano, S. Kuriki, M. Matsuda: Supercond. Sci. Technol. **15**, 1 (2002)
143. E. Dantsker, S. Tanaka, P. A. Nilsson, R. Kleiner, J. Clarke: Appl. Phys. Lett. **69**, 4099 (1996)
E. Dantsker, S. Tanaka, J. Clarke: Appl. Phys. Lett. **70**, 2037 (1997)
H. M. Cho, R. Mc Dermott, B. Oh, K. A. Kouznetsov, A. Kittel, J. H. Miller Jr., J. Clarke: IEEE Trans. Appl. Supercond. **9**, 3294 (1999)
144. K. K. Likharev: IEEE Trans. Magn. **13**, 242 (1977)
K. K. Likharev: ibid. **13**, 245 (1977)
145. M. B. Ketchen, W. J. Gallagher, A. W. Kleinsasser, S. Murphy, J. R. Clem: in H. D. Hahlbohm, H. Lubbig (Eds.), *SQUID '85* (de Gruyter, Berlin 1985) p. 865
146. S. Hirano, H. Oyama, M. Matsuda, T. Morooka, S. Nakayama, S. Kuriki: IEEE Trans. Appl. Supercond. **11**, 924 (2001)
147. T. P. Orlando, K. A. Delin: *Foundations of Applied Superconductivity* (Addison-Wesley, Reading 1991) Chap. 8
148. E. Zeldov, J. R. Clem, M. Mc Elfresh, M. Darwin: Phys. Rev. B **49**, 9802 (1994)
M. Benkraouda, J. R. Clem: Phys. Rev. B **53**, 5716 (1996)
149. S. Hirano, H. Oyama, T. Morooka, S. Nakayama, S. Kuriki: Physica C **357-360**, 1455 (2001)
150. S. Hirano, H. Oyama, S. Kuriki, T. Morooka, S. Nakayama: Ext. Abst. 8th Int. Supercond. Electron. Conf. (ISEC 2001) 325, (2001)
151. C. D. Tesche: J. Low Temp. Phys. **44**, 119 (1981)
152. P. Dutta, P. Dimon, P. M. Horn: Phys. Rev. Lett. **43**, 646 (1979)
P. Dutta, P. M. Horn: Rev. Mod. Phys. **53**, 497 (1981)
153. M. J. Ferrari, M. Johnson, F. C. Wellstood, J. Clarke, D. Mitzi, P. A. Rosenthal, C. B. Eom, T. H. Geballe, A. Kapitulnik, M. R. Beasley: Phys. Rev. Lett. **64**, 72 (1990)
154. M. J. Ferrari, J. J. Kingston, F. C. Wellstood, J. Clarke: Appl. Phys. Lett. **58**, 1106 (1991)
155. F. Ludwig, D. Drung: Appl. Phys. Lett. **75**, 2821 (1999)

Observation of Vortices

S. Ohshima[1], K. Tanabe[2], T. Morishita[2], and M. Tonouchi[3]

[1] Faculty of Engineering, Yamagata University
4-3-16 Jonan, Yonezawa, Yamagata 992-8510, Japan
ohshima@eie.yz.yamagata-u.ac.jp
[2] Superconductivity Research Center,
International Superconductivity Technology Center
1-10-13 shinonome, Kotoku, Tokyo 135-0062, Japan
{tanabe,mrst}@istec.or.jp
[3] Research Center for Superconductor Photonics, Osaka University
2-1 Yamadaoka, Suita, Osaka 565-0871, Japan
tonouchi@rcsuper.osaka-u.ac.jp

Abstract. Various kinds of special techniques have been developed to visualize quantized magnetic flux in superconductors, i.e. vortex. This chapter reviews the experimental techniques and discusses the vortex distributions in high-T_c superconductors under various conditions. As currently available techniques, the Bitter method, the magneto-optic method, the scanning SQUID microscope, and terahertz radiation imaging are used in the study.

1 Introduction

In 1957 *Abrikosov* pointed out the existence of a secondtype of superconductor [1]. He calculated that the energy of the interface between the normal state and the superconducting state became negative in some superconductors. That is, there is a superconducting state that takes a lower energy than the Meissner state. The London penetration λ and the coherence length ξ determine whether the superconductor is of the first or the second type. (The first type exhibits perfect diamagnetic behavior up to H_c.) When a magnetic flux penetrates into the superconductor the value of the flux is quantized, and it is called a vortex. Abrikosov also found that vortices arranged in a triangular lattice had the smallest free energy compared with any other arrangement of vortices. (First he pointed out that vortices arranged in a square lattice had the smallest free energy.)

What kind of structure does a vortex in a superconductor have? The conceptual scheme of the vortex is shown in Fig. 1. The nucleus of the normal conducting region is at the center and the shielding current J_0 flows around the circumference. The size of the nucleus is about 2ξ and the shielding current region is about 2λ. Therefore, the size of a vortex in a high-T_c superconductor (HTS) is several hundred nm when the vortex is determined by the shielding current, and it is several nm when determined by the nucleus of the central normal conducting region.

T. Kobayashi et al. (Eds.): Vortex Electronics and SQUIDs,
Topics Appl. Phys. **91**, 53–102 (2003)

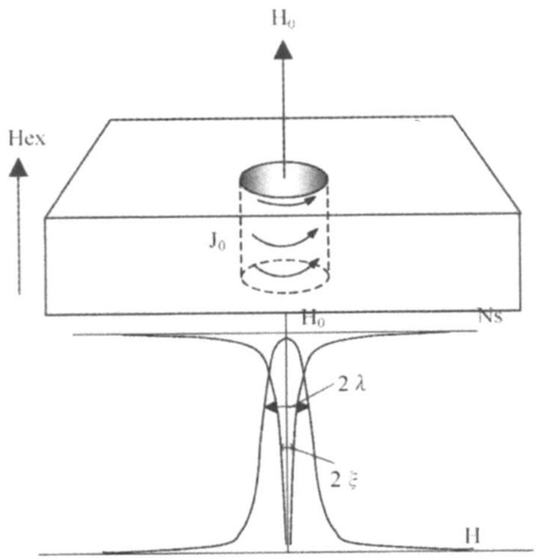

Fig. 1. Schematic drawing of a vortex in a superconductor. The nucleus of the normal conducting region, 2ξ is at the center, and the shielding current J_0 flows around the circumference

It was *Essmann* and *Träuble* who first observed vortices directly as shown in Fig. 2. They used the decoration by magnetic particles technique, which is used for observation of magnetic domains and is known as the Bitter method [2]. As the decoration was done by fine magnetic particles at very low temperature, they could not use the usual colloid used for magnetic-domain observation. They obtained the fine magnetic particles by deposition of iron in helium gas. *Essmann* and *Träuble* were also able to observe the intermediate state of superconducting lead [3]. Subsequently, the Bitter method was used as a powerful tool for the observation of vortices, and many experiments have been reported. When high temperature superconductors were discovered, the Bitter technique was already available for the observation of vortices.

Other techniques for observing vortices besides the Bitter method have also been reported. There are two kinds of technique; one is the observation of the shielding current and the other is the observation of the normal conducting nucleus.

The Lorentz microscope, SQUID microscope, magnetic force microscope (MFM), the magneto-optic method and the terahertz radiation effect are used for the former, while the scanning tunneling microscope (STM) is used for the latter. The Lorentz microscope was developed for magnetic domain observation. *Tonomura* et al. succeeded in the dynamic observation of vortices in HTS films [4]. They found that a hopping motion of the vortices occurred, and that the vortex was stable and formed a triangular lattice in the vicinity of T_c. MFM uses the tip of a cantilever coated with a ferromagnetic material. The space resolution for the MFM is very high. Therefore, it can be used to

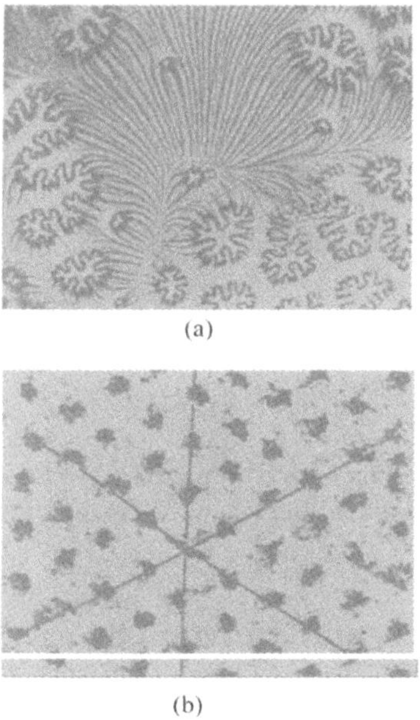

Fig. 2. Penetrated magnetic flux and vortex distribution of superconducting lead reported by *Essmann* and *Träuble* [1,2]. (a) Intermediate state of superconducting lead. *Black lines* show normal state regions. (b) A triangular lattice of superconducting lead. *Black points* are vortices

observe the structure of a vortex. It has been reported that a magnetic flux of $1\Phi_0$ and $1/2\Phi_0$ exists [5].

STM is also useful for observation of a vortex distribution. *Fisher* et al. were able to observe the vortex arrangement at less than 1.5 T [6]. The scanning SQUID microscope is also useful for observing vortex distributions in a low magnetic field [7]. Recently, terahertz radiation has been developed for the observation of vortices. *Tonouchi* et al. reported the observation of a vortex distribution by terahertz radiation imaging [8].

In this chapter, the experimental procedures and the results of vortex distributions measured by the Bitter method, the magneto-optic method, the scanning SQUID microscope and terahertz radiation imaging are described. The vortices in single crystalline $YBa_2Cu_3O_x$ (YBCO), YBCO thin films, single crystalline $Bi_2Sr_2CaCu_2O_x$ (BSCCO), quench-melt-growth (QMG) YBCO crystals, $Bi_2Sr_2Ca_2Cu_3O_x$ (Bi-2223) silver sheath tapes, polycrystalline MgB_2, and $NbSe_2$ single crystal are examined.

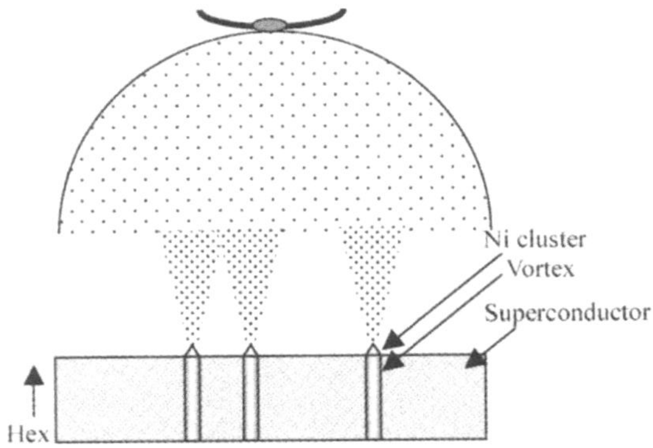

Fig. 3. Sketch drawing the main idea of the high-resolution Bitter method. Because a vortex behaves as a small magnet a hill is formed when fine magnetic clusters are condensed on it

2 Observation of a Vortex Pattern and Movement by the High-resolution Bitter Method

The main technique of the Bitter method is that fine magnetic particles are poured onto penetrated vortices as shown in Fig. 3. Because a vortex behaves as a small magnet, a hill is formed when fine magnetic particles are condensed on it. Therefore, the distribution and movement of the vortex can be traced by observing the pattern using a SEM, AFM or optical microscope.

2.1 Experimental Procedure

When observing a vortex by the high-resolution Bitter method it is necessary to pay careful attention to the next two points. (1) The fine magnetic particles should be as fine as possible. (2) The kinetic energy of the particles should be low enough to enable them to be caught by the vortices. Some schemes for doing this have been reported. *Gammell* et al. [9] used iron flash evaporation in helium gas at 1 atm and got fine magnetic particles with low kinetic energy. They succeeded in observing the vortex distribution of single crystal HTSs by this method. We reported that we could observe the vortex distribution and movement in HTS material by depositing nickel in a helium gas ambient. A schematic drawing of our equipment is shown in Fig. 4.

The collision of helium gas molecules with nickel particles produced fine magnetic particles with a low kinetic energy. The experimental conditions and procedures were as follows. (1) The chamber of the Bitter technique was evacuated, and then helium gases was introduced to a pressure of about 10 m Torr. (2) A fixed magnetic field was applied, and the sample was cooled by a mini-cryocooler. (3) The sample was cooled to a fixed temperature, and nickel was deposited. The nickel deposition time was 30 s. (4) After nickel

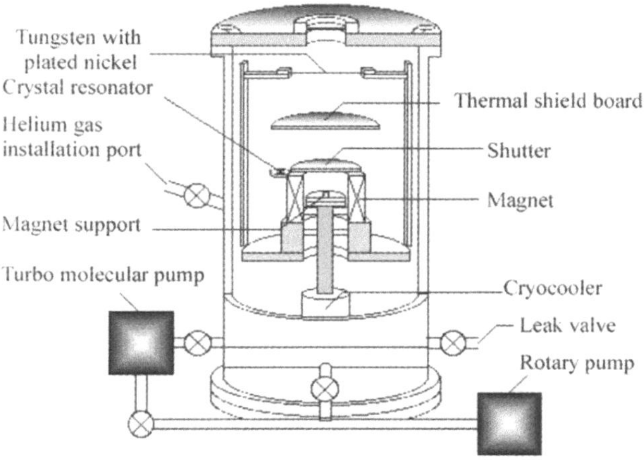

Fig. 4. Schematic drawing of high-resolution Bitter method equipment designed by *Ohshima* [14]

deposition, the applied magnetic field was reduced to zero and the sample was warmed to room temperature. Afterwards, the sample surface was observed by SEM, AFM and optical microscope.

In the following section, the vortex distribution and movement in HTS material observed by the high-resolution Bitter method are reported.

2.2 The Observation of a Static Vortex Distribution

2.2.1 YBCO Single Crystal

The high-resolution Bitter method is suitable for observation of the static vortex distribution. Therefore, many papers showing the vortex distribution of single crystal YBCO, BSCCO and $Tl_2Ba_2CaCu_2O_x$, etc, have been reported. Gammel reported the triangular lattice vortex distribution in single crystal YBCO by the high-resolution Bitter method, shown in Fig. 5a [9]. They found a nice triangular arrangement of vortices at 4.2 K. It has also been reported that a vortex chain in untwined single crystal YBCO was observed when the external magnetic field was tilted to about 70 degrees from the c axial direction (shown in Fig. 5b) [10]. *Vinnirov* et al. reported the vortex distribution in partly untwined single crystal YBCO [11].

They examined the question 'How do the twin boundaries work for the vortex distribution?' Figure 6 shows a polarized light microscope photograph of the vortex distribution of single crystal YBCO reported by *Vinnikuv* et al. [11]. In the polarized light photograph the oblique lines show the twin boundaries and the round points in the SEM photograph show the vortex. They found the twin boundaries acted as a potential well and the flux line lattice is parallel to the twin direction. The vortex density inside the

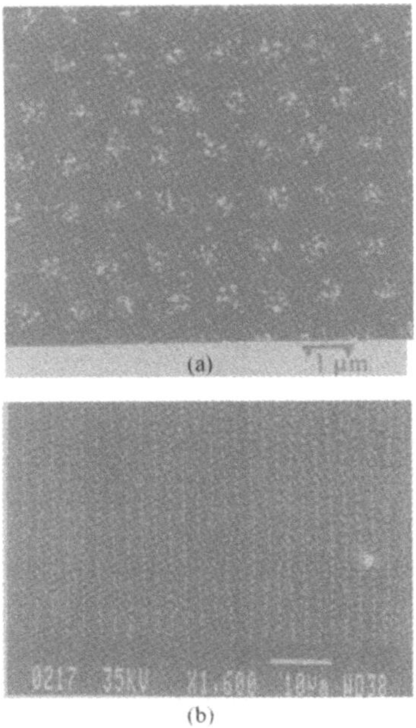

Fig. 5. Vortex distribution of YBCO single crystal reported by *Gammel* et al. (**a**) A trianglar lattice of YBCO single crystal with a field of 13 G at 4.2 K [9]. (**b**) SEM micrograph of the vortex chains with a field of 24.8 G applied at 70° from the *c*-axis of the crystal [10]

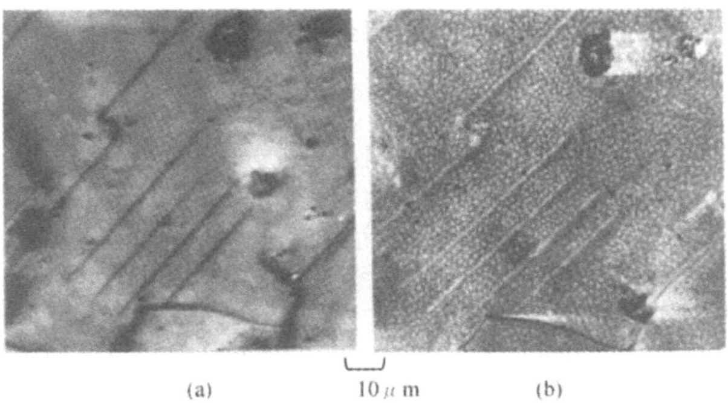

Fig. 6. Vortex distribution of partially untwined YBCO single crystal reported by *Vinnirov* et al. [11]. (**a**) A polarized light microscope photograph of YBCO single crystal. (**b**) Vortex distribution in the same section of the crystal. $T_d = 4.2\,K$, $H_{ex} = 10\,G$

twin boundaries was higher than in the other region. Since the twin boundary represents a potential well the distribution of the vortex differs from that of the other region. It is important from an engineering point of view to observe the distribution and motion of vortices at high temperature. Normally,

— 10 μm

(a)

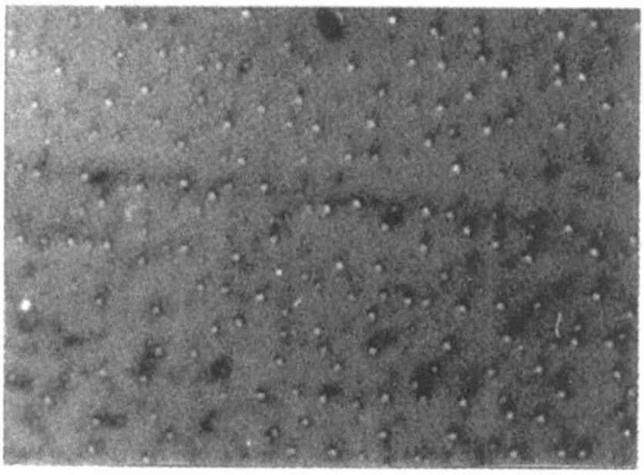

— 10 μm

(b)

Fig. 7. Polarizing microscope photograph and vortex distribution of the YBCO single crystal. (**a**) Polarizing microscope photograph: *oblique lines* show the twin boundaries. (**b**) Vortex distribution of the YBCO single crystal observed by the high-resolution Bitter method. $T_d = 40\,K$, $H_{ex} = 75\,G$

many twin boundaries exist in single crystal YBCO. We reported the vortex distribution of single crystal YBCO with a high-density twin boundary. A polarized light photograph and the vortex distribution of the sample for the same position are shown in Fig. 7. The oblique lines in the upper photograph show twin boundaries and the round points in the lower photograph show the vortex. It was confirmed that the grain boundaries had no influence on the vortex distribution, where the twin boundary density was high.

2.2.2 BSCCO Single Crystal

We carried out a systematic experiment on the distribution and movement of vortices at high temperature, and interesting results were obtained. The appearance of the vortex distribution of single crystal BSCCO at 50 K and 80 K is shown in Fig. 8. The surface after nickel decoration was observed by AFM. The bright spots and lines in the right hand picture are the places where the fine nickel particles have collected. From the experiment, they found the following result: the vortex did not move below 50 K, but it moved

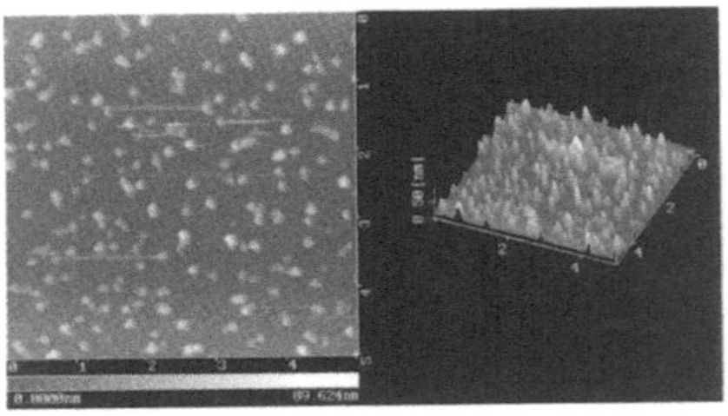

(a)

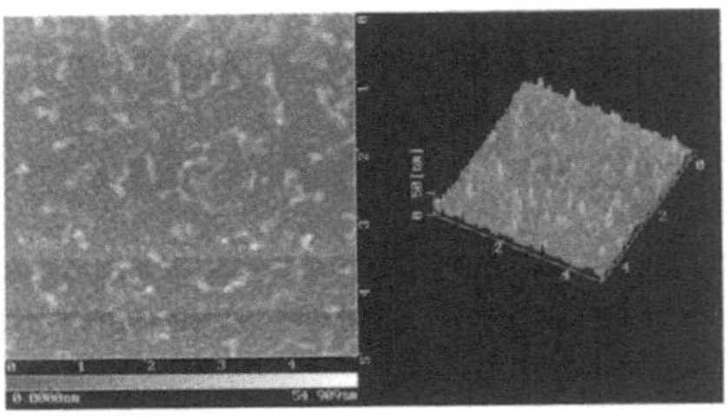

(b)

Fig. 8. Vortex distribution of a BSCCO single crystal observed by the high-resolution Bitter method. The surface of the sample was observed by AFM. (**a**): $T_{\rm d} = 55\,{\rm K}$, $H_{\rm ex} = 75\,{\rm G}$, (**b**) $T_{\rm d} = 80\,{\rm K}$, $H_{\rm ex} = 75\,{\rm G}$

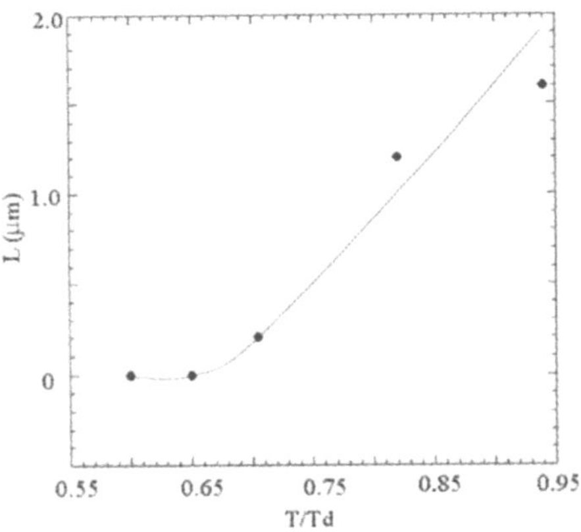

Fig. 9. The relationship between the migration length and $T_{\rm d}/T$ where $T_{\rm d}$ is the temperature at which fine nickel particles are deposited on the sample

randomly at a higher temperature of about 80 K. That is, vortices can easily be dislodged from pinning centers by thermal disturbance.

The vortex moved while the nickel was being deposited for 30 s. From this migration length it is possible to estimate the pinning potential of single crystal BSCCO. Figure 9 shows the relationship between the migration length and the observation temperature, $T_{\rm d}$. It was observed that the migration length of the vortex increased with increasing temperature. That is, the thermal energy released the vortex from the restriction of the pinning cen-

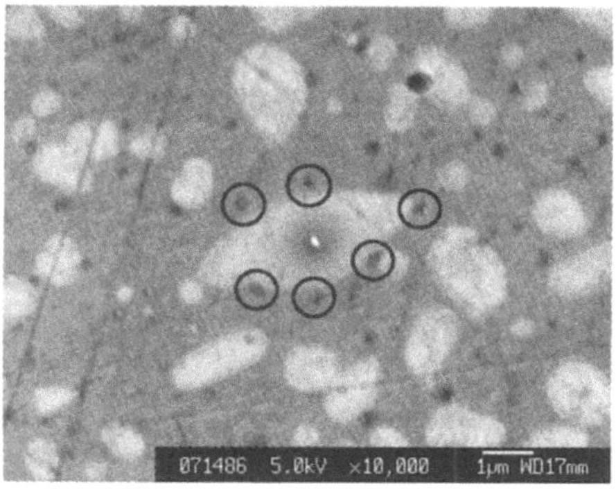

Fig. 10. Vortex distribution of a YBCO QMG sample observed by the high-resolution Bitter method. The surface was observed by SEM. White regions are the Y_2BaCuO_x (Y-211) phases and the *black points* show the vortices. Typical vortices around Y-211 particle are shown in *circles*

ter. The relationship between the pinning potential, $U_{\rm pin}$ and the migration length, L of the vortex is assumed to be expressed by the following equation:

$$L = A\exp(-U_{\rm pin}/k_{\rm B}T_{\rm d})\,, \qquad (1)$$

where $T_{\rm d}$ is the temperature at which fine nickel particles are deposited on the sample and $k_{\rm B}$ is Boltzmann's constant. We obtained a pinning potential of 41 meV for single crystal BSCCO.

2.2.3 YBCO QMG Sample

As the YBCO QMG (Quench-Melt-Growth) samples have a very strong pinning force, they are used in applications as superconducting permanent magnets and magnetic bearings, etc. The large pinning force is caused by the pinning centers of fine precipitated Y-211 particles in the YBCO phase. Figure 10 shows a photograph of the distribution of a vortex in a YBCO QMG sample observed by the high-resolution Bitter method. White regions in the picture are the Y-211 phases and the black points show the vortices. It is clear that the vortices collected at the boundary of the Y-211 and YBCO phases. This means that the Y-211 phase works as a pinning center.

2.2.4 Bi-2223 Silver Sheath Tape

As with the Y system, there are also many reports on the vortex distribution in the single crystal Bi system. In the Bi system of superconductors it is interesting to examine the vortex distribution and the temperature characteristic of Bi-2223 silver sheath tape. The vortex distribution of Bi-2223 silver sheath tape is shown in Fig. 11. The sample was cut from the silver

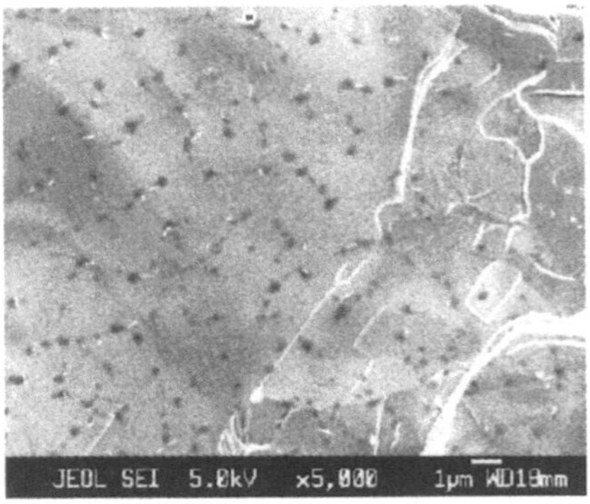

(a)

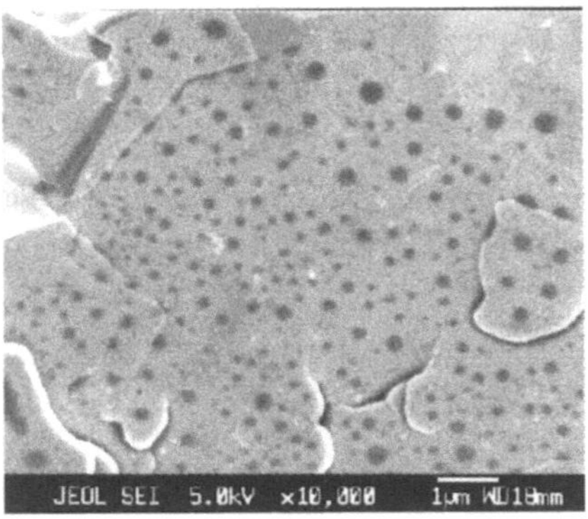

(b)

Fig. 11. Vortex distribution of Bi-2223 silver sheath tape observed by the high-resolution Bitter method. The surface of the sample was observed by SEM. *Black points* are vortices. (**a**) $T_\mathrm{d} = 25\,\mathrm{K}$, $H_\mathrm{ex} = 75\,\mathrm{G}$. (**b**) $T_\mathrm{d} = 50\,\mathrm{K}$, $H_\mathrm{ex} = 75\,\mathrm{G}$

sheath. At 25 K it was shown that the vortices developed into a heterogeneous arrangement. The vortices are seen distributed along the grain boundary of the silver sheath, which works as a weak pinning center. It was found that vortices were lined up relatively homogeneously at 50 K. That is, the surface pin potential produced by the silver sheath is not large compared with the thermal energy at 50 K.

2.3 The Observation of a Dynamic Vortex Behavior

2.3.1 Penetration of Vortices into the Sample

Though the high-resolution Bitter method is effective for observing the static vortex distribution, observation of the dynamic behavior of vortices is difficult. However, it was reported that the dynamic behavior of a vortex could

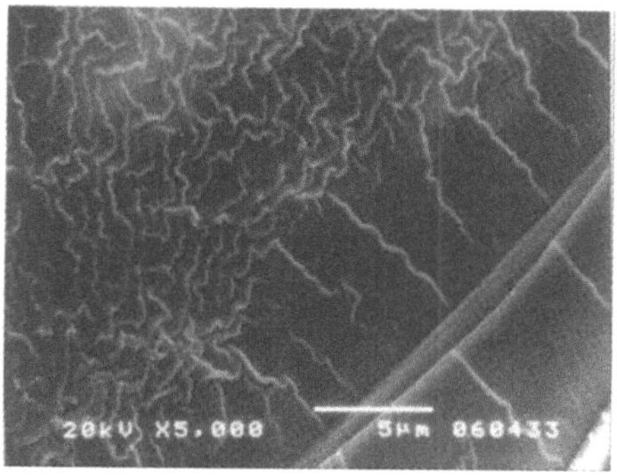

Fig. 12. Vortex movement of BSCCO single crystal when the vortices penetrated into the sample. The *white lines* are loci of vortex movement and two *oblique lines* show the step edge of the sample

also be observed by the high-resolution Bitter method under the right conditions. We have succeeded in observing the dynamic motion of a vortex using the equipment shown in Fig. 3 [12]. The success of the observation depends on controlling the feed rate and kinetic energy of fine magnetic particles. If these parameters can be controlled the trace of the vortex movement can be obtained. Figure 12 shows the vortex movement where the vortex penetrates the superconductor. The experimental procedure is as follows. The sample was cooled to 25 K and then a magnetic field was applied. After applying the magnetic field the fine magnetic particles were deposited immediately. The white line in the figure traces the movement of the vortex. The thick oblique white line on the right under the figure shows the sample edge and the two oblique lines show the step edge of the sample. At the sample edge the white lines are zigzagging white lines. That is, the vortex penetrates perpendicularly from the sample edge and moves randomly on the inside.

2.3.2 Vortex Movements under Application of a Transport Current

For observation of the dynamic behavior of the vortex motion there is another interesting study. It is very important to observe the vortex movement in the sample while applying transport current. Figure 13 shows the vortex distribution of single crystal BSCCO after applying a transport current observed by the high-resolution Bitter method. The transport current J_t was approximately 40 % of J_c for Fig. 13a and approximately equal to J_c for Fig. 13b. The directions of H_{ex} and J_t were normal to and parallel to the c plane, respectively. The direction of J_t is shown in the figure by an arrow. As shown in Fig. 13a, some elliptical spots were observed. We believe that these spots were caused by vortex movement during nickel deposition (30 s) and were not significantly influenced by the transport current. On the other hand, the vor-

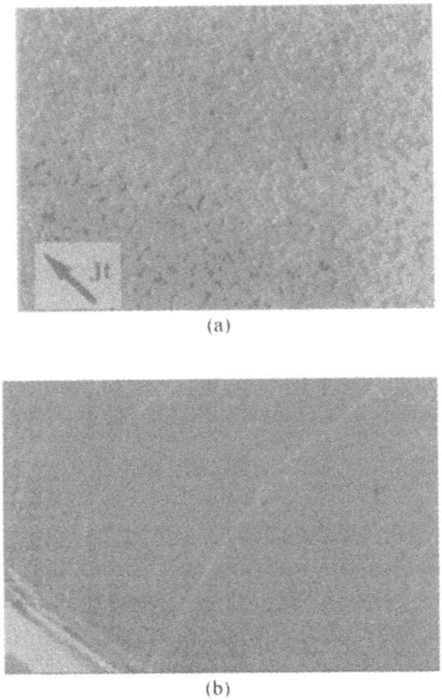

Fig. 13. Vortex movements of the BSCCO single crystal under application of a transport current. The *white* and *black points* in picture (**a**) show vortices, and *white lines* in picture (**b**) show vortices. (**a**) J_t; 40% J_c. (**b**) J_t; 100% J_c

tex distribution in Fig. 13b is quite different to that in Fig. 13a. We can see two different patterns of vortex arrangement. One is a chain-like distribution marked Fig. 13a and the other is a random distribution marked Fig 13b. The chain-like pattern consists of many vortices, which can be clearly seen in the AFM image shown in Fig. 14. The direction of the chains is approximately normal to the direction of the transport current. The AFM image of the nickel decoration clearly illustrates the hopping movement of the vortex. We can see the circle spots, which stand in a line, and the interval between the spots is not uniform. The loci of the vortices show a very sharp projection of nickel particles. It is also surprising that the height of the spots varies in different chains as shown in Fig. 14b. This indicates that the dwell time of the vortex for each pin potential is different and that the vortices in a narrow interval move more quickly.

In this section, we shall summarize the results obtained from the observation of vortex patterns and movement by the high-resolution Bitter method. The investigation of vortex distributions gave experimental evidence for the Abrikosov vortex lattice and the random arrangement caused by the random pinning centers. The vortex was able to move in the high temperature region by thermal energy. It was also clarified by observation of vortices that Y-211 precipitates in the YBCO crystal behave as pinning centers. The high-resolution Bitter method is also useful for observing dynamic vortex

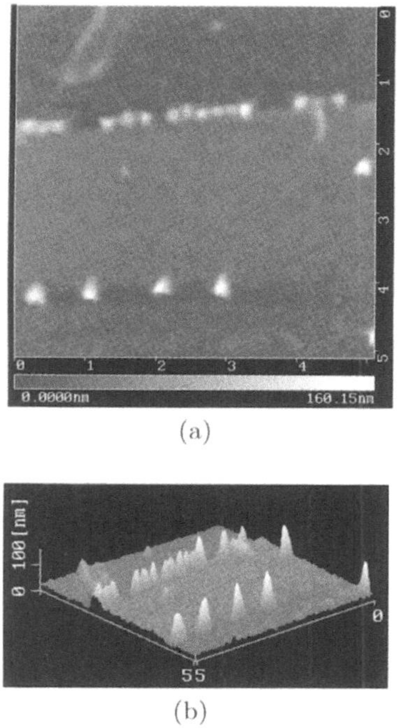

Fig. 14. Vortex movements in a BSCCO single crystal under application of a transport current. The nickel decoration pattern was observed by AFM and was the same section of the crystal shown in Fig. 12b

movement. Hopping vortex movement was observed by applying a transport current. Also the behavior of penetrating vortices could be observed.

3 Scanning SQUID Observation

The scanning SQUID microscope (SSM) system is one of the most powerful tools that enable direct observation of vortices in superconducting materials. SSM systems have a SQUID magnetometer integrated with a small pick-up loop and a scanning sample stage. Their spatial resolution is determined by the diameter of the pick-up loop and the so-called stand-off distance between the loop and the sample, and is typically 1–10 μm. Though this value is not as good as other observation methods such as the magnetic force microscope (MFM), the Bitter method, the magneto-optical (MO) method and the Hall probe microscope, SSM has much better field sensitivity than the other methods, as shown in Fig. 15. By using SSM, the local field strength can be accurately measured, which is another advantage over MFM and the Bitter method. Furthermore, in contrast to the Lorenz microscope, there is no restriction on the sample thickness and thus the distribution of magnetic field or current in electronic devices can be also measured by using SSM.

The first SSM with a Nb-based integrated SQUID magnetometer was developed in the early nineties [15] and applied to the observation of vortices

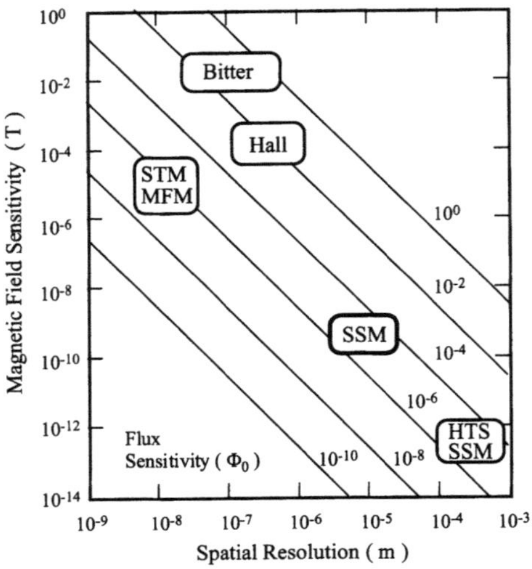

Fig. 15. Comparison of magnetic field sensitivity and spatial resolution for various vortex observation methods

trapped in a superconducting regular lattice in order to investigate flux dynamics [16]. An epoch-making application of SSM was the observation of half flux quanta trapped or generated at the tricrystal point in high-T_c cuprate thin films, which confirmed the *d*-wave symmetry of the order parameter in these "unusual" superconductors [17,18]. SSM was also applied to the observation of Josephson vortices along the CuO_2 planes in cuprate superconductors to examine the validity of the mechanism for high temperature superconductivity [19].

It is well known that the magnetic flux trapped during cooling through T_c in superconducting electronic devices causes serious problems such as malfunction of digital circuits or an increase of flux noise in high-T_c SQUIDs. The magnetic images of Nb-based digital circuits were first taken by *Jeffery* et al. using SSM [20], and the suppression of flux trapping in the circuit area surrounded by moats, i.e. artificially-made arrays of holes in superconducting thin films, was clearly shown. Recently, a commercial SSM system has been applied to the observation of high-T_c thin film patterns with various line widths [21], various configurations of moats [22,23] and grain boundaries [24] in order to investigate the flux trapping behavior in high-T_c thin films and electronic devices such as single flux quantum (SFQ) devices.

In the following subsections, the SQUID microscope observation technique and some recent results on vortex observation in high-T_c thin film patterns and devices are reviewed.

3.1 SSM System and Observation Technique

Figure 16 schematically shows the configuration of a commercial SSM system developed by Seiko Instruments [25]. In this system a Nb-based dc-SQUID

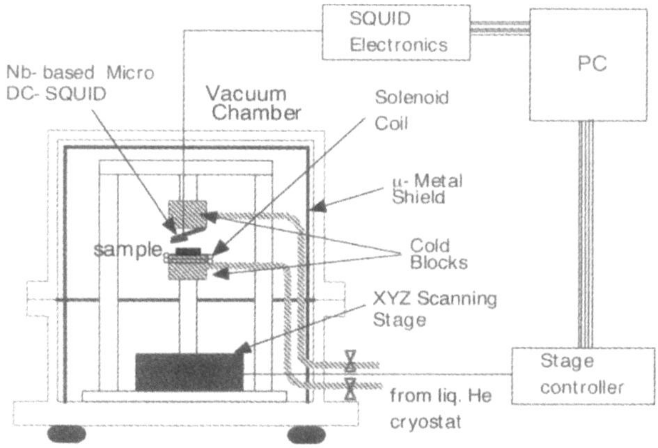

Fig. 16. Schematic illustration of a commercial SSM system developed by Seiko Instruments

magnetometer integrated with a one-turn pick-up loop and a reference magnetometer on a 3 mm square silicon chip is mounted on a Cu–Sn–P flexible cantilever. The outer and inner diameters of the pick-up loop are 12.5 and 7.5 μm, respectively. The separation between the pick-up loop and sample surface is several to ten μm. It was ensured experimentally that the minimum spatial resolution is approximately 5 μm [25]. In this system, the sample and the SQUID device are thermally attached to different cold blocks, which are cooled with evaporated He gas from a liquid He cryostat. The sample temperature can be varied in a wide range of 4–100 K while maintaining the temperature of the Nb-SQUID below 5 K. The sample cold block is mounted on an XYZ scanning stage which is moved by a stepping motor with a step of 50 nm. The motor and the SQUID electronics are controlled by a PC.

Figure 17a shows a typical 3D magnetic field image of an $YBa_2Cu_3O_y$ (YBCO) high-T_c thin film observed by using the SSM system. The sharp peak corresponds to a flux quantum Φ_0 ($= h/2e = 2.07 \times 10^{-15}$ Wb) trapped in the film. Figure 17b shows the flux profile of the trapped flux quantum shown in Fig. 17a. The full width at half maximum (FWHM) is estimated to be approximately 15 μm, which is larger than the diameter of the pick-up loop by a factor of 1.5. According to the simulation assuming a single magnetic dipole perpendicular to a thin film as a trapped single flux quantum [26], the distribution of the magnetic field coupled to the pick-up loop while scanning the loop over the dipole largely depends on the stand-off distance Z_{st} between the loop and the signal source. The distribution calculated for the case of $Z_{st} = D_p$ with D_p the diameter of the loop is shown in Fig. 18b, as an example. The ratio of FWHM to D_p is comparable to the result of Fig. 17b, indicating that the actual stand-off in the above experiment is approximately 10 μm. Even in this case, the integral of the magnetic field over the area sufficiently far from the dipole should be equal to Φ_0 [26]. Actually it is

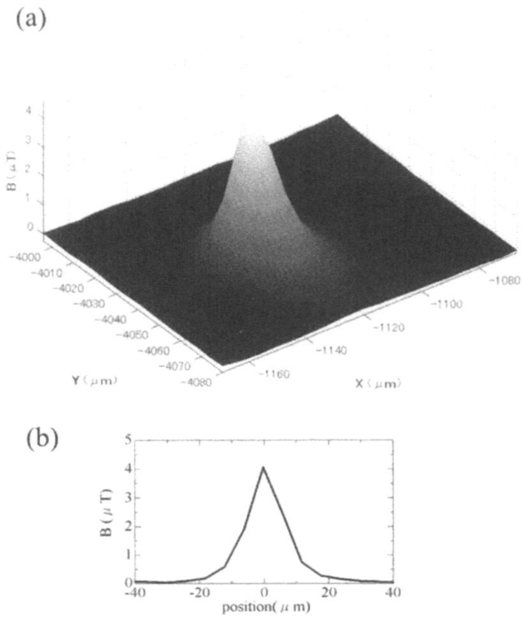

Fig. 17. 3D magnetic image (**a**) and the flux density profile along the Y-axis (**b**) for a single vortex trapped in a superconducting thin film

confirmed that the integral of the flux in Fig. 17a over the rectangular area almost equals the value of Φ_0.

Although the SSM system is equipped with μ-metal magnetic shields, a residual magnetic field less than 2 μT usually exists. Thus many trapped flux quanta are observed when a superconducting thin film is cooled to a low temperature. However, the residual magnetic field can be compensated with a solenoid coil surrounding the sample stage. The sample is heated above T_c and then a certain magnitude of opposite field is applied. After cooling the sample, a magnetic image is taken again. This procedure is repeated and finally no trapped flux is observed, indicating that the residual field is completely compensated. Considering this state as a reference, magnetic images are taken by applying additional fields and cooling the sample to low temperatures.

3.2 Flux Expulsion in Narrow High-T_c Thin Film Patterns

Use of high-T_c SQUIDs in unshielded environments is very important for biomagnetic applications such as magneto-cardiograms. However, a significant increase of the flux noise in field-cooled SQUIDs was previously observed and this was attributed to flux trapping in a SQUID washer coil and thermally activated flux hopping. With the aim of preventing flux trapping, SQUIDs

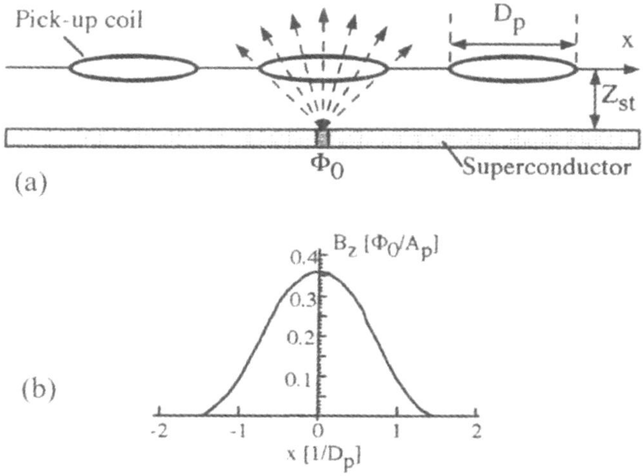

Fig. 18. A model of a single flux quantum (**a**) and the distribution of the magnetic field induced by a single flux quantum when $Z_{st} = D_p$ (**b**) [26]

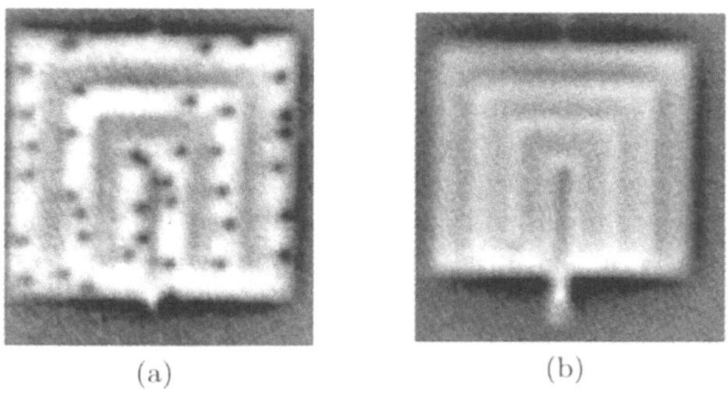

Fig. 19. Gray-scale magnetic field images of NBCO thin film patterns with line widths of (**a**) 40 μm and (**b**) 20 μm. The images were taken after cooling the patterns to a low temperature in a field of 4 μT [21]

with holes or slots in a washer have been employed [27]. The behavior of flux trapping in high-T_c thin film patterns with the shape of SQUID washers containing an array of slots has been investigated by using a SSM system [21].

150–200 nm thick $NdBa_2Cu_3O_y$ (NBCO) thin films were patterned by standard photolithography and ion milling into a series of four patterns with shapes of SQUID washers. Each washer pattern had an array of slots, and the line widths of the remaining thin film were 4, 10, 20 or 40 μm. These patterns had outer dimensions of about $300 \times 300\ \mu m^2$.

Figure 19 shows the gray-scale images at 5–7 K of NBCO patterns with line widths of (a) 40 μm and (b) 20 μm, which were cooled in a field of

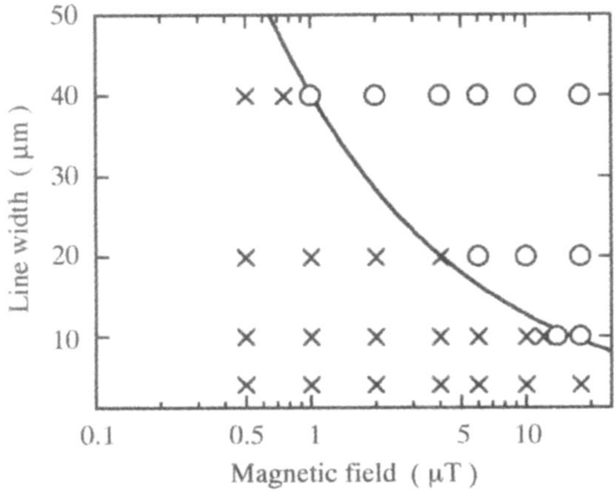

Fig. 20. Relation of the field to the line width giving rise to flux trapping (*open circles*) and flux expulsion (*crosses*) for a series of NBCO patterns fabricated on the same substrate. The *solid line* shows the threshold value predicted by *Clem* ($B_t = \pi\Phi_0/4w^2$) [21]

4 μT. The magnitude of the magnetic field is coded from the minimum value (bright) to the maximum value (dark) in these figures. The bright area corresponds to the NBCO thin film in the superconducting state, where the flux is excluded by the Meissner effect, and the slots are seen as the dark areas. Each of the dark spots seen in the film patterns corresponds to a trapped flux quantum Φ_0. Figure 19a shows trapped flux quanta, while Fig. 19b shows no trapped flux. This result clearly indicates that the flux trapping is suppressed by decreasing the line width.

The obtained results are summarized in Fig. 20. In this figure the circles and crosses indicate whether flux trapping or complete flux expulsion occurs, respectively. It was found that the density of trapped flux quanta decreases with decreasing the ambient field and no trapped flux is observed below certain threshold values. The threshold field for flux expulsion clearly increases with decreasing the line width w. The solid line represents the threshold field predicted by Clem ($B_t = \pi\Phi_0/4w^2$) [27] for lines with an infinite length. Clearly the results are very close to this theoretical curve.

This result is consistent with the results of $1/f$ noise measurements by *Dantsker* et al. [27] where a SQUID with many slots and a line width of 4 μm exhibited no significant increase in the noise for a field up to 100 μT. Thus it is clear that SSM observation provides useful information on flux behavior in practical devices.

3.3 Flux Trapping in High-T_c Thin Film Patterns with Moats

For the purpose of preventing flux trapping in superconducting electronic digital devices, moats in superconducting ground planes which surround dig-

(a) (b) (c)

Fig. 21. Moat configurations examined: (**a**) broken-line moats, (**b**) continuous moats with two gaps on the sides, and (**c**) continuous moats with four gaps at the corners. The moat-surrounded area is 160 × 160 μm^2 and the moat width is 4 μm [22]

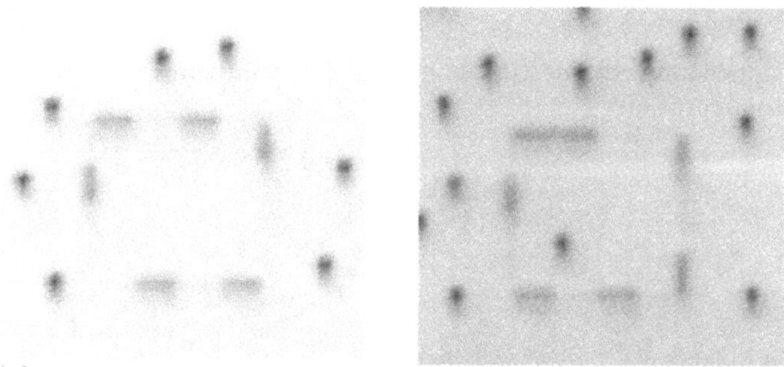

(a) (b)

Fig. 22. Gray-scale magnetic field images for an NBCO pattern with broken-line moats, where the moat length and the gap are 36 and 4 μm, respectively. These images were observed after cooling in fields of (**a**) 0.25–0.30 μT and (**b**) 0.30–0.35 μT

ital circuits have been employed [20]. By using a SSM system the threshold field for flux expulsion as a function of moat size for various moat configurations has been examined in detail [22,23]. As shown in Fig. 21, test patterns containing 4 μm wide moats or moat arrays with various configurations and lengths surrounding 160 μm square areas were fabricated on NBCO films. In the broken-line configuration shown in Fig. 21a, the square area is surrounded by sixteen equally spaced moats with the gap between moats of 4, 8, 16, 20, 24, 28 or 32 μm. In the configuration shown in Fig. 21b, the square area is surrounded by continuous moats with two gaps at the sides, where the gap between two moats is 8, 16, 24, 32 or 40 μm. In the configuration shown in Fig. 21c, the square area is surrounded by continuous straight moats with four gaps at the corners. The moat length is 148, 136, 124 or 112 μm.

Figure 22 shows gray-scale magnetic field images for an NBCO thin-film pattern with broken-line moats as shown in Fig. 21a, where the moat length and the gap are 36 and 4 μm, respectively. These images were observed after cooling in fields of (a) 0.25–0.30 μT and (b) 0.30–0.35 μT, respectively. Flux is trapped in some of the moats as well as the NBCO film. It is found that flux

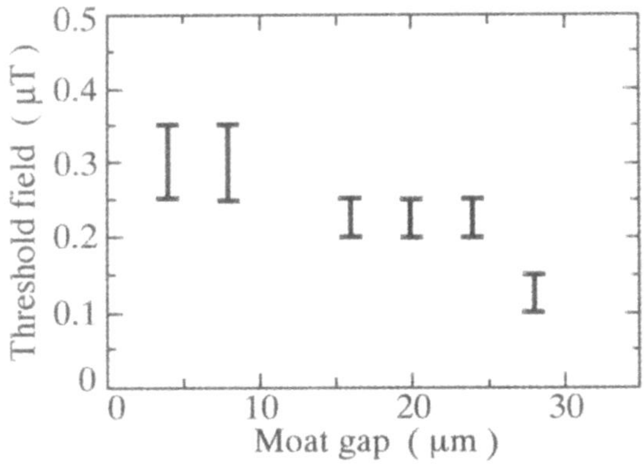

Fig. 23. Moat-gap dependence of the threshold field for complete flux expulsion for the broken-line configuration

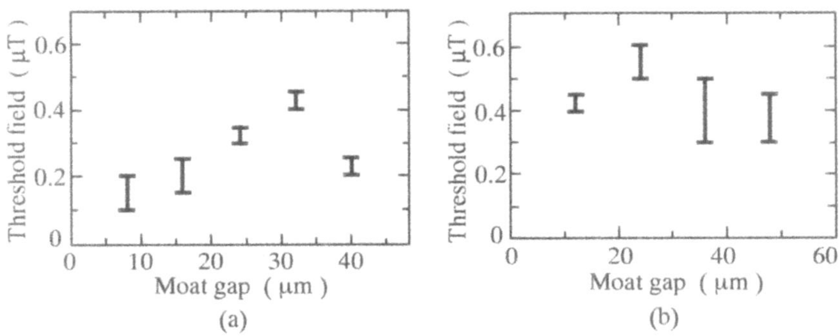

Fig. 24. Moat-gap dependence of the threshold field for complete flux expulsion for the NBCO patterns consisting of continuous moats (**a**) with two gaps on the sides and (**b**) with four gaps at the corners [23]

trapping in the moat-surrounded area is certainly suppressed by moats, and flux is completely excluded for fields less than a certain threshold field. From the result of Fig. 22, the threshold field is estimated to be 0.25–0.35 μT for this pattern. This value is approximately four times larger than the threshold field for an isolated square pattern with the same area, which is estimated based on a simple flux quantization condition, $B_t = \Phi_0/(160 \times 160\,\mu\text{m}^2) = 0.08\,\mu\text{T}$. Figure 23 shows the moat-gap dependence of the threshold field for the pattern in Fig. 21a. For the patterns with moat-gaps shorter than 28 μm, clear suppression of flux trapping in the moat-surrounded area was observed, while no complete flux expulsion was observed for the moat-gap of 32 μm. The threshold field clearly increases with decreasing the gap between the moats.

Figure 24a shows the moat-gap dependence of the threshold field for the side-gap patterns. The observed results indicate that the threshold field has a peak value for the moat gap around 32 μm. This result is clearly different from the case of broken-moat patterns, where the threshold field monotonically decreases with the moat gap. Figure 24b shows the moat-gap dependence of the threshold field for the corner-gap pattern. Here, the moat gap is defined as (160 μm − moat length). The obtained results indicate that the threshold field has a peak value for the moat gap around 24 μm. This tendency is similar to the case of the continuous moat configuration with side-gaps. The threshold field of 0.5–0.6 μT for the 136 μm length moat is the largest one in this study and is approximately seven times larger than the theoretical value of 0.08 μT for an isolated pattern.

It is deduced that there are two factors that determine the flux expulsion effect. The first one is the area which one moat hole can cover. It is easy to understand that the effect of moats is minimal if the gap between moat holes is much larger than the moat length. The threshold field is expected to increase with increasing the moat length or decreasing the moat gap. The second factor is the shielding current around the moat holes, which is schematically shown in Fig. 25. In the case of continuous moats with very narrow gaps, the shielding current may exceed the critical current in the gap regions at temperatures just below T_c. Then the situation becomes close to the case of isolated patterns, implying no enhancement of the threshold field. The peaks observed in continuous moat configurations are probably caused by the competition between these two factors. On the other hand, for the case of the broken-line configuration it seems very difficult to fill all the broken-line moats with Φ_0. In this case the flux exclusion effect is determined by the first factor alone and accordingly the threshold field has no peak.

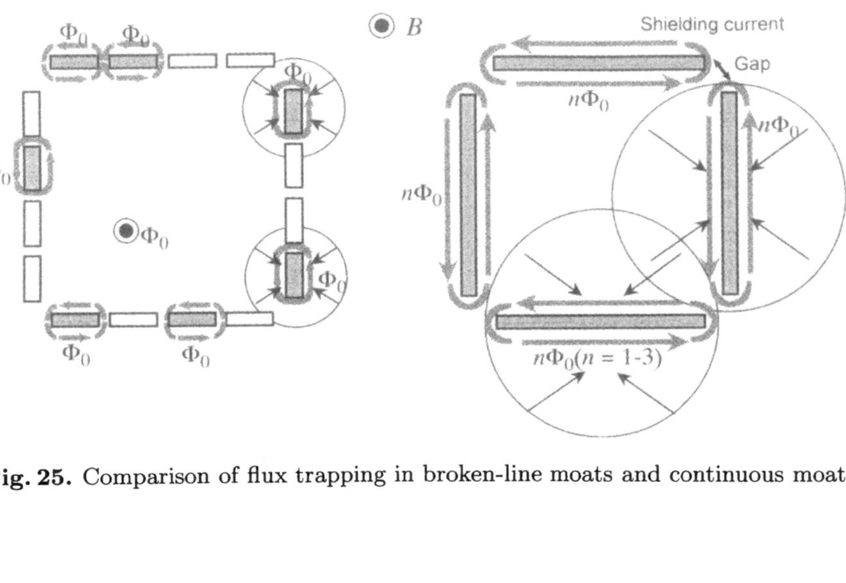

Fig. 25. Comparison of flux trapping in broken-line moats and continuous moats

3.4 Flux Trapping in High-T_c Films with a Bicrystal Grain Boundary

It is well known that large-angle grain boundaries in high-T_c cuprate superconductors behave as Josephson junctions or weak links. Thus, investigation of the properties of grain boundaries in high-T_c films is very important for applications to both electronic devices such as high-T_c SQUIDs and conductors. Moreover, it is also useful to clarify the physical properties peculiar to high-T_c materials such as *d*-wave symmetry of the order parameter [17,18,28]. The SSM observation of high-T_c thin film patterns with bicrystal grain boundaries, as schematically shown in Fig. 26, has also been tried [24].

Figure 27 shows gray-scale magnetic field images of a $200 \times 200\,\mu m^2$ NBCO thin film square pattern with a symmetric 36.8° [001]-tilt grain boundary, which was cooled in a field of 1.1 μT. Figure 27a was taken after cooling to a low temperature of 9 K. Trapped flux quanta are seen at the grain boundary as well as outside the grain boundary region. The peak height of the flux quanta at the boundary is reduced to about 80% of that for flux in the thin film, probably because of the Josephson penetration depth λ_J, comparable to the size of the pick-up loop.

In these measurements, the T_c values of the thin film and the grain boundary region were determined to be 83 K and 68–71 K, respectively, from the disappearance of the Meissner effect. Figure 27b–d was taken at 62, 71, and 82 K, respectively, by elevating the sample temperature. At 62 K, which is near the boundary T_c, the flux quantum near the boundary edge indicated by the lower arrow in Fig. 27a disappeared and the flux quantum indicated by the upper arrow moved. This implies that the flux pinning strength is weaker at the grain boundary. On the other hand, Fig. 27a–d shows that the number and the positions of the flux quanta trapped outside the grain boundary were unchanged even at 82 K, just below the film T_c, which is consistent with the strong pinning in RE-123 thin films [29].

Figure 28 shows gray-scale magnetic field images of the NBCO thin film cooled to low temperatures of 6–8 K in various magnetic fields. Figure 28a–d was taken after cooling in fields of 0.3, 2.1, 4.1 and 6.1 μT, respectively. As known from Fig. 28a–d, the amount of trapped flux increases with increasing ambient magnetic field outside the grain boundary. At the grain boundary, however, the number of trapped flux quanta does not increase with increasing field. Furthermore, Fig. 28d shows no flux at the grain boundary, while many trapped flux quanta are observed outside the grain boundary. This is probably due to the weak pinning at the boundary and repulsive interaction with the flux quanta outside the boundary. When the film is cooled through the boundary T_c, flux trapping in the grain boundary is probably suppressed by the repulsive interaction with the flux quanta outside the grain boundary, which were already trapped at the higher film T_c.

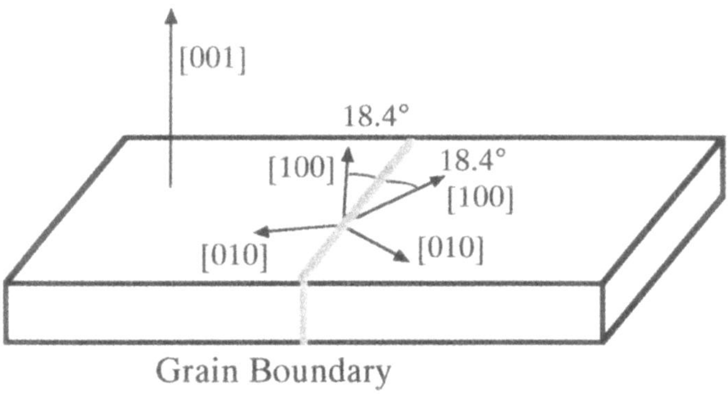

Fig. 26. Sketch of a symmetric 36.8° [001]-tilt grain boundary

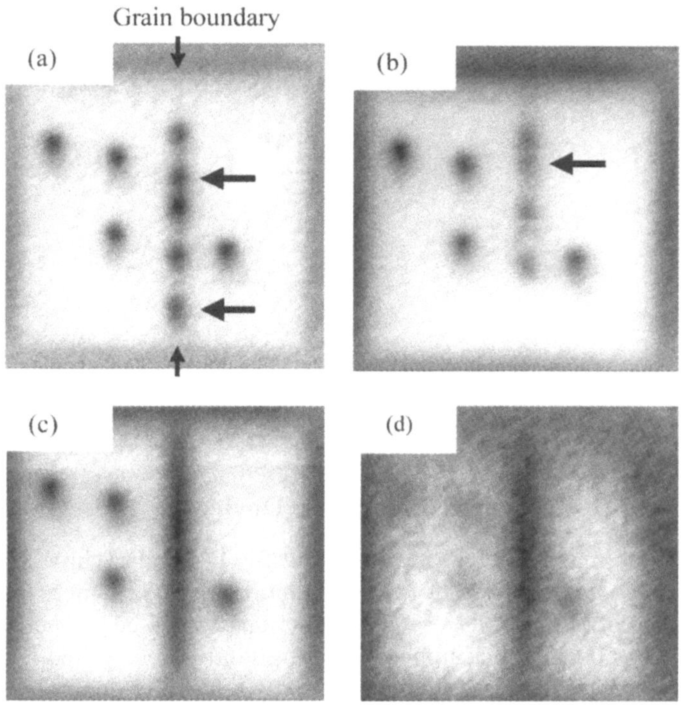

Fig. 27. Gray-scale magnetic field images of the NBCO thin-film with a symmetric 36.8° [001]-tilt grain boundary. The outer dimension of the film pattern is 200 × 200 μm^2. (**a**) was taken after cooling to 9 K in a field of 1.1 μT. (**b**)–(**d**) were taken by elevating the sample temperature to 62, 71 and 82 K, respectively [24]

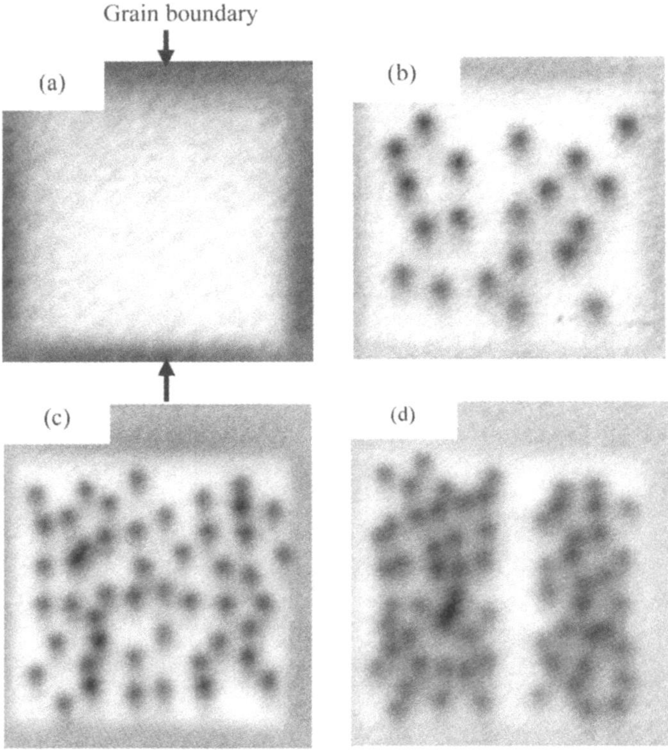

Fig. 28. Gray-scale magnetic-field images of the NBCO thin-film with a symmetric 36.8° [001]-tilt grain boundary. Images were taken after cooling in fields of (**a**) 0.3, (**b**) 2.1, (**c**) 4.1 and (**d**) 6.1 μT, respectively [24]

3.5 Observation of Multilayered Electronic Devices

Figure 29 shows SSM images at 5 K for a YBCO SFQ sampler chip [31] on $SrTiO_3$ substrate cooled in fields of (a) 0.4 μT and (b) 2 μT, which has a multilayer structure containing five Josephson junctions with edge geometry and an approximately 600 nm thick upper YBCO ground plane. When the circuit is cooled in a weaker field less than 0.4 μT, only a few flux quanta are trapped in the moat holes and outside the important region of the circuit. As shown in Fig. 29b, however, when the circuit is cooled in 2 μT, flux quanta are trapped in the regions of the ground plane without wiring thin film patterns of the circuit, while in the regions with wiring patterns and thus larger total thickness of superconducting layers, no trapped flux is observed. Another interesting fact is that in the narrow regions between the wiring patterns (upper left of Fig. 29b), the residual weak field penetrates to the ground plane. This suggests that the superconductivity of this region is rather weak possibly because of poor film quality and almost destroyed by the flux focusing effect.

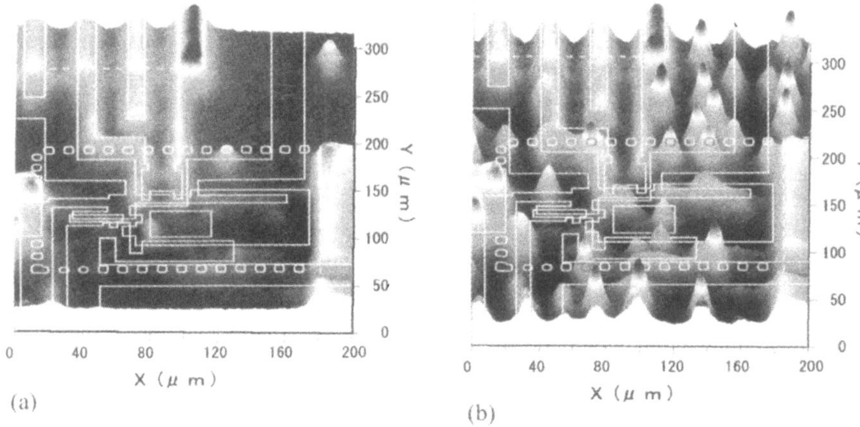

Fig. 29. Magnetic field images of a YBCO SFQ sampler chip on $SrTiO_3$ substrate cooled in fields of (**a**) 0.4 μT and (**b**) 2 μT [30]

3.6 Summary

The scanning SQUID microscope has advantages over other vortex observation techniques such as high field sensitivity, accurate determination of local field strength and no limitation on the sample thickness. In spite of the limited spatial resolution of 1–10 μm, it has been proven a powerful tool for nondestructive vortex observation in superconducting thin films and devices in a weak field region less than the earth's field which is important especially for electronic device applications.

By using a commercial scanning SQUID microscope system, the behavior of flux trapping in narrow high-T_c thin-film patterns has been investigated, and the threshold field for complete flux expulsion is found to exactly agree with Clem's criterion. The effects of moats on the suppression of flux trapping have been also examined using the same system, and two factors determining the threshold field for flux expulsion in moat-surrounded areas are clarified. Observation of a high-T_c thin-film pattern with a bicrystal grain boundary has clearly revealed weaker flux pinning at the boundary than that in thin-film regions. Furthermore, it has been demonstrated that SQUID microscope observation can be utilized in the diagnosis of practical electronic devices.

4 Magneto-Optical Imaging

A magneto-optical effect and polarizing microscopy have been used to observe a domain structure in a material with spontaneous magnetization [32]. Various techniques have been developed to investigate magnetic walls and their dynamics in bubble materials [33]. The discovery of high-temperature superconductors has stimulated dramatic growth in our understanding of the physics of quantized vortex lines. The mechanism of vortex motion strongly depends on the strength, density and distribution of pinning centers and the homogeneity of the superconducting materials. Thus the direct observation of magnetic flux distribution is of great interest. The magneto-optical image gives us total information including a demagnetizing effect and local inhomogeneity in the materials observed. The flux density profile measured with sufficient resolution allows comparison with a model calculation.

In this section we will explain magneto-optical imaging and review resent results mainly obtained for high-temperature superconductors.

4.1 Experimental Method

The magneto-optical technique is attractive in several respects. It is applicable to bulk samples of any material on which a suitable surface can be prepared. It is nondestructive to the surface. It is inherently a dynamic method and it can be used over a wide range of temperatures and applied fields. Both the magneto-optical Faraday and Kerr effects can be used to investigate a magnetic structure. In the former, a beam of plane polarized light passing through a material undergoes a nonreciprocal magnetic rotation of its plane of polarization. In the latter, plane polarized light reflected from a magnetic material acquires a rotation and an ellipticity. The Kerr effect is available to investigate an opaque material using light reflected on the surface. However the Kerr effect is very weak [32].

4.1.1 Faraday Effect

When linearly polarized light is incident normally on a magnetic material, the wave leaves the material remaining linearly polarized, but the plane of polarization is rotated in a sense depending on the direction of the magnetization normal to the material surface. The rotation angle is also proportional to the path length that the light traverses. The direction of the rotation is reversed when the light passes through the material in the opposite direction due to the antireciprocity of this effect [34]. If the light beam goes toward and comes back along the same path the total rotation of the plane of polarization is therefore double the value resulting from a single path.

Figure 30 is a schematic of the system for observing a domain structure with various magnetic orientations using difference in contrast [32,35]. When the analyzer is slightly uncrossed with respect to the polarizer, one domain appears brighter than the other. This is the most convenient way of visualizing the domain configurations. The light may be white or monochromatic, incoherent or coherent, continuous or pulsed depending on what kind of information is required. It is noted that the variation of absorption and the rotation with the wavelength results in a color as well as a contrast.

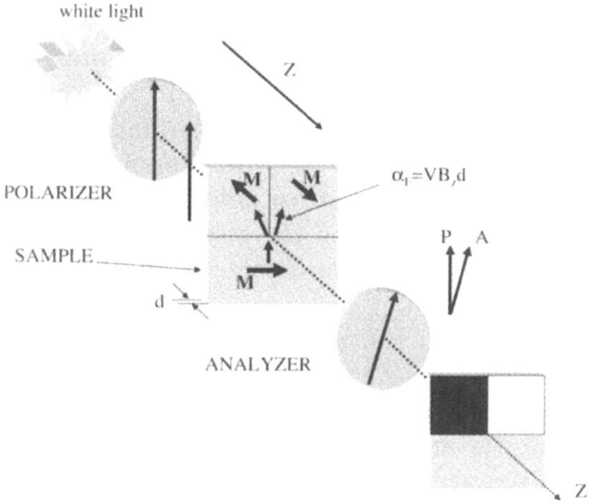

Fig. 30. Schematic drawing of the method in which contrast is made between magnetic domains of various orientation using Faraday rotation [35]

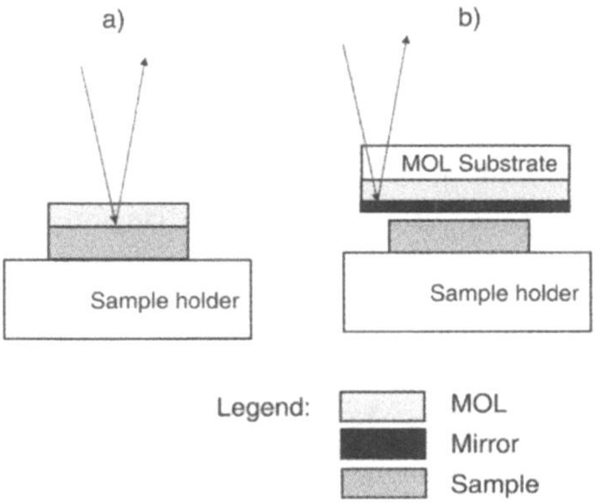

Fig. 31. Arrangement of the magneto-optically active layer, mirror layer and sample [36]

4.1.2 Magneto-optically Active Materials

The magnetic field distribution of a superconductor in the Shubnikov phase is imaged by observing a Faraday rotation in a magneto-optically active layer that is placed on the superconductor. A schematic drawing of the set-up for observing magnetic flux in a superconductor is shown in Fig. 31 [36]. The magneto-optically active layer deposited on the mirror layer is contacted to the sample. The polarizer and analyzer are in a cross-position. The mirror reflects the incident light after passing through the area where flux is present. Therefore a Faraday rotation angle is doubled at the analyzer.

The first magneto-optical observations of magnetic flux were performed using paramagnetic glasses of cerium metaphosphate or nitrate as a magneto-optical material [6,7]. This cerium compound was also used for observation of high temperature superconductors [39]. The paramagnetic glass presents a rather small Faraday rotation angle α which is proportional to the Verdet constant V and the thickness of the superconductor, d, according to $\alpha = VHd$, where H is a local magnetic field parallel to the direction of the light beam. The thickness of the glass was 0.25 to 3 mm to obtain a reasonable rotation angle. A large gap between the sample and the glass surface necessarily resulted in reduction of the spatial resolution of the magneto-optical image.

Adopting a thin-film technique to prepare the magneto-optical layer made a great improvement in the resolution [40]. The thin-film of a mixture of chalcogenides, ferromagnetic EuS and paramagnetic EuF_2 was used as a magneto-optical material. This material was applied for observing the magnetic flux in conventional [41] and high temperature [42] superconductors. Generally the mixture is troublesome to control its chemical composition in the process of deposition. Overcoming this, the single element EuSe film, which is still paramagnetic at 4.6 K, was used [43,44]. The drawback of chalcogenides for observing the behavior of flux is that the temperature is limited to below 20 K because the Verdet constant decreases with increasing temperature, and then a practical resolution is not enough for examining the distribution of vortices especially for high-temperature superconductors. Recently a thin film of ferromagnetic EuO with a Curie temperature of 69 K was successfully applied to the observation of the distribution of flux density in $NdBa_2Cu_3O_7$ single crystals up to 60 K and 7 μT [45].

4.1.3 Ferromagnetic Active Layers

In a ferrromagnet the Faraday effect is not so straightforward as in the case of paramagnets. The angle of Faraday rotation is proportional to the product of an intensity of an optical absorption and a splitting due to the spin-orbit coupling in an excited level. At temperatures lower than the Curie temperature, each magnetic ion possesses a ground level split due to the exchange field, resulting in spin polarization. Therefore the rotational angle is proportional to the magnetization and is much larger compared with paramagnets. An increase in Faraday rotation is found around the wavelength where an optical absorption is observed as in Fig. 32 [46]. In order to obtain a high signal-to-noise ratio in magneto-optical imaging, the figure of merit defined by a rotational angle divided by an intensity of optical absorption should be large.

4.1.4 Iron Garnets

Ferrimagnetic rare earth iron garnets ($R_3Fe_5O_{12}$: $R =$ Y, Lu, Gd) are optically transparent in the infrared region from 2000 to 7000 cm^{-1} (1.5 to 5 μm) while their Faraday rotation angles are fairly large. Substitution of bismuth

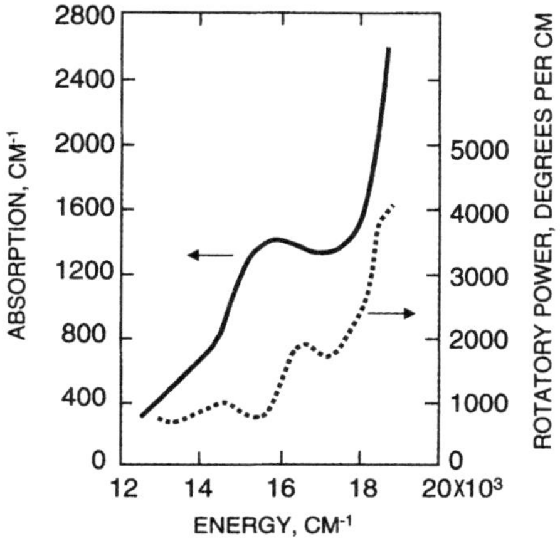

Fig. 32. Room temperature absorption and Faraday rotation of yttrium iron garnet [46]

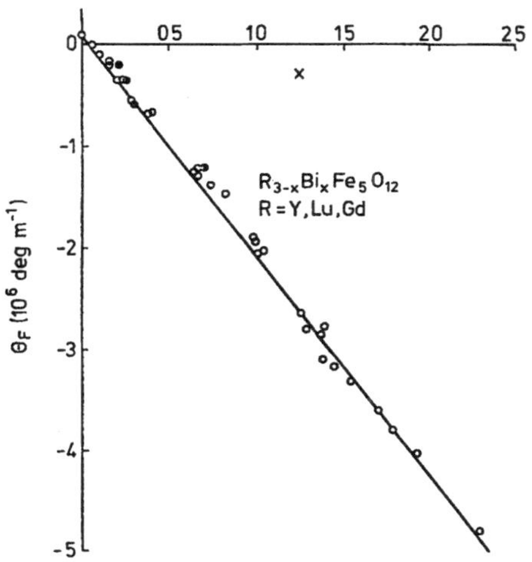

Fig. 33. Faraday rotation for bismuth-substituted iron garnet films vs bismuth content [47]

in iron garnets ($R_{3-x}Bi_xFe_5O_{12}$) strongly enhances the Faraday rotation as shown in Fig. 33 [47]. The figure of merit of $Gd_2BiFe_5O_{12}$, for example, is larger by one order of magnitude compared with $Y_3Fe_5O_{12}$ in the wavelength range in the visible and near infrared [48].

4.1.5 Resolution in Magneto-optical Images Using Garnets

The garnet films, so-called bubble films, have uniaxial anisotropy with easy axis perpendicular to the film, and have been successfully used for visualizing the distribution of flux density through a change of stripe domain structures. However, the change is produced only when the gradient of the field exceeds a critical value for wall motion. The size of the domains also limits the spatial resolution. A typical spatial resolution is 5 to 10 µm. Garnet films with an in-plane easy axis are used to overcome these problems in bubble films. In this case any out-of-plane deviation of the magnetic moment induced by an applied field results in Faraday rotation and then makes it possible to provide half-tone imaging suitable for visualizing the spread of magnetic vortices. Using $(Bi,Lu)_3(Fe,Ga)_5O_{12}$ a resolution of 0.37 µm was demonstrated [49].

4.1.6 Preparation of Garnet Active Layers

Bi-substituted iron garnet films are usually grown by liquid phase epitaxy on a substrate of single crystalline paramagnetic garnet ($Gd_3Ga_5O_{12}$) [50]. The growth parameters such as supercooling, melt composition, growth temperature and rate of substrate rotation are of primary importance for control of the magnetic properties, in particular concerning the uniaxial anisotropy. By varying the growth conditions a wide range of useful film characteristics is obtained. The garnet films have been developed as candidates for displays or bubble applications. Therefore, for a magneto-optical material used at low temperatures one must adjust the growth conditions so that the film has a suitable demagnetizing domain width and saturation field at working temperatures. These characteristics are satisfied for $Y_{2.7}Bi_{0.3}Fe_{3.7}Ga_{1.3}O_{12}$ at temperatures from 7 to 300 K, for example [51]. In order to obtain garnet films with an in-plane easy axis, the chemical composition is chosen so that the magnetic anisotropy is reduced almost to zero. The lattice constant of the film slightly larger than that of $Gd_3Ga_5O_{12}$ substrate results in compression in the film favoring in-plane magnetization.

4.1.7 Experimental Set-up

Figure 34 shows a standard apparatus for magneto-optical investigations at low temperatures, which was equipped with an optical cryostat operating below 10 K. The cryostat is installed directly on the stage of a polarizing microscope in reflection mode. The magneto-optical film and superconducting sample are fixed on a cooling block put in a system of coils to generate an external magnetic field. It is important that the magneto-optical film side (not substrate side) is placed in intimate contact with the surface of the sample for good thermal conduction and maximum spatial resolution. Although magneto-optical measurements provide local information on the magnetic flux density in a sample, one must calibrate the magneto-optical light intensity in terms of the flux density.

Using a magneto-optical film with perpendicular magnetization one can observe a strip domain pattern in the demagnetizing state. Figure 35 shows that the width of a domains increases with the field for the domain in the

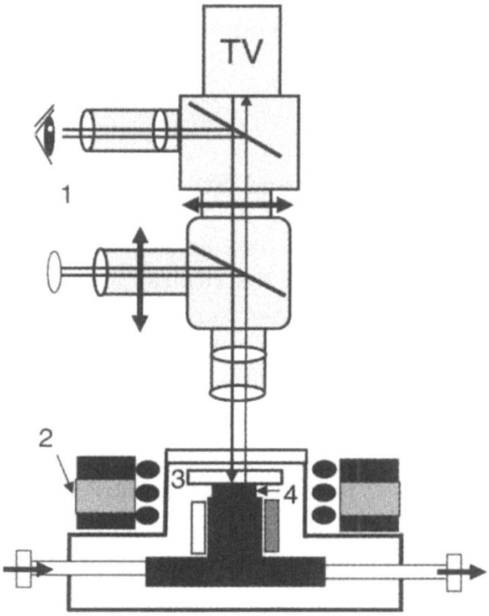

Fig. 34. Apparatus for magneto-optical imaging for cryogenic experiments: 1) polarizing microscope, 2) solenoids for external fields, 3) magneto-optical indicator, 4) sample on cooling stage [35]

Fig. 35. Schematic drawing of strip domains in different magnetic fields [33]

direction of an applied field. The dependence of the width on an applied field is strongly affected by the characteristics of a magneto-optical material. Usually a simple relation between the width and field is obtainable only for a limited range of temperature and magnetic field for each material [52].

For a magneto-optical indicator with in-plane magnetization, the calibration of the light intensity for the flux density is rather straightforward because the Faraday rotation angle depends linearly on the magnetic moment along the optical pass. However, for a detailed study of the flux distribution in a strip with transport current, the Faraday rotation is suppressed by a small magnetic moment perpendicular to the light beam as well as the in-plane anisotropy of magnetization [53].

4.2 Studies of Flux Density Profiles and Critical States

The flux density profile is a direct result obtained from magneto-optical measurements. Analyses of magneto-optical images of flux distributions are presented for a sample in the critical state [54], where the critical current flows parallel to the sample edges. Lines, so-called *d*-lines, form a hip-roof distribution of magnetic flux, along which the critical current is discontinuous. Since the vortices cannot cross the lines, the Meissner phase contracts along these lines and remains there up to a magnetic field at which the Meissner phase vanishes. One can expect to observe such behaviors by the magneto-optical image especially in thin samples.

Magnetic flux penetration in polycrystalline MgB_2 slabs was investigated by magneto-optical imaging [55]. Figure 36 shows typical magneto-optical images of the flux distributions for a MgB_2 slab in a perpendicular applied field. Figure 36a shows magnetic flux penetration in the zero field cooled sample at 20, 35 and 40 K, while Fig. 36b shows magneto-optical images of the trapped flux after an applied field of 160 mT was turned off. The images clearly exhibit the hip-roof patterns predicted by Schuster's analysis [53]. The images and corresponding flux density profiles across the sample in the field parallel to its long edge are shown in Fig. 37. In this configuration demagnetization effects are much reduced, which greatly simplifies the analysis of magneto-optical images. The field was first increased from 0 to 0.1 T and then reduced to zero at 37 K. It is worth noting that a constant flux gradient independent of an applied field indicates characteristics of the Bean critical state.

4.3 Dynamic Observations of Magnetic Flux

The magneto-optical imaging technique has a great advantage for the study of dynamic processes. The time resolution is limited by the image detector. Typically magneto-optical images are recorded with a digital charge coupled device camera and transferred to a computer for processing. Relaxation effects are noticeable in high-temperature superconductors due to high flux creep rates even in the absence of transport current [56]. In the presence of transport current the relaxation of flux and current density profiles is more complicated as shown in Fig. 38 [57]. The profiles were measured for $YBa_2Cu_3O_7$ at different times during a 5000 ms pulse. The relaxation tends to increase the flux density throughout the sample and to redistribute the current density more uniformly. Some extra current moves from the edges to the central region.

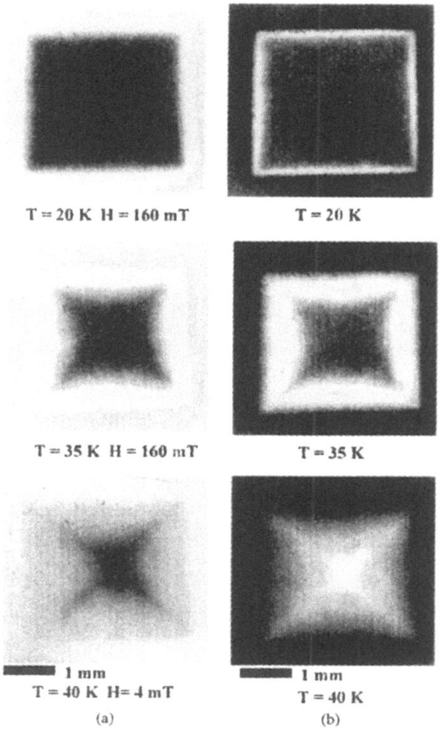

Fig. 36. Magnetic flux distribution in a MgB_2 slab: (**a**) flux penetration after zero field cooling: (**b**) distribution of trapped magnetic flux after turning off an applied field of 160 μTT [55]

4.4 Differential Magneto-Optical Technique

A magneto-optical image of good quality is obtained by averaging raw images from a charge coupled device camera. For further improvement in resolution, the differential scheme of imaging is adopted to compensate for images independent of magneto-optical signals, for example, the variations in the reflectivity and crystallographic defects in a magneto-optical indicator. The differential technique is also useful for recording the motion of flux in a sequence of magnetic fields or temperatures.

The melting process of a vortex lattice was studied as a model of a first-order phase transition using differential images for a small change of the field [58]. Figure 39 shows differential magneto-optical images during vortex lattice melting in $Bi_2Sr_2CaCu_2O_8$. The images were obtained by subtracting the image at an applied field from the image taken after an increase of the corresponding field by 0.1 mT. At 15.95 mT a liquid droplet appears, which is seen as a bright spot. As the field increases the droplet changes in shape to a ring-like domain, which is the solid-liquid interface region. These visualized

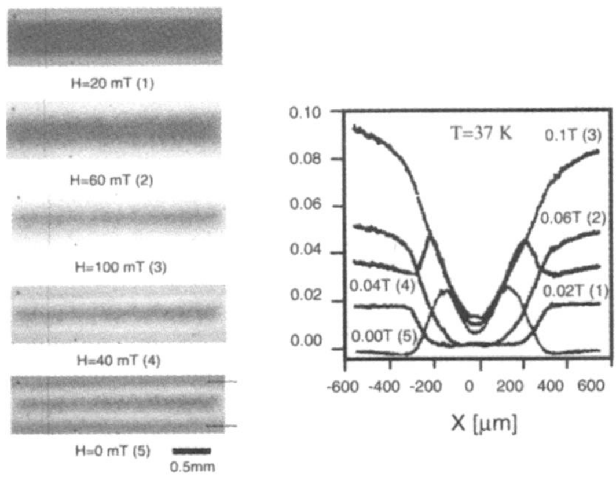

Fig. 37. Magneto-optical images (**a**) and corresponding flux density profiles for a MgB_2 slab in parallel fields at 37 K (**b**) [55]

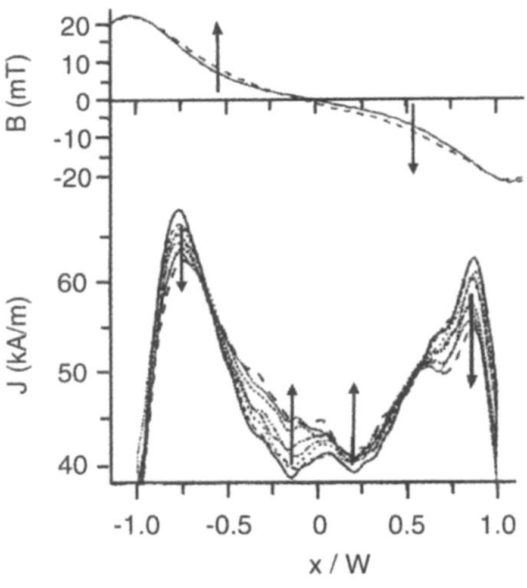

Fig. 38. Relaxation of flux and current density distributions in $YBa_2Cu_3O_7$ strip during a current pulse of 4.9 A. The profiles were observed at 40 and 1600 ms for flux density and at 40, 100, 200, 300, 450, 800, 1200 and 1600 ms for current density. The *arrows* show a sequence of increasing time [57]

melting processes make it possible to discuss the behavior of vortex lattice using the analogy of crystal growth.

The propagating flux front presents the instability in $YBa_2Cu_3O_7$ thin films when it is triggered by a local perturbation of the superconducting state. The dendritic penetration of the flux is observed after laser heating a spot using high speed magneto-optical imaging [59]. A similar dendritic behavior

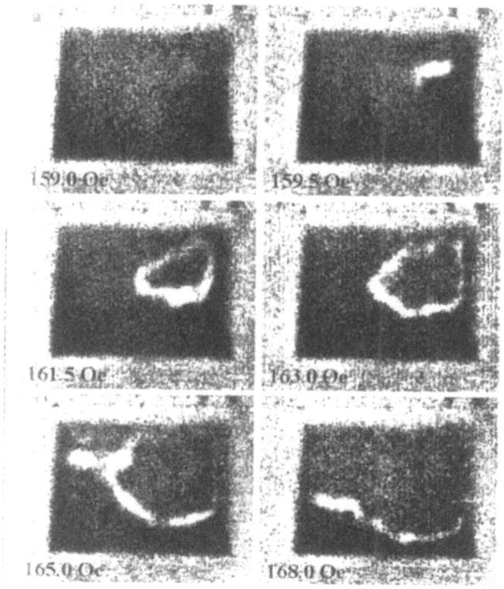

Fig. 39. Vortex lattice melting process in $Bi_2Sr_2CaCu_2O_8$ crystal at 60 K and applied fields along the *c*-axis [58]

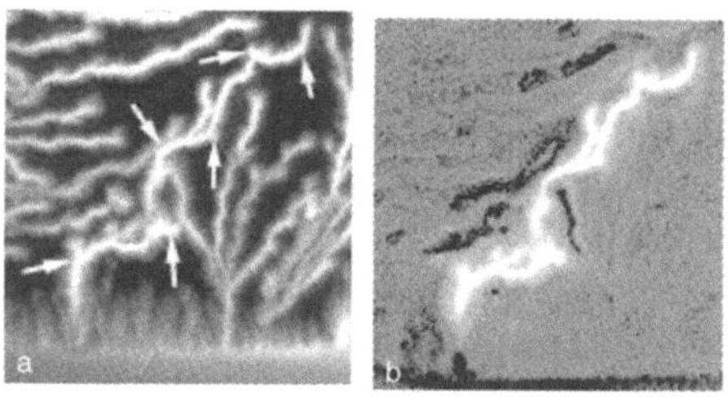

Fig. 40. Magneto-optical images of flux penetration into a zero field cooled MgB_2 film. (**a**) The dendrite changes its growth direction several times (indicated by *arrows*). (**b**) Differential image; the subtraction of (**a**) and the image taken before the last dendrite appears [60]

is observed in MgB_2 films [60]. Figure 40 shows the magneto-optical images of flux penetration into a zero field cooled MgB_2 film. The growth of each dendritic structure took place during less than 1 ms. The dendrites grown in a stronger applied field are seen to be brighter, i.e. including more flux than those formed in a weaker field. A newborn dendrite changes its direction of growth due to repulsion from existing dendrites as shown by arrows in Fig. 40a. In order to enhance the behavior of a single flux line the differential image (Fig. 40b) was composed from images before and after the last flux appeared. There are dark regions indicating a decrease in flux density near the newborn dendrite. It suggests that a new dendrite may consume some vortices in the existing dendrites.

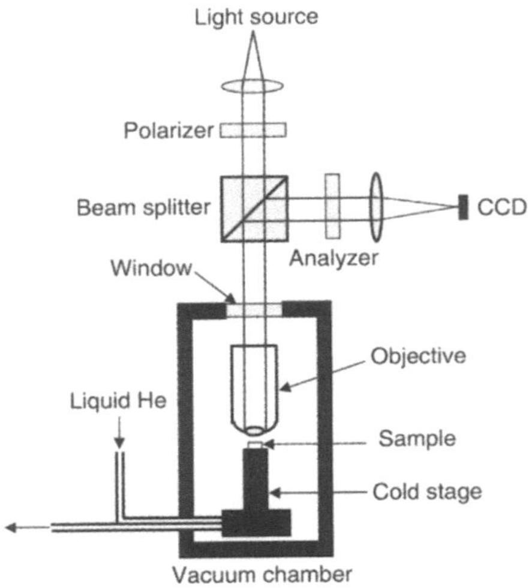

Fig. 41. Schematic drawing of the experimental set-up for observation of quantized magnetic vortex [61]

4.5 Real-Time Observations of Single Vortex

The visualization of individual vortices is the final goal of magneto-optical imaging. As compared with reported methods, magneto-optical imaging has the advantage of a high temporal resolution using a rather simple experimental set-up [61]. The set-up that makes it possible to observe a single vortex is shown in Fig. 41. The main improvements are to suppress a leakage magnetic field of vortices from between the sample and the magneto-optical detector and to minimize degradation of plane polarized light, which results in poor contrast of images. More exactly, in order to minimize the distance between the sample and the garnet film, they omitted the reflective layer and used the sample's surface itself as a mirror. The objective lens was put in the vacuum chamber so that the window was not between the sample and the objective.

Figure 42 shows the magneto-optical image of vortices in a $NbSe_2$ single crystal ($T_c = 7.2\,K$) at 4.0 K after cooling without magnetic shielding. The image of the vortex lattice was obtained by subtracting two raw images recorded with an analyzer rotated 2 degrees from the crossed position. This differential scheme increased the signal-to-noise ratio. The individual vortices were resolved up to 1 mT, corresponding to a distance between vortices of 1.4 µm.

4.6 Summary

The magneto-optical imaging technique has grown up enough to resolve individual vortices and to allow real-time investigation of flux dynamics. Possible

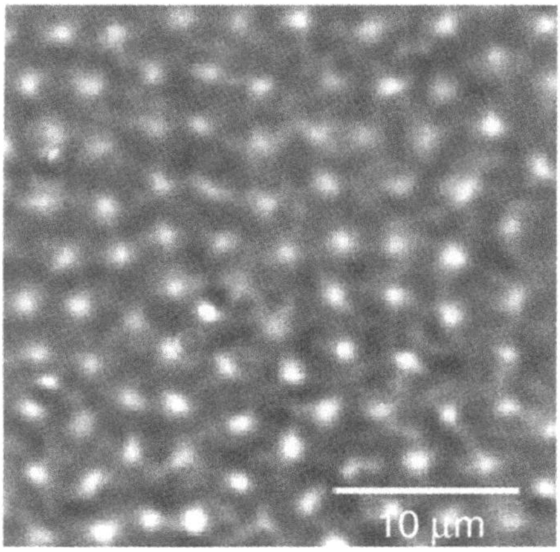

Fig. 42. Vortices in a $NbSe_2$ at 4.0 K. The scale bar is 10 μm [61]

studies in future include imaging of dynamical processes at the order-disorder transition in a vortex lattice, imaging of vortices interacting with pinning centers and imaging of rearrangement of the vortex distribution affected by transport current.

5 Terahertz Radiation Imaging

The terahertz (THz) radiation imaging technique can provide a unique and powerful tool to study supercurrent distributions in vortex-penetrated HTS thin films. Femtosecond (fs) optical pulses are employed to excite the pulsed electromagnetic (EM) radiation [62] with a THz spectral bandwidth (THz radiation) from superconductive thin films. The THz radiation mechanism is explained by partial supercurrent modulation without phase transition. Since the radiation amplitude is proportional to the local supercurrent density at the optically excited area [63], one can observe the supercurrent distributions by detecting the THz radiation images with scanning fs laser pulses on the sample surface. This THz radiation imaging technique has the following advantages [64,65,66,67,68,69,70]:

- Quantitative distributions can be obtained.
- Vector-map-distributions can be obtained.
- One can easily observe the temperature dependence of the distributions.
- The technique is a nondestructive, noncontact and free-space tool.

In this section the THz radiation imaging technique is reviewed. The supercurrent density distribution (SDD) in $YBa_2Cu_3O_{7-\delta}$ (YBCO) thin film strips is observed under various conditions and the vortex penetration is discussed.

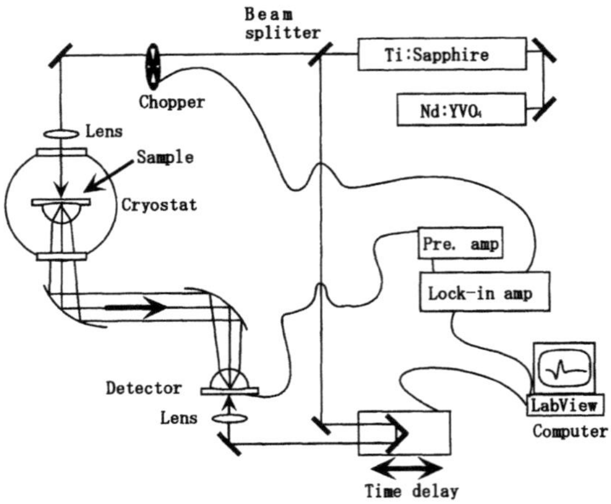

Fig. 43. Schematic view of the experimental setup

5.1 THz Radiaton Imaging System

Figure 43 shows a schematic configuration of the femtosecond time-resolved EM pulse generation and detection system [67]. A mode-locked Ti:sapphire laser operating at a repetition rate of 82 MHz is used to produce 50 fs pulses with a center wavelength of about 800 nm. The fs light pulses are focused by an objective lens onto the sample mounted on a cold finger of a closed-cycle helium cryostat. THz radiation emitted from the opposite side is collimated and focused by a pair of off-axis paraboloidal mirrors onto the detector through a Si hemispherical lens. An Au/Ge/Ni alloy bow-tie antenna fabricated on low-temperature grown GaAs (LT-GaAs) is used as a photoconductive detector. The detector is triggered by fs pulses separated from the main beam by a beam splitter. The integrated photocurrent is lock-in detected after being amplified. The EM pulse waveforms in the time domain are monitored by changing a delay time between the main and trigger pulses. After fixing the delay time at the maximum amplitude position, the laser beam is scanned to visualize the THz radiation images by translating the sample. The movable sample holder is connected to a cold base of the refrigerator using a metal mesh and also to an exterior X-Z stage controlled by LabView using a thermally insulating polyethylene rod. The spatial resolution is limited by the laser spot size. Using the beam profiler, the minimum full-width at half maximum (FWHM) of the beam diameter was measured to be 25 μm.

YBCO films grown on 0.5-mm-thick MgO substrates with a thickness of 100 nm are patterned into a stripline structure or other structures using conventional photolithographic techniques and Ar ion milling. A hemispherical MgO lens with a diameter of 3 mm is attached to the back side of the substrate with vacuum grease to increase the collection efficiency of the THz radiation when we observe small current. Although the MgO lens enhances

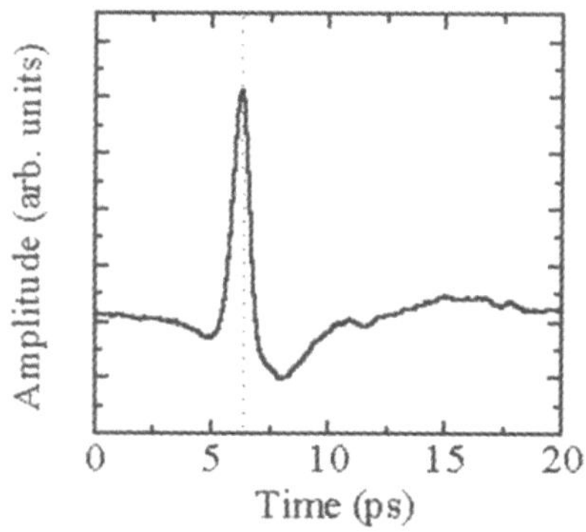

Fig. 44. Typical waveform of the THz radiation emitted from YBCO thin film strip

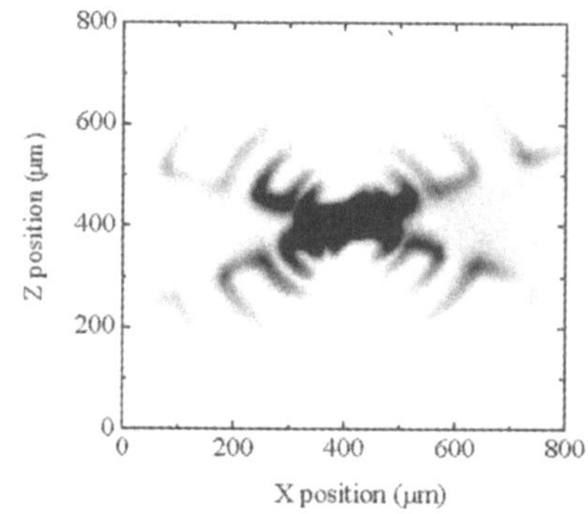

Fig. 45. Two-dimensional plot of the THz radiation amplitude observed from YBCO thin film log-periodic antenna

the collection efficiency about 10 times more than that without the lens, the observable area is limited to about 200 μm in diameter.

Figure 44 shows a typical THz radiation waveform emitted from a YBCO strip. Since we use the bow-tie antenna detector to increase the detection efficiency [71], the temporal waveform is integrated in the time domain compared to the original emission waveform traced by a dipole antenna detector [72]. The waveform shows that the radiation has its maximum amplitude at around 6.3 ps. This amplitude is proportional to the local current density at the optically excited area [63]. After fixing the time delay at 6.3 ps as indicated by the dashed line in Fig. 44, the sample is scanned along the x and z directions. Figure 45 shows an example of the THz radiation peak amplitude distribution. A supercurrent of 150 mA was applied to the log-periodic antenna which has a 30 μm wide bridge at the center at 16 K. The LT-GaAs bowtie antenna mainly detects the radiation polarized in the x direction as expected from the antenna structure so that only the x component of the supercurrent distribution is shown.

The quatitative local current density distribution in the stripline struture is estimated from the detected THz radiation amplitude in the following manner. A constant supercurrent is applied to the strips and the distribution is observed. We first obtain the conversion function to change the amplitude distribution mirror-symmetric at the center of the bridge to avoid a slight distortion in the images due to misalignment of the system. We then estimate the current density by calibrating the distribution so that the integrated amplitude becomes identical with the bias current. We apply the calibration function from the observed THz radiation amplitude images to the current density distribution. Since the radiation efficiency depends on the sample temperature, the calibration function is corrected by taking the temperature dependence into account.

The observation is easily affected by optical influence. In the present experiments, the laser power P_L is less than 5 mW. Typically 3 mW is used for the THz excitation. The optical excitation with P_L larger than 10 mW induces redistribution of the vorties in the remanent state.

The observed images are the results of the convolution between the laser beam profile and real supercurrent distribution. The theoretical distribution under various condition in the superconductive thin films has been intensively studied by many researchers. *Zeldov* et al., and *Blandt* et al. have derived basic formulas for the supercurrent distributions for these films [73,74]. For instance, the distribution in the remenent state is expressed by the following equations.

$$J_x(z) = \begin{cases} \frac{2J_c}{\pi}\left[\arctan\left(\sqrt{\frac{W^2-a_0^2}{a_0^2-z^2}}\right) \right. \\ \qquad \left. -\,2\arctan\left(\sqrt{\frac{W^2-a^2}{a^2-z^2}}\right)\right], & -a_0 \le z \le a_0 \\ J_c\left[1-\frac{4}{\pi}\arctan\left(\sqrt{\frac{W^2-a^2}{a^2-z^2}}\right)\right], & a_0 < |z| \le a \\ -J_c, & a \le |z| < W \end{cases} \tag{2}$$

where J_c and 2W are the critical current density and the width of the strips, respectively. a_0 and a correspond to the field-free region at the edge and are given by

$$a_0 = W\sqrt{1-\left(\frac{I_B}{I_c}\right)^2}, \qquad a = W\sqrt{1-\left(\frac{I_B}{2I_c}\right)^2}, \tag{3}$$

where I_c is the critical current of the strip and I_B is the bias current.

The real distribution is discussed by fitting the convolution curves between the theoretical distribution and the observed beam profile to the data. Only I_c is taken as a parameter to calculate the fits to the data. The vortex penetration due to the transport current or external magnetic fields is examined. For the case of vortex penetration due to the external magnetic field, a field B_{ext} of 0.9 mT or 25 mT is applied perpendicular to the sample surface by placing the permanent magnet outside the cryostat after the sample being cooled without an intentional external field, which corresponds to a zero field cool (ZFC) state.

5.2 Vortex Penetration Due to Transport Supercurrent

Supercurrent density distributions (SDDs) along the z axis in the 114 µm wide YBCO thin film strip are shown in Fig. 46a at various bias currents. The distributions indicate that the supercurrent flows near the strip edges, which is qualitatively explained by Bean's critical state model for thin films [73,74]. Vortices penetrate into the superconductive thin film strips only when the bias current reaches the critical current at the edges. Since the entry field for the thin films is very small [75,76] they easily penetrate from the edges and are trapped inside the film when the transport current is removed. Figure 46b shows the SDDs in the remnant state after removing the transport current. The distributions indicate that a vortex trap near the strip edges due to the self-field occurs easily in the strip with antidots. The clear vortex trap is observed after removing the transport currents of 250 mA.

The solid lines in Fig. 47 are the fits calculated by convolution between the laser profile and the $J_x(z)$ given by (2). Close agreement between observed and calculated distributions was obtained with an I_c of 7 A. This value corresponds to a J_c of 6×10^7 A/ cm^2 at 17 K. The convolution fits suggest that the real supercurrent is confined in the very thin layer near the edges and

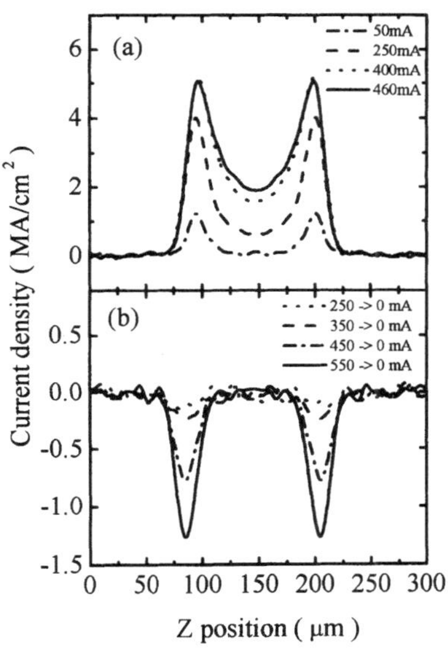

Fig. 46. Supercurrent density distributions along z direction in the 114 µm wide strip (**a**) at various transport currents, and (**b**) in the remnant state. Measured at 17 K

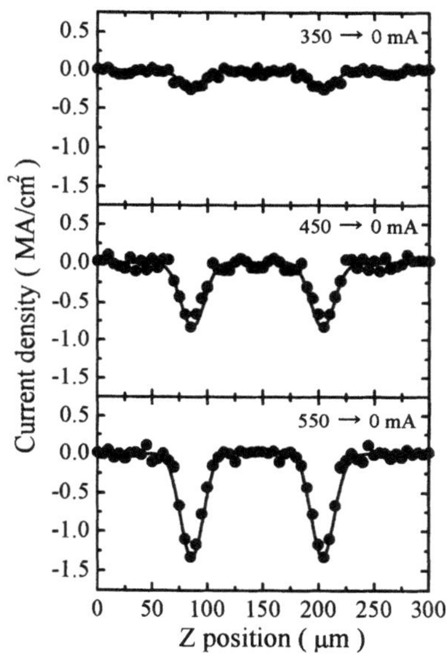

Fig. 47. Supercurrent density distributions after removal of the bias current. *Solid lines* are the fits to the data calculated by convolution between the observed laser beam profile and the theoretically expected distribution on the assumption of an I_c of 4 A

the vortex penetration depth is of the order of sub-microns. Since the imaging resolution is much larger than the values, the estimated distribution may contain uncertainty. However, we see no strong disagreement between the theory and observed data within our experimental conditions.

5.3 Vortex Entry at Weak Magnetic Field

Figure 48a shows the SDDs at a B_{ext} of 0.9 mT and in its remnant state. The supercurrent under the magnetic field flows in opposite directions along each edge to shield the field, and the direction is reversed after removal of the field. This indicates that the field of 0.9 mT was large enough to penetrate into the inner part of the strip. The vortices were trapped near the edge of the strips in the remnant state. The formula for the distributions under the magnetic field and in the remnant state, similar to (2), quantitatively explains both distributions by assuming the same I_{c} of 4 A as indicated by the solid lines in Fig. 48a. (The sample is different from that used in Figs. 4 and 5.)

The corresponding magnetic flux density distributions (MFDDs) are estimated by solving Biot-Savart's law as illustrated in Fig. 48b. The field is

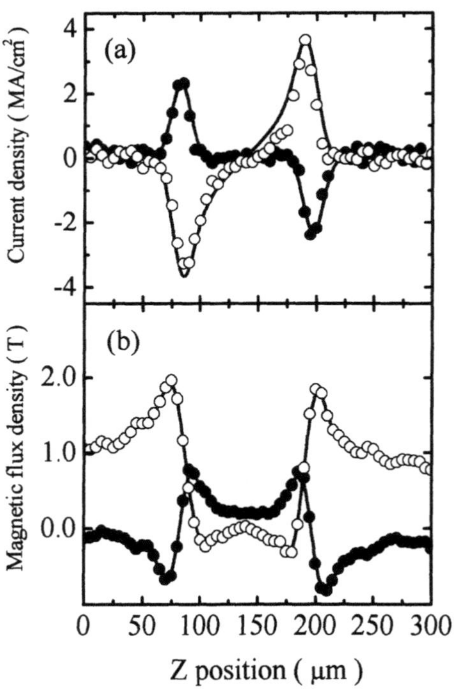

Fig. 48. (**a**) Supercurrent density distributions under B_{ext} of 0.9 mT (*open symbols*) and its remnant state (*closed symbols*). The *solid lines* are the calculated fits. (**b**) Corresponding magnetic field calculated from the current distribution. The *solid lines* are drawn as a guide to the eye

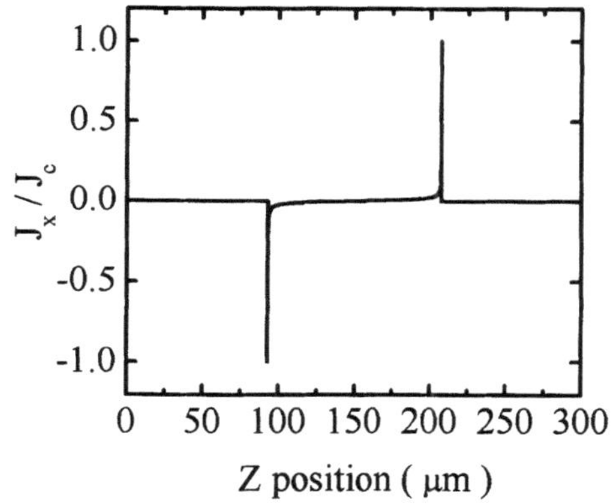

Fig. 49. Theoretical supercurrent distribution normalized by J_c. The function was used to calculated the curve in Fig. 48a under a magnetic field

compressed near the strip edge and no flux penetrates into the center of the strip. Since the magnetic flux density at the center of the strip and far outside the strip is estimated at zero and 1 mT, the results indicate that the calibration from the detected THz radiation amplitude of the supercurrent density is quantitatively correct within our experimental resolution.

Since the obtained distributions are visualized by the convolution between the laser beam profile and the supercurrent distribution, the real distribution is expected to be confined very near the edges. The curve in Fig. 49 is a function to obtain the fit in Fig. 48a under the field. This suggests that the supercurrent distribution is simply explained by an edge current with a width less than 1 μm, and the width decreases after removal of the field. The observed laser beam profile coincides almost exactly with the supercurrent distribution in the strip in the remnant state. Although expressions are not available in the present case that has a very narrow vortex penetration depth compared to the London penetration depth, the results at least indicate that vortex penetration at a B_{ext} of 0.9 mT is limited to a very narrow region near the edges, e.g., less than 0.5 μm.

5.4 Vortex Penetration Due to a Strong Magnetic Field

In the presence of a strong magnetic field vortices penetrate into the deep part of the superconductor. Figure 50a shows the observed SDD in the remnant state after removal of a B_{ext} of 25 mT. The solid line is a fit to the data assuming an I_c of 2 A for the same strip examined in Fig. 49. The observed SDD is quite different from the calculated one. The corresponding MFDD suggests that the vortex is distributed less near the center of strip as depicted in Fig. 50b. In order to discuss this phenomenologically, an artificial function for MFDD is defined by multiplying the original by the function $f(z) = \frac{3}{4W^2}z^2 + \frac{1}{4}$ as displayed in Fig. 51a. The convolution between the

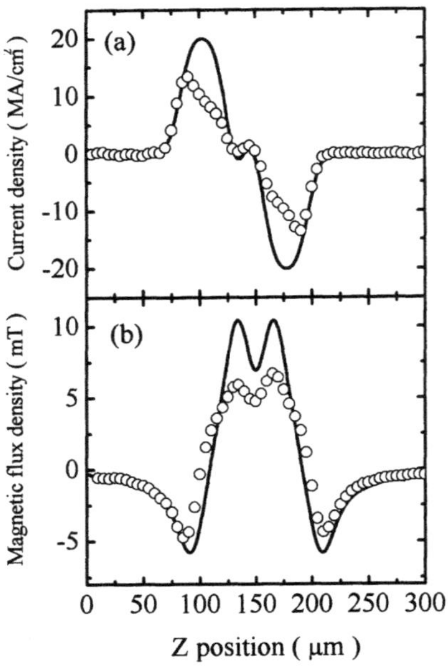

Fig. 50. (**a**) Supercurrent density distribution in the remnant state after removal of a $B_{\rm ext}$ of 25 mT. (**b**) Corresponding magnetic field calculated from the current distribution. The *solid lines* are the calculated curve assuming an I_c of 2 A

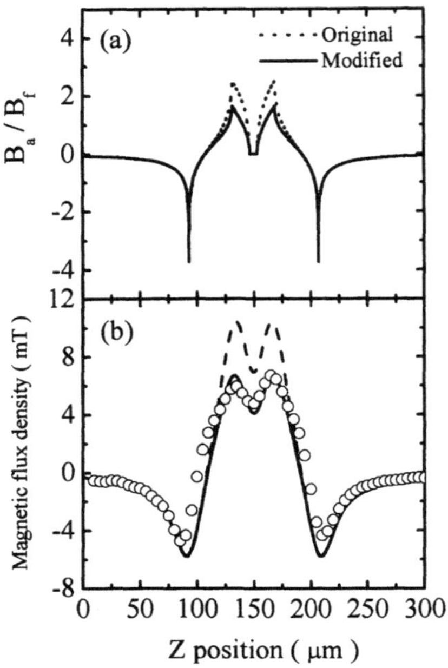

Fig. 51. (**a**) Modified theoretical magnetic flux density distribution (*solid line*) in comparison with the original curve (*dashed line*). (**b**) Calculated convolution (*solid line*). The *open circles* represent the same data in Fig. 50b

modified MFDD and the laser beam profile qualitatively agrees with the observed MFDD as depicted in Fig 51b. This suggests that near the deep part of the strip vortices exist less compared to the theoretically expected distribution. This is presumably attributed to pairing of vortex and anti-vortex near the center, which are depinned due to the transport current and/or thermal excitation.

5.5 Temperature Dependence of the Trapped-Vortex Behavior

The critical current density decreases with increasing temperature, which should affect the vortex penetration effects. Figure 52 shows the SDD curves in the remnant state after removal of a $B_{\rm ext}$ of 25 mT at various temperatures. The supercurrent decreases with increasing temperature monotonically, but the rate of decrease is smaller near the center than near the edges. This is naturally acceptable becasue the trapped vortices escape from the edges.

The temperature dependence of the maximum current density at the stripline edge for the various conditions is summarized in Fig. 53. The strips under

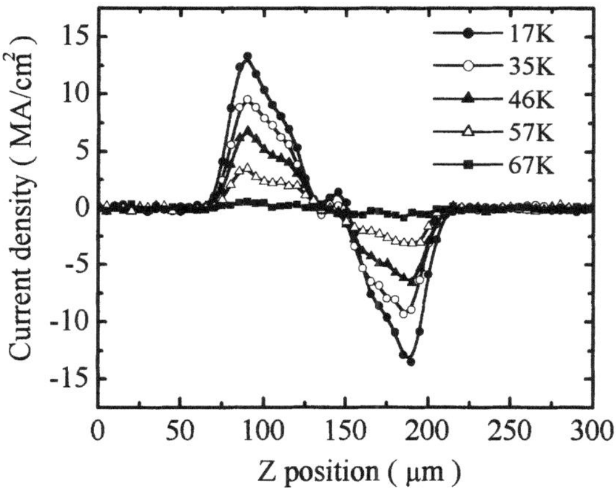

Fig. 52. Supercurrent density distributions in the remnant state at various temperatures. The ZFC remnant state iwas generated by removal of a $B_{\rm ext}$ of 25 mT at 17 K

a $B_{\rm ext}$ of 0.9 mT and in its remnant state after removal of $B_{\rm ext}$ show a weak temperature dependence of the maximum supercurrent density below 60 K. However, above this value the supercurrent density rapidly decreases with increasing temperature. The rate of decrease of the supercurrent density in the strip under the field above 60 K is larger than that in the vortex-trapped strip. This would be attributed to optical excitation effects. By neglecting these effects, one can expect similar behavior in both cases. Namely, the supercurrent distribution and the trapped-vortex distribution are not affected by the thermal activation below around 60 K. On the other hand, the temperature dependence in the strip with trapped vortices under a $B_{\rm ext}$ of 25 mT differs remarkably from the above dependencies. The rate of decrease of the maximum supercurrent density below 60 K is observed to be much larger than that in the former cases. This thermal effect was confirmed by measuring the maximum current density at temperatures of 16.5 K, 35 K and 16.5 K, in that order. We were able to obtain almost the same current density in the last two cases. It is also confirmed that the maximum current densities in the strip before and after the sample temperature is elevated to 35 K from the initial flux-trapped state and then cooled down to 16.5 K without imaging at 35 K are identical to the first and the last, respectively. The results indicate that the decrease in the maximum current density with increasing temperature below 60 K originates in the thermal activation effect.

One may expect that the thermal effect can be attributed to the temperature dependence of the critical current density $J_{\rm c}(T)$ in the YBCO thin film. Here, we discuss the temperature dependence of the maximum current density in the strip in the remnant state trapped with a $B_{\rm ext}$ of 25 mT. At temperatures below the near-temperature regime $J_{\rm c}(T)$ has been explained

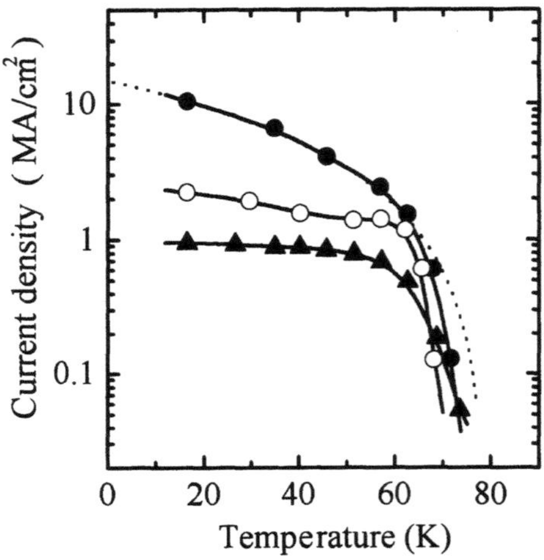

Fig. 53. Temperature dependence of the maximum supercurrent density. *Open circles*, *closed triangles*, and *closed circles* represent the data measured under the conditions (**a**) applied with B_{ext} of 0.9 mT, (**b**) in the remnant state after removal of B_{ext} of 0.9 mT, and (**c**) in the remnant state after removal of B_{ext} of 25 mT, respectively. The *solid lines* are drawn as a guide to the eye; the *dashed line* represents the fit to the data with a power law

by a nearly linear dependence [77,78], or by a flux-creep-controlled transport current of the form [79,80]

$$J_c(T) = J_0 \left(1 - \alpha\frac{T}{T_c} - \beta\frac{T}{T_c}^2\right) , \tag{4}$$

where α and β are coefficients and J_0 is the critical current density at 0 K. The linear dependence cannot explain the observed temperature dependence. Curve fitting to the relationship given by (4) gives $\alpha = 1.5$ and $\beta = -0.45$. Since α and β are usually selected to be close to 1 and positive the flux creep controlled model also cannot explain such dependence. *Baziljevich* et al. explained the temperature dependence under a weak magnetic field of 60 mT by an exponential term of the form [81]

$$J_c(T) = J_0 \exp\left(-A\sqrt{\frac{T}{T_c}}\right) , \tag{5}$$

where A is a constant. However, this relationship also failed to explain the dependence. We then tried to explain it by a power-law dependence. The dashed line in Fig. 53 is a fit to the data below 65 K with a power law of the form

$$J_c(T) = J_0 \left(1 - \frac{T}{T_c}\right)^n \tag{6}$$

which yields $n = 1.5$ and $J_0 = 15\ \mathrm{MA/cm^2}$. The nonlinear behavior with $n = 1.5$ is explained by the Ginzburg-Landau theory [82]. In HTS thin films, however, this dependence is usually observed only near T_c. Therefore, the observed temperature dependence of the maximum current density in the remnant state cannot be simply explained by the temperature dependence of the critical current density. *Cukauskus* et al. reported that composite thin films of Au and YBCO prepared on various substrates have large n, which strongly depends on the substrates and the deposition conditions. They suggested that the grain boundary plays an important role in determining the temperature dependence [83]. As seen in Fig. 52 the vortex behavior near the film edges differs from that inside the film. The rate of decrease of the maximum current density with increasing temperature at the edges is larger than that in the inner part of the strip. This suggests that the edge has a weaker pinning force than the inner part of the strip, and may originated in process damage, which would affect the coupling properties between the grains. The coexistence of this edge effect and the intrinsic one in the strip would account for the observed temperature dependence. Above around 60 K we observed an abrupt deviation from the power law dependence in the maximum current density. *Cao* et al. found a logarithmic decrease of the critical current density with increasing temperature near T_c, and explained it by the pinned vortex liquid [84]. The weak trapped magnetic field in the remnant state might be responsible for the pinned vortex liquid regime formation.

5.6 Summary

THz radiation imaging technique has proven to be a powerful tool for nondestructive, noncontact and free-space evaluation to study vortex penetration into superconductive thin film strips for practical use. The distributions in the remnant state after vortex penetration due to the transport current or the external magnetic field can be quantitatively observable.

The observed distributions in comparison with the theoretical ones indicate that though the supercurrent distributions in the remnant state where the vortices are mainly trapped only near the edges are well described by the conventional model, the vortex distributions after a large amount of vortex entry show discrepancies with the theory. The temperature dependence of the distributions also reveals that, below 60 K, thermal activation produces no significant effects on the penetrated vortices at a B_{ext} of 0.9 mT. However, the vortices in the remnant state after removal of the field of 25 mT are strongly affected by thermal activation.

References

1. A. A. Abrikosov: J. Exp. Theor. Phys. **32**, 144 (1957) (in Russian) [Soviet Phys. JETP **5**, 1174 (1957)]
2. H. Träuble, U. Essmann: Phys. Status Solidi **25**, 373 (1968) (in German)
3. H. Träuble, U. Essmann: Phys. Status Solidi **18**, 813 (1966) (in German)
4. K. Harada, T. Matsuda, J. Bonevich, E. Yoshida, U. Kawabe, A. Tonomura: Phys. Rev. Lett. **71**, 3371 (1993)
5. H. Hallen, R. Seshadri, A. Chand, H. Hess: Phys. Rev. Lett. **71**, 3007 (1993)
6. C. Renner, A.D. Kent, P. Niedermann, O. Fischer: Phys. Rev. Lett. **67**, 1650 (1991)
7. P.E. Goa, H. Haublin, M. Bazilievich, E. Il'yashenko, P.L. Gammel, T.H. Johansen: Supercond. Sci. Technol. **14**, 729 (2001)
8. M. Tonouchi, S. Shikii, M. Yamashita, K. Shikita, T. Kondo, O. Morikawa, M. Hangyo: Jpn. J. Appl. Phys. **37**, L1301 (1998)
9. P.L. Gammel, D.J. Bishop, D.J. Dolan, J.R. Kwo, C.A. Murray, L.F. Shneemeyer, J.V. Waszczak: Phys. Rev. Lett. **59**, 2592 (1987)
10. P.L. Gammel, D.J. Bishop: Phys. Rev. Lett. **68**, 3343 (1992)
11. L. Ya. Vinnikov, I.V. Grigovieva, L.A. Gurevich, A.E. Koshelev, Yu. A. Ossipyan: J. Less-Common Met. **164**, 1271 (1990)
12. S. Ohshima, T. Koseki, A. Kamimura: Jpn. J. Appl. Phys. **37**, L375 (1998)
13. S. Kanno, T. Kawai, S. Satou, K. Aizawa, M. Kusunoki, M. Mukaida, S. Ohshima, S. Hayashi, T. Nishizaki, N. Kobayashi: Physica C **357–360**, 625 (2001)
14. S. Ohshima, A. Kamimura, K. Okuyama: *Advances in Superconductivity IX* (Springer, Berlin, Heidelberg 1997) p. 479
15. L.N. Vu, D.J. Van Harlingen, IEEE Trans. Appl. Supercond. **3**, 1918 (1993)
16. J.R. Kirtley, IEEE Spectrum, Dec. **41** (1996)
17. J.R. Kirtley, C.C. Tsuei, J.Z. Sun, C.C. Chi, Lock See Yu-Jahnes, A. Gupta, M. Rupp, M.B. Ketchen: Nature **373**, 225 (1995)
18. C.C. Tsuei, J. Kirtley: Rev. Mod. Phys. **72**, 969 (2000)
19. J.R. Kirtley, K.A. Moler, G. Villard, A. Maignan: Phys. Rev. Lett. **81**, 2140 (1998)
20. M. Jeffery, T. Van Duzer, J.R. Kirtley, M.B. Ketchen: Appl. Phys. Lett. **67**, 1769 (1995)
21. K. Suzuki, Y. Li, T. Utagawa, K. Tanabe: Appl. Phys. Lett. **76**, 3615 (2000)
22. K. Suzuki, S. Adachi, Yijie Li, T. Utagawa, K. Tanabe: IEEE Trans. Appl. Supercond. **11**, 238 (2001)
23. K. Suzuki, H. Suzuki, S. Adachi, T. Utagawa, U. Kawabe, K. Tanabe: Physica C **357–360**, 1469 (2001)
24. K. Suzuki, T. Takagi, S. Adachi, T. Utagawa, N. Koshizuka, K. Tanabe: Physica C **357–360**, 1469 (2001)
25. T. Morooka, S. Nakayama, A. Odawara, M. Ikeda, S. Tanaka, K. Chinone: IEEE Appl. Supercond. **9**, 3491 (1999)
26. T. Morooka, S. Nakayama, A. Odawara, K. Chinone: Jpn. J. Appl. Phys. **38**, L119 (1999)
27. E. Dantsker, S. Tanaka, John Clarke: Appl. Phys. Lett. **70**, 2037 (1997)
28. J. Mannhart, H. Hilgenkamp, B. Mayer, Ch. Gerber, J.R. Kirtley, K.A. Moler, M. Sigrist: Phys. Rev. Lett. **77**, 2782 (1996)

29. Yijie Li, K. Tanabe: Physica C **324**, 198 (1999)
30. K. Tanabe, K. Suzuki, Yijie Li, T. Utagawa, N. Inoue, T. Sugano, Y. Honami, U. Kawabe, M. Hidaka, T. Satoh, S. Tahara: Physica C **341–348**, 2691 (2000)
31. M. Hidaka, T. Satoh, M. Koike, S. Tahara: IEEE Trans. Appl. Supercond. **9**, 4081 (1999)
32. J. F. Dillon: *Magnetism 3* (Academic, New York 1963) p. 415
33. A. P. Malozemoff, J. C. Slonczewski: *Magnetic Domain Walls in Bubble Materials* (Academic, New York 1979) p. 41
34. L. D. Landau, E. M. Lifshitz: *Electrodynamics of Continuous Media* (Pergamon, New York 1959) Chap. 83
35. A. A. Polyanskii, X. Y. Cai, D. M. Feldmann, D. C. Larbalestier: *Nano-Crystalline and Thin Film Magnetic Oxides* (Kluwer Academic, Dordrecht 1999) p. 353
36. M. R. Koblischka, R. J. Wijngaarken: Supercond. Sci. Technol. **8**, 199 (1995)
37. P. B. Alers: Phys. Rev. **105**, 104 (1957)
38. W. de Sorbo: Phys. Rev. Lett. **4**, 406 (1960)
39. E. Batalla, E. G. Zwartz, R. Goudreault, S. L. Wright: Rev. Sci. Instrum. **61**, 2194 (1990)
40. H. Kirchner: Phys. Lett. A **30**, 437 (1969)
41. H. U. Habermeier, R. Aoki, H. Kronmuller: Phys. Status Solidi A **50**, 187 (1978)
42. M. R. Koblischka, N. Moser, B. Gegenheimer, H. Kronmuller: Physica C **166**, 36 (1990)
43. B. Dutoit, L. Rinderer: Jpn. J. Appl. Phys. **26**, 1661 (1987)
44. T. Tamegai, T. Shibauchi, S. Ooi, H. Nakagawa, T. Kumeno: Adv. Supercond. **7** (Springer, Berlin, Heidelberg 1995) p. 419
45. N. Iwata, T. Takagi, T. Machi, T. Morishita, K. Kohn: Adv. Supercond. **12** (Springer, Berlin, Heidelberg 2000) p. 350
46. J. F. Dillon, Jr.: J. Appl. Phys. **29**, 539 (1958)
47. P. Hansen, J.-P. Krumme: Thin Solid Films **114**, 69 (1984)
48. H. Takeuchi, K. Shinagawa, S. Taniguchi: Jpn. J. Appl. Phys. **12**, 465 (1973)
49. R. M. Grechishkin, M. Yu. Goosev, S. E. Il'yashenko, N. S. Neustroev: J. Magn. Magn. Mater. **157/158**, 305 (1996)
50. W. Tolksdorf, C.-P. Klages: Thin Solid Films **114**, 33 (1984)
51. Y. Yokoyama, Y. Hasumi, H. Obara, Y. Suzuki, T. Katayama, S. Gotoh, N. Koshizuka: Jpn. J. Appl. Phys. **30**, L714 (1991)
52. S. Gotoh, N. Koshizuka: Physica C **176**, 300 (1991)
53. T. H. Johansen, M. Baziljevich, H. Bratsberg, Y. Galperin, P. E. Lindelof, Y. Shen, P. Vase: Phys. Rev. B **54**, 16265 (1996)
54. Th. Schuster, M. V. Indenbom, M. R. Koblischka, H. Kuhn, H. Kronmuller: Phys. Rev. **49**, 3443 (1994)
55. A. A. Polyanskii, A. Gurevich, J. Jiang, D. C. Larbalestier, S. L. Bud'ko, D. K. Finnemore, G. Lapertot, P. C. Canfield: Supercond. Sci. Technol. **14**, 811 (2001)
56. Y. Yeshurun, A. P. Malozemoff, A. Shaulov: Rev. Mod. Phys. **68**, 911 (1996)
57. A. V. Bobyl, D. V. Shantev, Y. M. Galperin, T. H. Johansen, M. Baziljevich, S. K. Karmanenko: LANL preprint cond-mat/0108047
58. A. Soibel, E. Zeldov, M. Rappaport, Y. Myasoedov, T. Tamegai, S. Ooi, M. Konczykowski, V. B. Geshkenbein: Nature **406**, 282 (2000)
59. P. Leiderer, J. Boneberg, P. Brull, V. Bujok, S. Herminghaus: Phys. Rev. Lett. **71**, 2646 (1993)

60. T. H. Johansen, M. Baziljevich, D. V. Shantsev, P. E. Goa, Y. M. Galperin, W. N. Kang, H. J. Kim, E. M. Choi, M-S. Kim, S. I. Lee: Supercond. Sci. Technol. **14**, 726 (2001)
61. P. E. Goa, H. Hauglin, M. Baziljevich, E. Il'yashenko, P. L. Gammel, T. H. Johansen: Supercond. Sci. Technol. **14**, 729 (2001)
62. M. Tonouchi, M. Tani, Z. Wang, K. Sakai, S. Tomozawa, M. Hangyo, Y. Murakami, S. Nakashima: Jpn. J. Appl. Phys. **35**, 2624 (1996)
63. M. Tani, M. Tonouchi, M. Hangyo, Z. Wang, N. Onodera, K. Sakai: Jpn. J. Appl. Phys. **36**, 1984 (1997)
64. S. Shikii, T. Kondo, M. Yamashita, M. Tonouchi, M. Hangyo: Appl. Phys. Lett. **74**, 1317 (1999)
65. M. Tonouchi, S. Shikii, M. Yamashita, K. Shikita, T. Kondo, O. Morikawa, M. Hangyo: Jpn. J. Appl. Phys. **37**, L1301–L1303 (1998)
66. O. Morikawa, M. Yamashita, H. Saijo, M. Morimoto, M. Tonouchi, M. Hangyo: Appl. Phys. Lett. **75**, 3387–3389(1999)
67. M. Tonouchi, M. Yamashita, M. Hangyo: J. Appl. Phys. **87**, 7366–7375 (2000)
68. A. Moto, M. Hangyo, M. Tonouchi: IEICE Trans. Electron. **E84-C**, 67–73 (2001)
69. M. Tonouchi, A. Moto, M. Yamashita, M. Hangyo: IEEE Trans. Appl. Supercond. **11**, 3230 (2001)
70. A. Moto, M. Tonouchi: Physica C **357–360**, 1603 (2001)
71. M. Tonouchi, M. Tani, Z. Wang, K. Sakai, S. M. Hangyo, N. Wada, Y. Murakami: IEEE Trans. Appl. Supercond. **7**, 2913 (1997)
72. M. van Exter, Ch. Fattinger, D. Grischkowsky: Appl. Phys. Lett. **55**, 337 (1989)
73. E. Zeldov, J. R. Clem, M. McElfresh, M. Darwin: Phys. Rev. B **49**, 9802 (1994)
74. E. H. Brandt: Phys. Rev. B **52**, 15442 (1995)
75. J. Z. Sun, W. J. Gallagher, R. H. Koch: Phys. Rev. B **50**, 13664 (1994)
76. A. V. Kunznetsou, D. V. Ermenko, V. N. Trofimov: Phys. Rev. B **59**, 1507 (1999)
77. C. Ferdeghini, E. Giannini, G. Grassano, D. Marre, I. Pallecchi, A. S. Siri: Physica C **294**, 233 (1994)
78. J. Fang, X. J. Xu, J. Z. Shi, X. W. Cao: Physica C **246**, 119 (1995)
79. J. Mannhart, P. Chaudhari, D. Dimos, C. C. Tsuei, T. R. McGuire: Phys. Rev. Lett. **61**, 2476 (1988)
80. L. W. Song, E. Narumi, F. Yang, D. T. Shaw, H. M. Shao, Y. H. Kao: J. Appl. Phys. **69**, 1538 (1991)
81. M. Baziljevich, T. H. Johansen, H. Bratsberg, Y. Galperin, P. E. Lindelof, Y. Shen, P. Vase: Physica C **266**, 127 (1996)
82. M. Thinkam: *Introduction to Superconductivity* (MaGraw-Hill, New York 1975)
83. E. J. Cukauskas, L. H. Allen: Physica C **313**, 11 (1999)
84. X. W. Cao, Z. H. Wang, K. B. Li: Physica C **305**, 68 (1998)

New Aspect of Vortex in HTSC

M. Tonouchi[1], G. Oya[2], Y. Matsuda[3], and K. Kumagai[4]

[1] Research Center for Superconductor Photonics, Osaka University
2-1 Yamadaoka, Suita, Osaka 565-0871, Japan
tonouchi@rcsuper.osaka-u.ac.jp
[2] Department of Electrical and Electronic Engineering, Utsunomiya University
7-1-2 Yoto, Utsunomiya 321-8585, Japan
ohya@cc.utsunomiya-u.ac.jp
[3] Institute for Solid State Physics, The University of Tokyo
Kashiwanoha, Kashiwa-shi, Chiba 277-8581, Japan
ym@issp.u-tokyo.ac.jp
[4] Division of Physics, Graduate School of Science, Hokkaido University
Kita-ku, Sapporo 060-0810, Japan
kumagai@phys.sci.hokudai.ac.jp

Abstract. Some topics related to a new aspect of the vortices observed in high-T_c superconductors (HTSCs) are reviewed in this chapter. One of the most important physical properties of vortices created in type-II superconductors is that a vortex line can support a magnetic flux with a flux quantum. Recently the possibility that a vortex can accumulate a finite electric charge as well has come to be realized. The sign and magnitude of the vortex charge is not only closely related to the microscopic electronic structure of the vortex, but also strongly affects the dynamical properties of the vortex.

1 Introduction

Section 2 describes the discussion on the electrostatics and the charge distribution in the vortex state, demonstrating that a vortex in high-T_c superconductors indeed traps a finite electronic charge, using the high resolution measurements of nuclear quadrupole frequencies. The vortex dynamics associated with the vortex charge, our attention on the vortex Hall effect, is also discussed.

Section 3 describes a basic model for a stack of Josephson junctions, the basic properties of intrinsic Josephson junctions in $Bi_2Sr_2CaCu_2O_y$ (BSCCO) single crystals, some properties of vortices in the intrinsic Josephson junctions and the behavior of vortices in stacks of Josephson junctions. Strongly anisotropic layered high-T_c superconductors such as $Bi_2Sr_2Ca_nCu_{n+1}O_{2n+6}$ ($n = 1, 2$) represent natural stacks of atomic scale intrinsic Josephson junctions, in which metallic and superconducting CuO_2 single, double or triple layers alternating with semiconducting or insulating other layers are stacked in the c-axis direction, and moreover every pair of adjacent supeconducting layers, together with the intervening nonmetallic layers, is weakly coupled

T. Kobayashi et al. (Eds.): Vortex Electronics and SQUIDs,
Topics Appl. Phys. **91**, 103–139 (2003)

by the Josephson tunnel effect. When a magnetic field is applied parallel to the layers of the stacked junctions in the superconductors, Josephson type vortices (fluxons) can be induced between the superconducting layers of the junctions, with circulating currents both along and across the layers. In this case, strong inductive coupling between the junctions in the stack can occur, and hence Josephson type vortices in the stacked junctions can exhibit many static and dynamic behaviors different from those in a single Josephson junction. Thus, the behavior of vortices in stacked junctions is an interesting object not only for fundamental studies on nonlinear dynamics but also for electronics applications to a variety of devices, such as high frequency oscillators and detectors in the range of GHz–THz, and further voltage standards and high-speed switches.

One of the most important aspect of the HTSC materials is the low reflection coefficient of the visible light. The effective photon absorption in HTSC induces a new type of nonequilibrium state. As one of the attractive fields, one can effectively study ultrafast phenomena in quantized system by means of the femtosecond fs optical excitation. Excitation and observation of ultrafast phenomena in the solid state are an essential interest in the field of condensed matter physics. Superconductivity can provide an excellent platform to open a new field such as transient quantum physics. The coherent states quantizing macroscopically are based on the minimum free energy of the ensemble system. Fs-laser illumination on the system could bring out strong local perturbation, which would propagate in both the time and spatial domains, and result in the generation of the new quantized states. In Sect. 3, as an example of such effects, optical vortex generation effects in high-T_c superconductors by means of optical supercurrent modulation with a single shot fs optical pulse is demonstrated.

2 Electrostatics and Charge Distribution in the Vortex State of Type-II Superconductors

A vortex in a type-II superconductor can be viewed as comprised of two parts: the inner core characterized by the coherence length ξ and the circulating superfluid electrons outside characterized by the magnetic penetration length λ. The superconducting energy gap Δ is suppressed in the core region and goes to zero at the center. Perhaps one of the most fundamental physical properties of the vortex created in type-II superconductors is that a vortex line can support a magnetic flux with a flux quantum $\Phi_0 = hc/2e$ ($= 2.07 \times 10^{-7}\,\mathrm{Oe{\cdot}cm^2}$) [1]. This fact was confirmed experimentally more than 4 decades ago. On the other hand, it is only very recently that another prominent feature, namely the possibility that a vortex of the superconductor can accumulate a finite electric charge as well, has been proposed [2,3,4,5,6]. The vortex charge appears as a result of the chemical potential difference between the vortex core and the region away from the core. If electrons are expelled from the vortex core, the core is positively charged and hence a hole-like vortex appears, while if electrons (holes) are trapped inside (expelled from) the core, the core is negatively charged and an electron-like vortex appears. The sign and magnitude of the vortex charge reflect the fundamental nature of the superfluid electrons and the low energy excitation out of the condensate. Moreover, it has also been pointed out recently that the vortex charge strongly affects the dynamical properties of the vortex [7,8,9]. For instance, the origin of the vortex Hall anomaly, in which the Hall sign change its sign in the superconducting state, has been attributed to the vortex charge. Clarification of the issue of the vortex charge serves as an important test of the predictions for the vortex electronic structure and dynamics.

Here we study the electrostatics in the vortex state of high temperature superconductors (HTSC) by high resolution measurements of the nuclear quadrupole frequency ν_{Q}, which is very sensitive to the local charge distribution [10,11]. We show that a vortex in HTSC indeed traps a small but finite electronic charge as well. In slightly overdoped $YBa_2Cu_3O_7$ the vortex is negatively charged, while in underdoped $YBa_2Cu_4O_8$ it is positively charged. The sign of the trapped charge is opposite to the sign predicted by the conventional BCS theory. Moreover, in both materials the accumulated charge is much larger than expected in ordinary superconductors. We discuss several possible origins of these discrepancies.

2.1 Electrostatics of the Vortices

2.1.1 Vortex Charge

We first briefly introduce the electrostatics of type-II superconductors. In the cores of conventional superconductors the distances between each discrete energy levels of the quasiparticles in the Andréev bound states are merely of the order of a few mK. Thus it is sufficient to view the energy levels as forming a continuous spectrum, just like in a normal metallic state [1,12]. Generally the chemical potential μ in the superconducting state differs from that in the normal state if an electron-hole asymmetry is present. Assuming therefore that the vortex core is a normal metallic region surrounded by superconducting materials, this difference in μ is expected to arise and should lead to redistribution of the electrons. In order to maintain the same electrochemical potential on both sides, charge transfer occurs between the core and the outside.

In the framework of the BCS theory taking into account the metallic screening effect, the charge accumulated within the vortex core Q_ξ per layer normal to the magnetic field is given as

$$Q_\xi \approx \frac{2ek_{\mathrm{F}}s}{\pi^3}\left(\frac{\lambda_{\mathrm{TF}}}{\xi}\right)^2\left(\frac{d\ln T_{\mathrm{c}}}{d\ln\mu}\right), \tag{1}$$

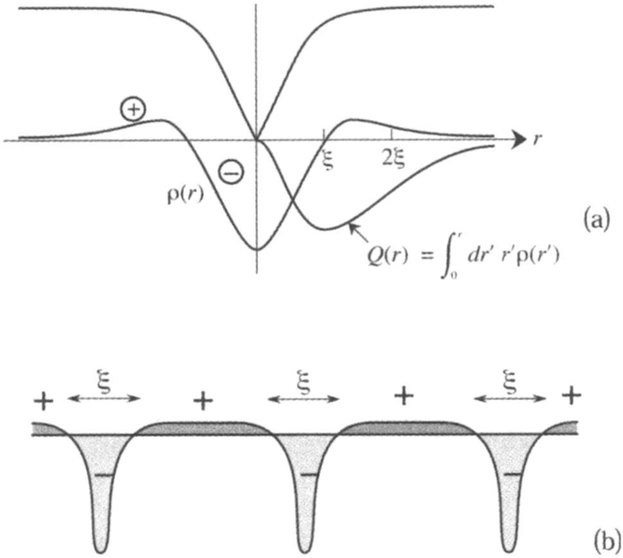

Fig. 1. (**a**) Schematic figure of the charge distribution around a single vortex core when the electrons are trapped within the core (negatively charged core). $\Delta(r)$ is the superconducting energy gap. $\rho(r)$ is the charge density. The charge accumulated inside the core is screened by charges with opposite sign. $\rho(r)$ decays gradually as r^{-4} well outside the core region. $Q(r)$ is the total charge within the distance r. $Q(r)$ goes to zero as $r \to \infty$ due to the requirement of overall charge neutrality. (**b**) Charge density modulation in the strong magnetic field ($H_{c1} \ll H \ll H_{c2}$) where each vortices overlaps

where λ_{TF} is the Thomas-Fermi screening length, s is the interlayer distance, μ is the chemical potential and $e(>0)$ is the electron charge [3]. The sign of the core charge is determined by $d\ln T_c/d\ln\mu$, which represents the electron-hole asymmetry. Outside the core, the charges with opposite sign screen the core charge, similar to a charged particle in a metal. Far outside the core, this screening charges decay gradually with a power law dependence as r^{-4} (Fig. 1a) [3,9]. In strong fields ($H_{c1} \ll H \ll H_{c2}$) each vortex overlaps with its neighborhood. Then the charge density outside the core is nearly constant and periodic modulation of the charge density appears for the periodic vortex lattice (Fig. 1b). In ordinary superconductors, $|Q_\xi|$ is estimated to be $\sim 10^{-5}$–$10^{-6}e$ using $k_F \sim 1\,\text{Å}^{-1}$, $\lambda_{TF} \sim k_F^{-1} \sim 1\,\text{Å}$, $\xi \sim 100\,\text{Å}$ and $|d\ln T_c/d\ln\mu| \approx \ln(\hbar\omega_D/k_BT_c) \sim 1$–$10$, where ω_D is the Debye frequency. Thus $|Q_\xi|$ is negligibly small and is very difficult to observe.

However, the situation in the case of HTSC seems to be promising because ξ is extremely short compared to that of conventional superconductors. Moreover, the strong electron correlation effects and d-wave pairing symmetry of HTSC are expected to change the electronic structure of the vortex dramatically [13,14,15,16,17,18,19,20,21]. In fact, recent STM measurements revealed that the vortices of HTSC are very different from those of conventional superconductors [22,23]. These unusual features of HTSC are expected to enhance the charging. We will discuss these issues later.

The vortex charge also plays an important role for the vortex dynamics. When a vortex moves in the superfluid electrons, the core plays a key role in dissipation processes [24]. One of the most striking phenomena is the vortex Hall anomaly, namely sign reversal of the flux flow Hall effect below T_c, which is observed in most HTSC [25,26]. This Hall sign reversal indicates that the vortices move upstream against the superfluid flow. Such an unusual motion has never been observed in any other fluid including superfluid helium and cannot be explained in the framework of classical hydrodynamic theory [27]. Recently, this phenomena has been discussed in terms of the vortex charge which produces an additional force acting on the vortices [7,8,9].

2.1.2 Nuclear Quadrupole Frequency

To elucidate the vortex charge, a direct observation of the local carrier density is strongly required. It has been pointed out by many authors that ν_Q in HTSC is very sensitive to the local hole density [28,29,30,31]. In this section we briefly review the nuclear quadrupole frequency. A nucleus with spin $I > 1/2$ has a quadrupole momentum due to the inhomogeneous distribution of the charged particles. In a solid, an electron distribution with spherical asymmetry such as an unclosed 3*d* shell and noncubic surrounding ions induces a local electric field gradient (EFG) in the vicinity of the nuclei. This EFG lifts the degeneracy of the nuclear spin levels interacting with the nuclear quadrupole moment, and ν_Q corresponds to the nuclear spin energy difference. The relevant information is obtained from nuclear quadrupole resonance (NQR) in zero magnetic field and nuclear magnetic resonance (NMR) in a finite magnetic field (Fig. 2). For ^{63}Cu nuclei with spin $I = 3/2$, the NQR resonance frequency ν_Q^{NQR} is expressed as

$$\nu_Q^{\mathrm{NQR}} = \frac{e^2 Q q_{zz}}{2h}\sqrt{1+\frac{\eta^2}{3}} = \nu_Q \sqrt{\left(1+\frac{\eta^2}{3}\right)}, \tag{2}$$

where eq_{zz} is the largest principle (z-axis) component of EFG at the nuclear site and Q ($= -0.211$ barn) is the quadrupole moment of the copper nuclei. The asymmetry parameter η of the EFG defined as $\eta = |(q_{xx} - q_{yy})/q_{zz}|$ is close to zero for the Cu site in the two dimensional CuO_2 planes (Cu(2) site). In a strong magnetic field when the Zeeman energy is much larger than the quadrupole energy, each Zeeman level is shifted by the quadrupole interactions and thus two satellite peaks ($\pm 3/2 \leftrightarrow \pm 1/2$) appear on both sides of the central ($\pm 1/2 \leftrightarrow \pm 1/2$) resonance peak. The frequency difference between the upper and the lower satellites exactly coincides with $2\nu_Q$ for $H \parallel z$. It

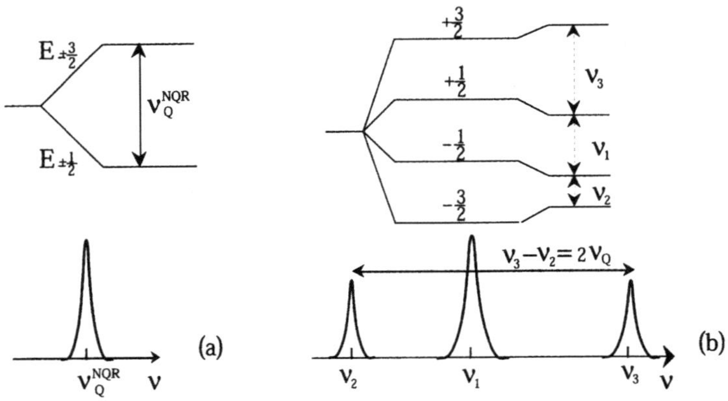

Fig. 2. (**a**) Energy splitting of ^{63}Cu nuclei with spin $I = 3/2$ in zero field. EFG lifts the the degeneracy of the nuclear spin levels and resonance occurs at $\nu = \nu_Q^{NQR}$. (**b**) In a strong magnetic field when the Zeeman energy is much larger than the quadrupole energy each Zeeman level is shifted by the quadrupole interactions and two satellite peaks ν_3 $(+3/2 \leftrightarrow +1/2)$ and ν_2 $(-3/2 \leftrightarrow -1/2)$ appear on both sides of the central ν_1 $(\pm 1/2 \leftrightarrow \pm 1/2)$ resonance peak. ν_Q is obtained from the difference between the upper and lower resonance frequencies; $\nu_3 - \nu_2 = 2\nu_Q$. In HTSC, ν_Q almost exactly coincides with ν_Q^{NQR} due to very small η

should be noted that this procedure for obtaining ν_Q is essentially free from the influence of the change of magnetic shift (or Knight shift) [32]. Moreover, the magnetic effect of the asymmetric broadening due to the vortex lattice is exactly cancelled out in the process determining ν_Q. Thus, we obtained the ν_Q values simply from the difference of the peak frequencies of the two satellite lines.

Generally the EFG originates from two different sources, namely from the on-site distributions $\nu_{\text{on-site}}$ of the electrons and from the surrounding ions ν_{ion}, $\nu_Q = \nu_{\text{on-site}} + \nu_{\text{ion}}$. Recent analysis of ν_Q on the Cu(2) site suggests that $\nu_{\text{on-site}}$ is mainly composed of the Cu $4p$ and $3d$ shell terms [28]. In HTSC the holes in the Cu $3d_{x^2-y^2}$ orbital play the most important role for the onset of superconductivity. Figure 3 shows the doping dependence of ν_Q on the Cu(2) site for $YBa_2Cu_3O_{7-\delta}$[29], $La_{2-x}Sr_xCuO_4$ [30] and $HaBa_2CuO_{4+\delta}$ [31]. In all materials, ν_Q increases linearly with the number of holes in the planes and can be written as

$$\nu_Q = A n_{\text{hole}} + C \,, \tag{3}$$

where n_{hole} is the number of holes per Cu(2) atom and A and C are constants. Although C is strongly material dependent, reflecting the difference in ν_{ion}, $A \approx 20$–30 MHz per hole per Cu(2) atom is essentially material independent. Thus measurement of ν_Q makes possible accurate determination of the local hole number.

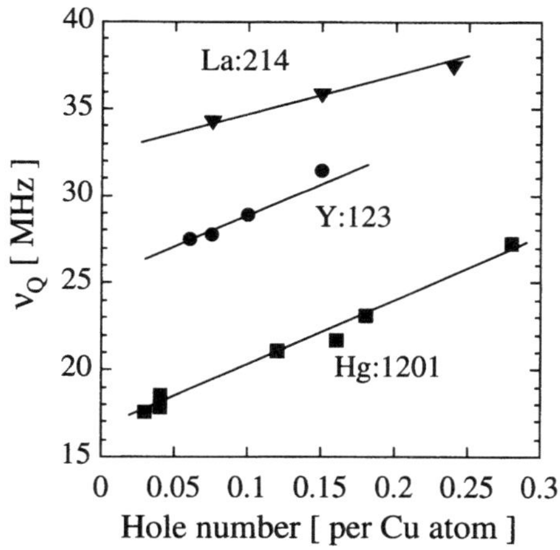

Fig. 3. Doping dependence of ν_Q on the Cu(2) site for $YBa_2Cu_3O_{7-\delta}$ [29], $La_{2-x}Sr_xCuO_4$ [30] and $HaBa_2CuO_{4+\delta}$ [31]. ν_Q is proportional to the hole number

2.1.3 Experiment

The principle of the experiment is the following. In the measurement, only the resonance of the ^{63}Cu(2) nuclei *outside the vortex core* is detected. This is because the applied field is much less than H_{c2} and hence the core region occupies a smaller area in the sample. If the vortex core traps (expels) a finite amount of electrons, the electron density outside the core should decrease (increase) from that in zero field where the electron distribution is uniform, as shown in Fig. 1b. We are able to detect the change of carrier density through the change of the value of ν_Q.

In the present measurements we used slightly overdoped $YBa_2Cu_3O_7$ and underdoped $YBa_2Cu_4O_8$ in which the NQR and NMR spectra are very sharp compared to those of other HTSC. The NMR spectra are obtained for fine powdered samples (the grains are less than 33 μm) with uni-axial alignment. Each grain aligns to an easy axis (the *c*-axis) in a high external field at room temperature, and is then fixed with epoxy (Stycast 1266).

The Cu NQR and NMR spectra were obtained by a conventional pulse spectrometer. The NMR experiments were performed in the field cooling condition under a constant field of 9.4 T by using a highly homogeneous superconducting magnet which was stabilized to less than 1 ppm during the experiment. The present measurement in the vortex state is made in the so-called Bragg glass phase in which the quasi-long-range order of the vortex lattice is preserved. We see a slightly asymmetric shape of the NMR satellite spectra due to the somehow miss-alignment of some grains. Line broadening of the NQR spectra in zero external field is not observed even below T_c in both $YBa_2Cu_3O_7$ and $YBa_2Cu_4O_8$. The width of the NMR spectra becomes broader with decreasing T below T_c partly due to the magnetic field inho-

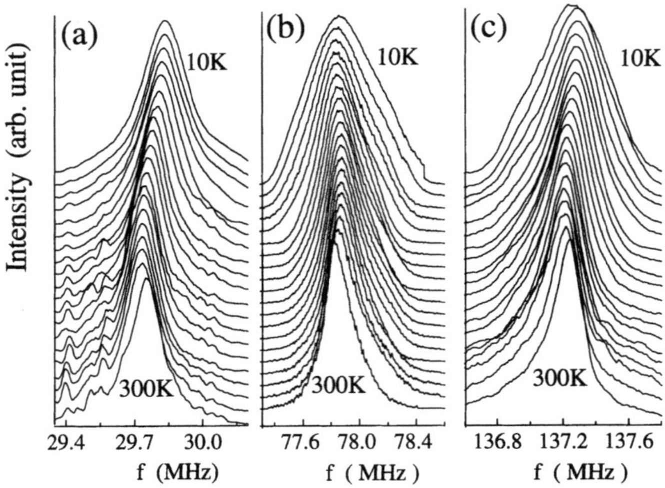

Fig. 4. The NQR in zero field (**a**) and lower (**b**) and upper (**c**) NMR-satellite spectra at 9.4 T for the ^{63}Cu(2) site of $YBa_2Cu_4O_8$ for various temperatures (300 K, 220 K, 200 K, 180 K and 160 K to 10 K by 10 K steps

mogeneity caused by the introduction of the vortices and partly due to the spatial distribution of carrier density, which we will discuss later.

2.1.4 Charge Profile in Vortices

Figure 4a–c shows also the NQR and NMR spectra of the ^{63}Cu(2) site of $YBa_2Cu_4O_8$. All data were taken by utilizing a phase coherent spin echo spectrometer. The spectra are obtained with a superposed method of Fourier Transform spectra of the spin echo measured in a certain frequency interval. In Fig. 5 we plot the difference between ν_Q in zero field and in the vortex state, $\Delta\,\nu_Q = \nu_Q(0) - \nu_Q(H)$, for $YBa_2Cu_3O_7$ and $YBa_2Cu_4O_8$. It should be noted that the procedure for obtaining ν_Q from the frequency difference between the upper and the lower satellites is essentially free from the influence of the change of magnetic shift (or Knight shift). Moreover, the magnetic effect of the asymmetric broadening (so-called Redfield pattern) due to the vortex lattice in the superconducting state is exactly cancelled out in the process determining ν_Q from NMR. Thus, we obtained the ν_Q values simply from the difference of the peak frequencies of the two satellite lines. The ν_Q in zero field for NQR is obtained after correction by the factor of $\sqrt{1+\eta^2/3}$ in (2), although this factor is at most 0.03 % of ν_Q for $\eta \approx 0.04$ for the present materials. In both materials $\Delta\,\nu_Q$ is essentially zero above T_c, indicating no modulation of the carrier density. Meanwhile a nonvanishing $\Delta\,\nu_Q$ is clearly observed below T_c in both materials. While $\Delta\,\nu_Q \sim -25$ kHz is of negative sign in $YBa_2Cu_3O_7$, $\Delta\,\nu_Q \sim 50$ kHz is of positive sign in $YBa_2Cu_4O_8$ at $T = 0$.

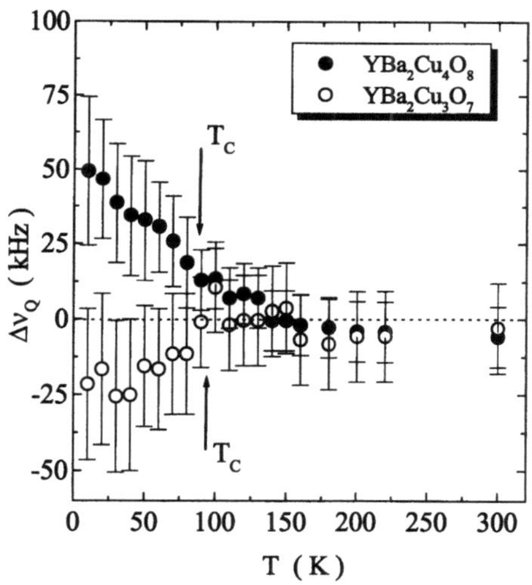

Fig. 5. T dependence of $\Delta\,\nu_Q = \nu_Q(0) - \nu_Q(H = 9.4\ \mathrm{T})$ for $YBa_2Cu_3O_7$ and $YBa_2Cu_4O_8$. In both materials nonzero $\Delta\,\nu_Q$ is clearly observed below T_c, showing that the electron density outside the core differs from that in zero field

We discuss here several possible origins for the nonzero $\Delta\,\nu_Q$. We first point out that magnetostriction cannot be the origin of the nonzero $\Delta\,\nu_Q$. In fact, both the magnetostriction and ultrasonic absorption measurements showed that the local lattice distortion $\Delta\,\ell$ caused by the magnetostriction is negligibly small under the field cooling condition: $\Delta\,\ell/\ell < 10^{-8}$ below 10 T where ℓ is the lattice constant [33]. Moreover there is no evidence of charge modulation due to CDW or stripe formation in zero field in our crystals. It is also quite unlikely that the magnetic field induces the CDW or stripe formations. Having ruled out these various possibilities, we conclude that the nonzero values for $\Delta\,\nu_Q$ naturally lead to the fact that *the electron density outside the vortex core is different from that in zero field.*

2.1.5 Sign and Magnitude of the Vortex Charge

We now discuss the issue of the sign and magnitude of the accumulated charge. The negative $\Delta\,\nu_Q$ in $YBa_2Cu_3O_7$ indicates the increment of the hole density outside the core. This excess density of holes is nothing but the holes expelled from the core. Therefore the accumulated charge in the core of $YBa_2Cu_3O_7$ is negative. By the same reasoning, the positive $\Delta\,\nu_Q$ in $YBa_2Cu_4O_8$ indicates a positive accumulated charge. Meanwhile, since the chemical potential decreases monotonically with doping holes, (1), $\mathrm{sgn}Q_\xi = \mathrm{sgn}(d\ln T_c/d\ln\mu)$ predicts that $Q_\xi > 0$ in the underdoped regime while $Q_\xi < 0$ in the overdoped regime. This is strikingly in contrast to the sign determined by the present experiment. The deviation of the magnitude of the charge from theory is also noteworthy. The magnitude of charge per pancake vortex, which is roughly estimated using $Q_\xi \approx \Delta\nu_Q H_{c2}/AH$ assuming $H_{c2} \sim 200\,\mathrm{T}$ are $Q_\xi \sim -0.005e$ to $-0.02e$ for $YBa_2Cu_3O_7$ and $Q_\xi \sim 0.01e$ to $0.05e$ for $YBa_2Cu_4O_8$. However, according to (1), Q_ξ is estimated to be $\sim 10^{-4} - 10^{-5}e$ where we assumed $\xi \sim 30\,\text{Å}$. Therefore $|Q_\xi|$ determined by the present experiments are still one or two orders of magnitude larger than expected by (1). Thus *the BCS theory not only predicts the wrong sign of the charge but also underestimates* $|Q_\xi|$ *significantly.*

There are several intriguing possible origins for these discrepancies. For example, because of the extremely short ξ, the vortex core may be in the quantum limit $k_F\xi \sim 1$ where k_F is the Fermi wave number. In this limit, the description of the quasiparticles in terms of semiclassical wavepackets breaks down in contrast to conventional superconductors [16]. Furthermore, as suggested by recent theories of the vortex core based on e.g. the t-J or SO(5) models, the antiferromagnetic (AF) state may be energetically preferable to the metallic state in the vortex core of HTSC [17,18,21]. If this is indeed so, the AF correlation is expected to enhance the charging effect because it causes a large shift of μ by changing the density of states of the electrons inside the core dramatically. We note here that the present results exclude the possibility of the SO(5) insulating AF core [17,21] in which holes should be expelled from the core and the accumulated charges are always negative; the present result yields the opposite sign for $YBa_2Cu_4O_8$ in the underdoped regime where the AF correlation is important. Therefore a detailed microscopic calculation is needed to evaluate the accumulated charge quantitatively including the sign.

2.2 Hall Anomaly and Vortex Charge

2.2.1 Vortex Hall Effect

In this section we shift to the subject of the dynamical properties of the vortex. The vortex motion in the superfluid electrons has presented a persistent problem in the superconducting state of type-II superconductors. Here we focus our attention on the vortex Hall effect. The Hall sign is determined by the topology of the Fermi surface in the normal state, while it is determined by the vortex motion in the superconducting state. In the latter, the vortex motion generates an electric field which results in longitudinal and Hall resistivities. This electric field is given by the Josephson relation

$$\boldsymbol{E} = \boldsymbol{B} \times \boldsymbol{v}_{\mathbf{s}}\,, \tag{4}$$

where $\boldsymbol{E}$ is the measured electric field and $\boldsymbol{v}_{\mathbf{s}}$ is the vortex velocity. From (4) it follows that vortex motion perpendicular to the transport current corresponds to a dissipative longitudinal electric field E_x while parallel motion corresponds to a Hall electric field E_y. Knowledge of the Hall effect enables

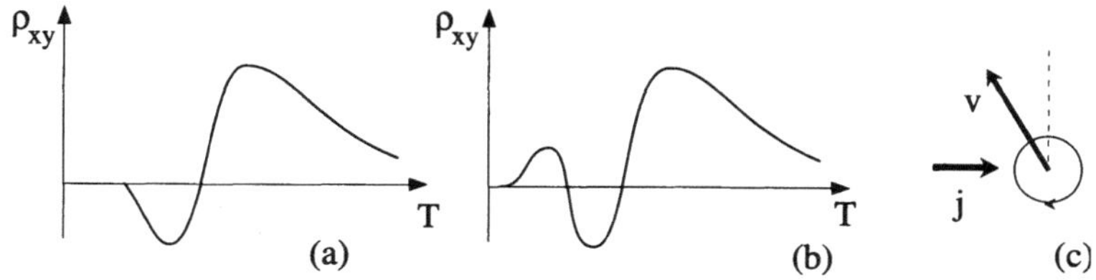

Fig. 6. Schematic figures of the vortex Hall anomaly. (**a**) The Hall resistivity ρ_{xy} changes the sign below T_c from the normal state. (**b**) A second sign change occurs in very anisotropic Bi or Tl based materials. (**c**) The negative Hall sign indicates that the vortices move upstream the superfluid flow $\boldsymbol{j}$

us to obtain important information on the problem of the energy dissipation process, which in turn reflects the electronic structure of the vortex.

A most puzzling and controversial phenomenon is the sign change that has been observed in the Hall effect in the superconducting state in most HTSC and some conventional superconductors [25,26] (Fig. 6a). Recent measurements revealed that the second Hall sign reversal occurs at still lower temperature (Fig. 6b). The occurrence of the sign reversal in one unit cell thick ultrathin $YBa_2Cu_3O_{7-\delta}$ film demonstrates that the Hall anomaly occurs within a two-dimensional CuO_2 plane [34] and is not ascribable to a three dimensional interlayer effect. When the vortices move, the Magnus force acts on them. This Hall sign reversal indicates that the vortices move upstream against the superfluid flow (Fig. 6c). This unusual motion has never been observed in any other fluid and cannot be explained in the framework of classical hydrodynamic theory [27]. In fact, the classical theories of vortex motion predict that the superconducting and normal states will have the same Hall sign. Thus, in spite of the various proposals made to explain the anomaly, e.g., as a vortex pinning induced phenomenon [35] or as collective vortex phenomena [36], the issue is still controversial. Recently, the vortex charging effect has been invoked to account for the vortex Hall anomaly. In this model the charged vortex produces an additional force which gives rise to vortex motion upstream to $\boldsymbol{v}_\mathrm{s}$ [2,7,8,9]. However, even if the charged vortices were established, the influence of the charging effect on the Hall conductivity is still controversial. For example, the results in [7,8] are opposite in sign to [2,36]. Therefore it is very desirable to have a clarification of the relation between the vortex charge and the vortex Hall anomaly.

2.2.2 Time Dependent Ginzburg–Landau Theory

Recent theories based on the time dependent Ginzburg–Landau (TDGL) equation have shed light on describing the Hall effect in the superconducting state [37,38]. According to the TDGL theory the vortex Hall conductivity $\sigma_{xy}(=\rho_{xy}/(\rho_{xx}^2+\rho_{xy}^2))$ in the flux flow state can be expressed as the sum of two contributions,

$$\sigma_{xy} = \sigma_{xy}^{\mathrm{V}} + \sigma_{xy}^{\mathrm{N}} , \tag{5}$$

where $\sigma_{xy}^{\mathrm{V}} (\propto 1/H)$ is the vortex Hall term arising from the hydrodynamic contribution while $\sigma_{xy}^{\mathrm{N}} \propto H$ is the contribution of the quasiparticle inside the vortex core. Accordingly, the Hall sign reversal occurs at low field when σ_{xy}^{V} has a sign opposite that of the normal state Hall effect. In Fig. 7 we summarize the doping dependence of the Hall sign in the superconducting state of HTSC [26]. In this figure, T_{c} is normalized by T_{c} at optimal doping, $T_{\mathrm{c}}^{\mathrm{opt}}$, for the corresponding system. The filled circles depict the electron-like Hall sign (Hall anomaly), and the open circles depict the hole-like Hall sign (no Hall anomaly). Figure 7 demonstrates that the Hall anomaly always occurs in the underdoped and slightly overdoped regimes. The Hall anomaly is sample dependent near $T_{\mathrm{c}}/T_{\mathrm{c}}^{\mathrm{opt}} \sim 0.9$ in the slightly overdoped region, but no Hall anomaly is observed beyond this regime. This behavior is observed in the crystals with monolayer, double layer, or triple layer structures. Moreover, the Hall sign depends neither on the magnitude of anisotropy nor on the pinning strength of the materials. We therefore conclude that the doping dependence shown in Fig. 7 is a universal transport property of HTSC, showing that *the*

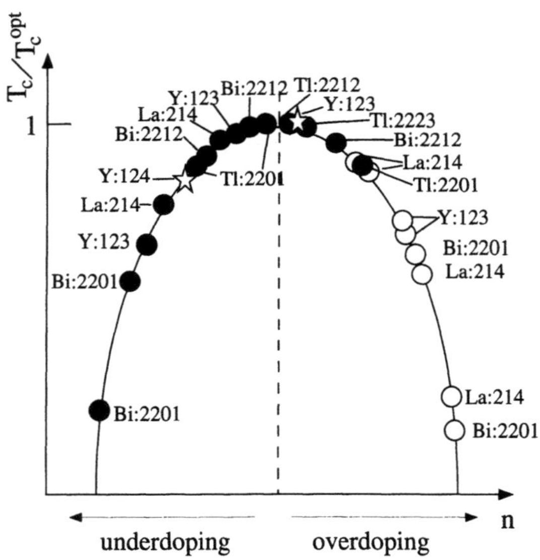

Fig. 7. Doping dependence of the Hall anomaly in the superconducting state for various high-T_{c} cuprates including La, Y, Bi, and Tl based compounds. *Filled circles* denote the presence of and *open circles* the absence of Hall anomaly. n represents the carrier number and T_{c} are normalized by $T_{\mathrm{c}}^{\mathrm{opt}}$. The Hall anomaly is present in the underdoped and slightly overdoped materials while it is absent in the overdoped materials. The *stars* show slightly overdoped Y:123 and underdoped Y:124 which are used for the present study

Hall sign in the superconducting state is closely related with the characteristic electronic structure determined by the doping.

Within the framework of the BCS theory, several authors have calculated σ_{xy}^{V} and emphasized the importance of the electronic structure of the materials for understanding the Hall effect. It has been shown that the imaginary part of the complex relaxation time ($\gamma = \gamma_1 + \mathrm{i}\gamma_2$) in the TDGL equation gives rise to σ_{xy}^{V} ($\sigma_{xy}^{\mathrm{V}} \propto -\gamma_2$) [37,38]. Accordingly the sign of σ_{xy}^{V} is given by the sign of $\mathrm{sgn}(e) d \ln T_c / d\mu$ [39]. From the effective action for the vortex motion based on the BCS Hamiltonian, σ_{xy}^{V} has been microscopically derived and it has been pointed out that σ_{xy}^{V} can be interpreted as the vortex charging effect [7]. Interestingly, the sign of the vortex Hall effect is opposite to the sign of the accumulated charge inside the core. All these calculations remain valid for s-wave weak coupling superconductors regardless of the nature of the interaction.

The present results show that the relation between the sign of the vortex charge and the sign of the vortex Hall effect is not simple. According to the photoemission experiment, μ decreases monotonically with doping holes [40]. Thus the BCS theory leads to the conclusion that the Hall sign is positive in the underdoped region where $\mathrm{sgn}(e) d \ln T_c / d\mu > 0$ and is negative in the overdoped regime where $\mathrm{sgn}(e) d \ln T_c / d\mu < 0$; the sign reversal should occur in the overdoped region and be absent in the underdoped region. This is in striking contrast to the results displayed in Fig. 7. Moreover, comparing the sign of the vortex charge determined by NMR and the vortex Hall sign, the situation turns out to be more complicated. The positively charged vortex in $YBa_2Cu_4O_8$ probed by NMR appears to be consistent with the negative Hall sign reported in [41]. On the other hand, the negatively charged vortex in $YBa_2Cu_3O_7$ probed by NMR leads to positive σ_{xy}^{V}, namely no Hall anomaly. However, a clear Hall anomaly is observed in $YBa_2Cu_3O_7$.

Therefore the prediction of the conventional TDGL equation based on the BCS theory fails to evaluate the hydrodynamic force acting on the vortices of HTSC and yields the wrong Hall sign. This discrepancy in Hall sign seems to suggest that the terms which have not been taken into consideration in the conventional TDGL theory give additional contributions to σ_{xy}^{V} [42]. It has been shown that the term arising from the requirement of the charge neutrality which has not been incorporated into the conventional theory does not change the sign of σ_{xy}^{V} from the conventional theory [43]. Quite recently it has been suggested that σ_{xy}^{V} can be opposite in sign to the conventional result of TDGL theory due to the effect of the local electric field at the vortex core, if the charging and screening effects are taken into account [9]. Thus, summarizing this situation, we conclude that the vortex Hall anomaly is closely related to the electronic structure of the vortex state, but the precise relation between the vortex charge and the vortex Hall effect still remains to be resolved.

2.3 Summary

In this subchapter we demonstrated that a vortex in HTSC indeed traps a finite electric charge, using high resolution measurements of the nuclear quadrupole frequencies. We showed that the sign of the trapped charge is opposite to the sign predicted by the conventional BCS theory and deviation of the magnitude of the charge from the theory is also significant. These unexpected features observed in the electrostatics of the vortex can be attributed to the novel electronic structure of the vortex in HTSC. We also discussed the vortex Hall anomaly. We showed that the sign of the vortex Hall effect is closely related to the electronic structure of the vortex state and again contradicts the prediction of the BCS theory. In spite of the close relationship between the vortex Hall anomaly and the electronic structure in the vortex state, the precise relation between the vortex charge and the vortex Hall effect still remains to be resolved.

3 Vortices in Intrinsic Josephson Junctions

High-T_c superconducting cuprates generally present anisotropic characteristics in directions parallel and perpendicular to the crystallographic c-axis of the materials. This originates in the two dimensional nature of the electronic state of the superconductors which are composed of a stack of metallic and superconducting CuO_2 single, double or triple layers alternating with non or weakly metallic layers as charge reservoirs donating carriers to the CuO_2 layers [44]. This structure is regarded as a natural multilayer on an atomic scale, or a natural superlattice.

$Bi_2Sr_2Ca_nCu_{n+1}O_{2n+6}$ $(n = 1, 2)$ and $Tl_2Ba_2Ca_nCu_{n+1}O_{2n+6}$ $(n = 1, 2)$ are extremely anisotropic high-T_c superconductors. In such superconductors the coherence length ξ along the c-axis (ξ_c) is comparable to or shorter than the thickness of the superconducting CuO_2 layer, but ξ along the direction perpendicular to the c-axis (ξ_{ab}) is somewhat longer [44], so that the superconducting order parameter is modulated periodically along the c-axis. Especially, $Bi_2Sr_2CaCu_2O_8$ (BSCCO) has a ξ_c of ~ 0.1 nm [45], being built of superconducting CuO_2 double layers of 0.3 nm thickness separated by insulating or semiconducting BiO and SrO layers of 1.2 nm thickness, as shown in Fig. 8 [46]. In it every pair of adjacent superconducting layers, together with the intervening nonsuperconducting layers, is weakly coupled by the Josephson tunneling effect, and hence forms an intrinsic Josephson junction on an atomic scale. Actually, this behavior was first observed in c-axis transport measurements for BSCCO by *Kleiner* et al. [47] and for similar $(Bi,Pb)_2Sr_2CaCu_2O_8$ (BPSCCO) by *Oya* et al. [48] in 1991. Since then, intrinsic Josephson junctions have been studied extensively and also observed in $Tl_2Ba_2Ca_2Cu_3O_{10}$, $Tl_2Ba_2CaCu_2O_8$, $(Bi,Pb)_2Sr_2Ca_2Cu_3O_{10}$, $(PrCe)_2CuO_4$, and underdoped $YBa_2Cu_3O_{7-\delta}$ [49,50,51,52,53,54]. Thus,

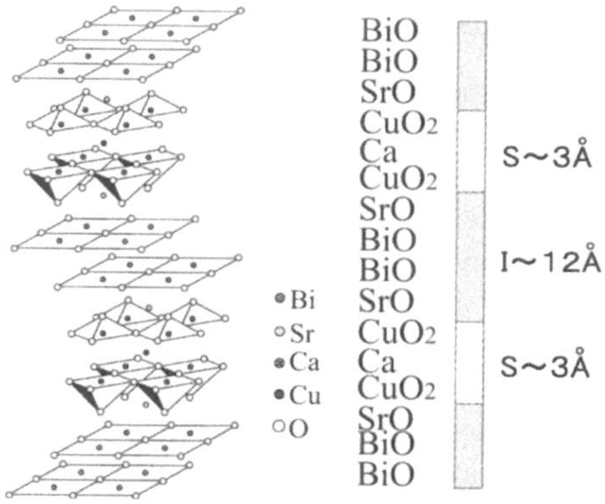

Fig. 8. Crystal structure of BSCCO and the model of stacked intrinsic Josephson junctions

strongly anisotropic high-T_c superconductors are composed of natural stacks of atomic scale intrinsic Josephson junctions as their common feature.

Moreover, such high-T_c superconductors belong to type-II superconductors. However, their magnetic properties are qualitatively different from those of ordinary type II superconductors, owing to the stacking of intrinsic Josephson junctions with the superconducting layers much thinner than the London penetration depth λ. In BSCCO λ along the direction perpendicular to the layers (λ_{ab}) is $\sim$ 150 nm and λ along the layers (λ_c) is $\sim$ 100 μm under an external magnetic field parallel to the layers [55]. Therefore, when a magnetic field is applied parallel to the layers of the stacked junctions in the superconductors, Josephson type vortices (fluxons) can be induced between the superconducting layers of the junctions, having circulating currents both along and across the layers. In this case, a strong inductive coupling between the junctions in the stack can occur, and hence Josephson type vortices in the stacked junctions can exhibit many static and dynamic behaviors different from those in a single Josephson junction. An interesting example is the collective motion of vortices moving with an interaction along different junctions in a stack, which can excite cavity resonance (Fiske resonance) or a Josephson plasma in the stack under appropriate conditions, together with the phase locking of ac Josephson oscillations of junctions in the stack. Thus, the behavior of vortices in stacked junctions is an interesting object not only for fundamental studies on nonlinear dynamics but also for electronics applications to a variety of devices, such as high frequency oscillators and detectors in the range of GHz–THz, and further voltage standards and high-speed switches [49,50,51,52,53,54].

From this point of view, a basic model for a stack of Josephson junctions, the basic properties of intrinsic Josephson junctions in BSCCO single crystals, some properties of vortices in the intrinsic Josephson junctions, and the behavior of vortices in stacks of Josephson junctions are described in this section.

3.1 Basic Model for a Stack of Josephson Junctions

Figure 9 shows a schematic drawing of a stack of N Josephson junctions ($N \gg 1$), which consists of $N+1$ superconducting layers (S layer) separated by N insulating layers (I layer), i.e., a Josephson coupled $SISI \cdots IS$ layered structure. Usually, the static and dynamic phenomena of this system are well described by the inductive coupling model on the basis of the coupled sine-Gordon equations, which has been recently developed by *Sakai* et al. [56]. This model is briefly introduced in the following.

In Fig. 9 the stack has a length L in the x direction and a width W in the y direction. The nth S layer in it has a thickness t_n and a London penetration depth λ_n and the thickness of an I layer between S layers n and $n-1$ is $d_{n,n-1}$. When an external magnetic field B is applied along y and a bias current I_b is fed into the outmost S layers labeled 0 and N, the density of the Josephson current $J^z_{n,n-1}$ flowing in the z direction across the nth Josephson junction will be described by the two Josephson equations

$$J^z_{n,n-1} = J_{\mathrm{c}n,n-1} \sin \varphi_{n,n-1} \,, \tag{6}$$

$$\hbar \dot{\varphi}_{n,n-1} = 2eV_{n,n-1} \,, \tag{7}$$

where

$$\varphi_{n,n-1} = \theta_n - \theta_{n-1} + \frac{2e}{\hbar} \int_{Q_1}^{Q_2} \boldsymbol{A} \cdot \boldsymbol{ds} \,, \tag{8}$$

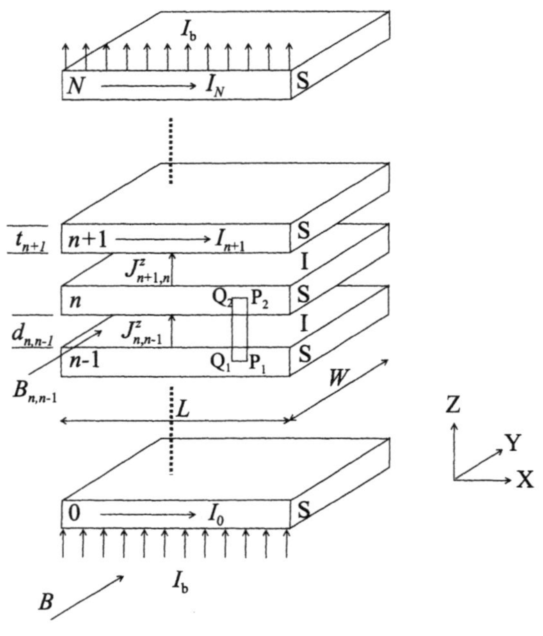

Fig. 9. Schematic drawing of a stack of N Josephson junctions

is the gauge-invariant phase difference of the order parameter between the S layers n and $n-1$ of the nth junction, $J_{cn,n-1}$ is the dc maximum Josephson current density of the junction and $V_{n,n-1}$ is the voltage across the junction. Here the phase changes to the z direction in each S layer can be ignored. Then the flux quantum can exist in the I layers as a Josephson vortex, and hence the magnetic flux Φ through the area $P_1P_2Q_2Q_1$ in Fig. 9 is given by

$$\Phi = \int B_{n,n-1} dS = \oint \boldsymbol{A} \cdot \boldsymbol{ds}\,, \tag{9}$$

where $B_{n,n-1}$ is the magnetic-flux density between S layers n and $n-1$, and $\boldsymbol{A}$ is the vector potential. Moreover, inside each S layer the magnetic-flux density obeys the London equation.

In this case, $\varphi_{n,n-1}$ can satisfy the equation

$$-\frac{\hbar}{2e}\frac{\partial \varphi_{n,n-1}}{\partial x} = s_{n-1}B_{n-1,n-2} + d'_{n,n-1}B_{n,n-1} + s_n B_{n+1,n}\,, \tag{10}$$

where $d'_{n,n-1}$ is the effective magnetic thickness, which is defined by

$$d'_{n,n-1} = d_{n,n-1} + \lambda_n \coth\left(\frac{t_n}{\lambda_n}\right) + \lambda_{n-1} \coth\left(\frac{t_{n-1}}{\lambda_{n-1}}\right)\,, \tag{11}$$

and the parameter s_n describing the coupling between layers is defined by

$$s_n = -\frac{\lambda_n}{\sinh\left(t_n/\lambda_n\right)}\,. \tag{12}$$

Moreover, from the usual Josephson junction model introducing the capacitive, resistive, and Josephson current between S layers n and $n-1$ of the nth junction, the total current is given by

$$J^z_{n,n-1} \equiv \frac{\hbar}{2e}C_{n,n-1}\frac{\partial^2 \varphi_{n,n-1}}{\partial t^2} + \frac{\hbar}{2e}G_{n,n-1}\frac{\partial \varphi_{n,n-1}}{\partial t} + J_{cn,n-1}\sin\varphi_{n,n-1}\,, \tag{13}$$

where $C_{n,n-1}$ is the unit area capacitance and $G_{n,n-1}$ the unit area conductivity of the junction. Combining these equations together with Maxwell's equations, and adding the density of bias current J_b yield the perturbed coupled sine-Gordon equation describing the dynamics of stacked Josephson junctions,

$$\frac{\hbar}{2e\mu_0}\frac{\partial^2}{\partial x^2}\begin{bmatrix} \varphi_{1,0} \\ \vdots \\ \varphi_{n,n-1} \\ \vdots \\ \varphi_{N,N-1} \end{bmatrix} =$$

$$= \begin{bmatrix} d'_{1,0} & s_1 & 0 & \cdots & & & \\ s_1 & d'_{2,1} & s_2 & 0 & \cdots & & \\ & & \ddots & & & & \\ & 0 & s_{n-1} & d'_{n,n-1} & s_n & 0 & \\ & & & & \ddots & & \\ & & & 0 & s_{N-2} & d'_{N-1,N-2} & s_{N-1} \\ & & & & 0 & s_{N-1} & d'_{N,N-1} \end{bmatrix} \begin{bmatrix} J^z_{1,0} - J_b \\ \vdots \\ J^z_{n,n-1} - J_b \\ \vdots \\ J^z_{N,N-1} - J_b \end{bmatrix} . \quad (14)$$

This inductive coupling model accounted very well for experimental observations of dynamical phenomena of Josephson vortices such as flux-flow and Fiske resonance in stacks of low-T_c Josephson junctions. It is now believed that this model is also available for the interpretation of the observed dynamic behavior of stacks of intrinsic Josephson junctions in high-T_c superconductors.

3.2 Basic Properties of Intrinsic Josephson Junctions

Various properties of the intrinsic Josephson junctions were examined by transport measurements in the *c*-axis direction of single crystals, whiskers or epitaxial thin films of the highly anisotropic high-T_c superconductors mentioned above. For the measurements several configurations of devices with dimensions down to submicron scale were fabricated by using existing and well-developed photolithography, electron beam lithography, and chemical or Ar-ion milling, focused-ion-beam etching or Si-ion implantation techniques [49,50,51,52,53,54]. In the following we focus on and briefly summarize the basic properties of intrinsic Josephson junctions in BSCCO ($T_c = 80$–85 K) as a basic material, on the basis of the current-voltage (I-V) characteristics of its small-sized single crystal mesas ($\leq 5 \times 5\,\mu\mathrm{m}^2$).

Figure 10a shows a typical overall I-V characteristic for current flow in the *c*-axis direction of a rectangular mesa of a single crystal BSCCO with an area as small as $2 \times 2\,\mu\mathrm{m}^2$ at 4.2 K without an external magnetic field. The mesa structure is shown in the inset. This mesa is composed of a stack of 6 intrinsic Josephson junctions. In this figure a clear gap structure appears at ~ 320 mV for 6 junctions, without any negative differential resistance as the effect of self-heating. Figure 10b shows the corresponding I-V characteristic on expanded scales, which was obtained by multiple sweep up and down of the current. This characteristic exhibits a superconducting current and multiple hysteresis resistive branches, whose number is equal to that of the intrinsic Josephson junctions in the mesa, each of which corresponds to individual junctions switching to the quasiparticle state as the bias current

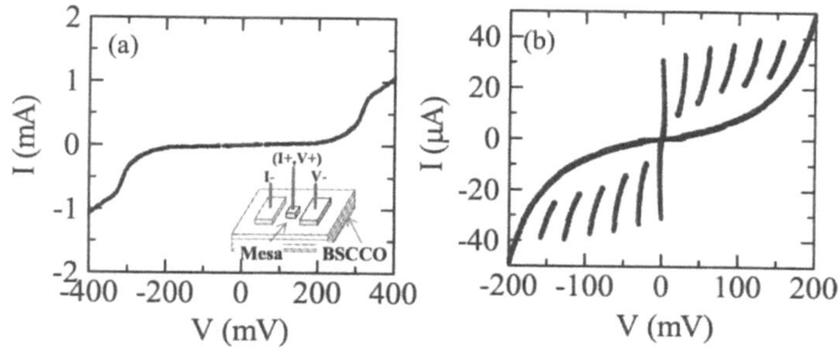

Fig. 10. (**a**) Typical overall *I-V* characteristic of a single-crystal mesa of BSCCO with an area as small as $2 \times 2\,\mu\mathrm{m}^2$, and (**b**) the corresponding one on expanded scales at 4.2 K. The *inset* shows a schematic view of the mesa structure fabricated on the single crystal

exceeds the corresponding critical current I_c. From this observation the inherent properties of intrinsic Josephson junctions in BSCCO are characterized as follows [57]:

- Each intrinsic Josephson junction in the mesa acts individually as a *SIS* Josephson tunnel junction with a critical current density J_c of $\sim$1 kA/cm^2 and a product of the critical current and the normal state resistance $I_c R_n$ of $\sim$ 2.5 mV at 4.2 K.
- The energy gap 2Δ (4.2 K) of the superconducting layer is $\sim$ 50 meV, with $2\Delta(0)/k_B T_c \sim 7$.

In addition the following properties have been also found [53,54]:

- The normalized superconducting energy gap $\Delta(T)/\Delta(0)$ has a BCS-like temperature dependence below T_c.
- A pseudogap larger than the superconducting gap appears between $\sim$ 70 K and $\sim$ 150 K.
- Fine subgap structures appear below the superconducting gap, which are interpreted as due to the interference of active optical phonons with oscillating Josephson current.
- The temperature dependence of I_c and hence J_c of junctions larger than $1 \times 1\,\mu\mathrm{m}^2$ approximately obeys but slightly deviates from the Ambegaokar–Baratoff relation, $I_c(T) = \pi\Delta(T)/2eR_n \times \tanh[\Delta(T)/2k_B T]$, reflecting the *d*-wave symmetry of the superconducting order parameter below T_c.
- As the junction area becomes less than $1\,\mu\mathrm{m}^2$, J_c decreases abruptly and then a periodic structure of current peaks appears due to the Coulomb blockade effect even at 4.2 K.
- Shapiro steps appear as the response of the ac Josephson effect to irradiation by microwaves, in particular up to 2.5 THz for the stacked junctions integrated with a bow-tie antenna.

3.3 Properties of Vortices in Intrinsic Josephson Junctions

One of the most characteristic manifestations of the Josephson effect is the Fraunhofer-type magnetic field B dependence of the maximum supercurrent I_c across a small rectangular Josephson junction, i.e., the dc Josephson effect. For a layered stack of identical Josephson junctions the behavior of $I_c(B)$ was theoretically predicted as $I_c(B) = I_c(0)|\sin(\pi B/B_0)/(\pi B/B_0)|$, with $B_0 = \Phi_0/L(t+d)$ [58]. Here, the external magnetic field is oriented parallel to the junction barrier, Φ_0 is the flux quantum, t is the thickness of the S layer and d is the thickness of the I layer. However, in order to achieve a Fraunhofer-like pattern the junction dimensions L and W should not exceed the Josephson penetration depth λ_J and the junction should be homogeneous. According to the inductive coupling model, λ_J for a single junction is given by $\lambda_J = (\hbar/2e\mu_0 d' J_c)^{1/2}$. Hence, for intrinsic Josephson junctions in BSCCO λ_J is estimated to be only $\sim$ 0.5–1 μm. This implies that small sized stacks consisting of as few as possible junctions are required to observe the dc Josephson effect in BSCCO.

Recently the Fraunhofer-like pattern has been successfully observed in small-sized intrinsic Josephson junctions of BSCCO with lateral dimensions close to λ_J [59]. Figure 11 shows the magnetic field dependence of the normalized I_c of 12 intrinsic Josephson junctions in a $1.2 \times 1.2\,\mu\text{m}^2$ mesa of BSCCO at 4.2 K. It is evidently observed that the behavior of $I_c(B)$ for all junctions in this stack is close to the Fraunhofer pattern with a B_0 of $\sim$ 1.1 T, so that individual junctions show the dc Josephson effect.

In contrast, with increasing L of the junctions to $\sim$ 20 μm the first minimum of $I_c(B)$ shifts to a lower field according to $\Phi_0/L(d+t)$, and furthermore for L much larger than λ_J the behavior of $I_c(B)$ is considerably disturbed by vortices entering the junctions at fields $B > B_{c1}$ [49,60]. An example of the magnetic field dependence of I_c for a $320 \times 40\,\mu\text{m}^2$ mesa of BSCCO at

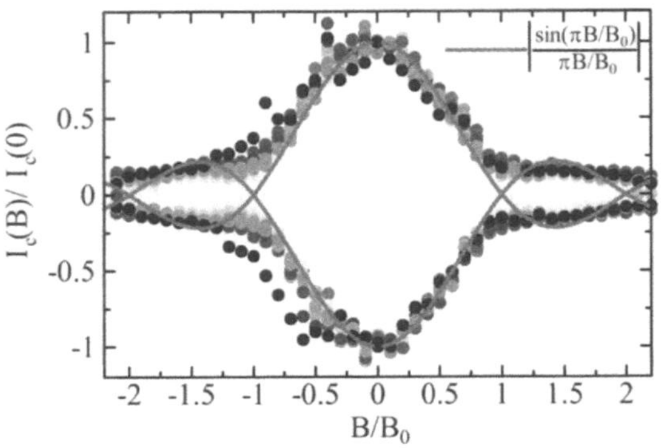

Fig. 11. Magnetic field dependence of the normalized I_c of the 12 intrinsic Josephson junctions of a $1.2 \times 1.2\,\mu\text{m}^2$ mesa of BSCCO at 4.2 K. $\lambda_J \approx 1\,\mu\text{m}$. *Lines* correspond to Fraunhofer pattern $B_0 = 1.1$ T

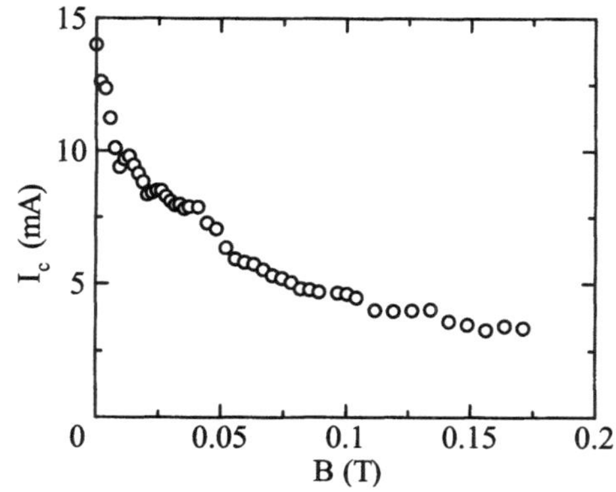

Fig. 12. Magnetic field dependence of I_c for a $320 \times 40\,\mu m^2$ mesa of BSCCO at 4.2 K

4.2 K is shown in Fig. 12 [61]. I_c is so sensitive to the magnetic field that it decreases abruptly with initial low B and then decreases slowly accompanied by several modulations with increasing B.

Thus, a stack of intrinsic Josephson junctions much longer than λ_J allows penetration by vortices in the presence of an external magnetic field. Then, under the influence of a bias current a Lorentz force acts on the Josephson vortices and gives rise to vortex flow in the stack. And, according to further investigation of (14) [62,63], in the stack of N Josephson junctions N different electromagnetic modes and their corresponding velocities c_n exist. They are given by

$$c_n = \omega_{pl}\lambda_J \frac{1}{\sqrt{1+2s\cos[\pi n/(N+1)]/d'}}\,, \; n = 1, 2, \cdots, N\,, \tag{15}$$

where $\omega_{pl} = (2edJ_c/\varepsilon)^{1/2}$ is the Josephson plasma frequency, ε is the dielectric constant and n is the mode number. With increasing n, c_n decreases. The highest mode velocity corresponds to $n = 1$, the lowest one corresponds to $n = N$. These also characterize the vortex motion in the stack. In the following we show some dynamic properties of vortices in stacks of intrinsic Josephson junctions of BSCCO.

Usually, in the vortex-flow regime the I-V characteristics of Josephson junctions exhibit the vortex-flow branch. Figure 13 shows the typical I-V characteristics of a $320 \times 40\,\mu m^2$ mesa of BSCCO at 4.2 K as a function of B. On increasing B to 0.185 T vortex-flow branches appear on the $V = 0$ branch and on the other quasiparticle branches, and they shift toward higher voltages. The maximum voltage of the vortex-flow branch V_f is given by $V_f = n_J(t+d)v_f B$ where n_J is the number of junctions in a vortex-flow state and v_f is the velocity of vortex flow. Using this relation, the vortex flow was found to occur partly in only several junctions of this mesa in that magnetic field range and the corresponding v_f was estimated to be $\sim 3 \times 10^6$ m/s [61]. On the other hand, at 77 K vortex flow was observed to occur at lower v_f in all

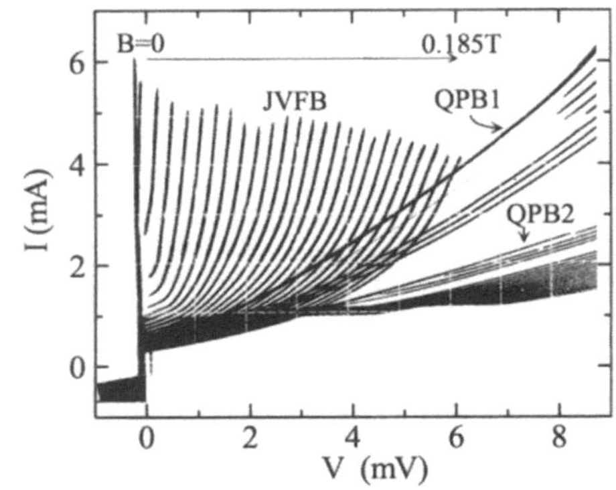

Fig. 13. *I-V* characteristics of a 320 × 40 μm² mesa of BSCCO in magnetic fields in the range of 0 T ≤ B ≤ 0.185 T at 4.2 K

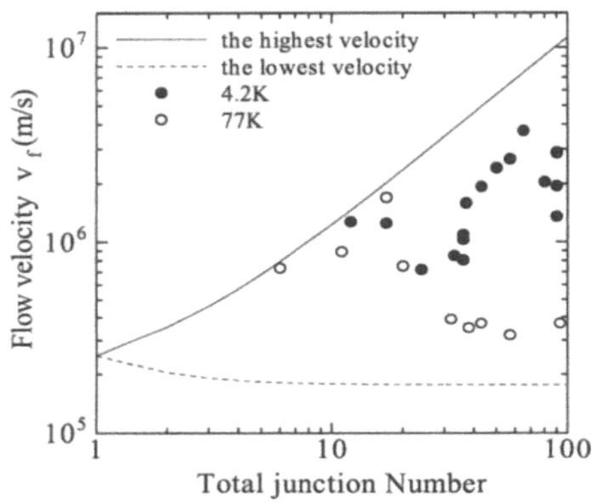

Fig. 14. Flow velocity v_f vs. the number N of intrinsic Josephson junctions in mesas at 4.2 and 77 K

junctions of the mesa with a decrease in I_c in the same field range [61]. The v_f observed for some mesas at 4.2 and 77 K is shown in Fig. 14, as a function of the total number N of intrinsic Josephson junctions in the mesas [61]. Here, solid and dashed lines indicate the highest and lowest velocities of the electromagnetic waves in the mesas, respectively, which are obtained from (15) using typical parameters of BSCCO. The v_f at 4.2 K is close to the highest velocity, but that at 77 K becomes lower with increasing N. This may be due to a collective motion of vortices with higher density and stronger interaction between them in the mesas at 77 K [61,64].

Furthermore, a Fiske resonance with vortex flow was observed in a 37 × 30 μm² mesa at 11.3 K [65]. When the external magnetic field B was increased from zero, the I_c of the superconducting branch first decreased, and then for $B =\sim 0.06$ T Fiske resonant steps with a voltage spacing ΔV_{FS} of ~ 43 μV appeared together with vortex-flow branches on the *I-V* characteristics, as shown in Fig. 15. This implies the superposition of vortex-flow branches and Fiske steps, which are excited by the vortex motion in intrinsic Josephson junctions in the mesa. The phase velocity of the electromagnetic wave in the junctions was estimated to be $\bar{c} = 1.5 \times 10^6$ m/s from the Fiske resonance,

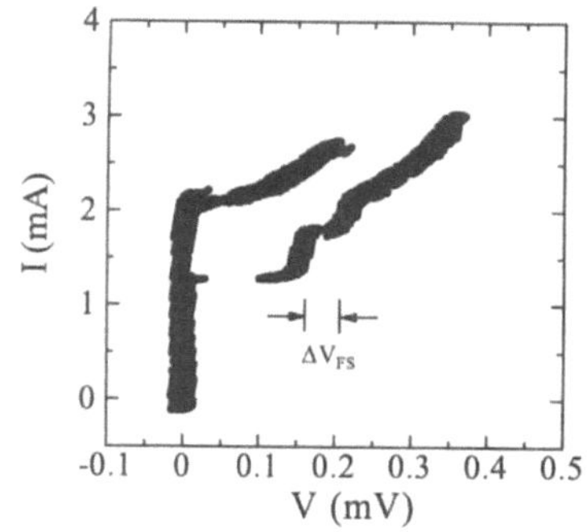

Fig. 15. *I*-*V* characteristics of a $37 \times 30\,\mu m^2$ mesa in B of 0.06 T at 11.3 K

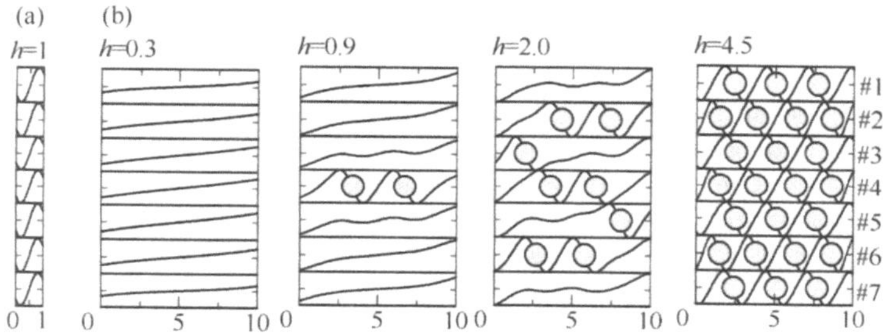

Fig. 16. Simulation of the static states of 7-fold stacks of Josephson junctions with (**a**) $\ell = 1$ and (**b**) $\ell = 10$

using the relation $\bar{c} = 2L\Delta V_{FS}/\Phi_0$. On the other hand, from the vortex-flow branches the velocity of vortex-flow was estimated to be $v_f = 1.8 \times 10^6$ m/s. These values almost agree with each other. This indicates that the Fiske step results from the highest velocity mode of vortex motion in the mesa.

Thus the vortex flow in the stack of intrinsic Josephson junctions can generate an oscillation higher than several 10 GHz.

In addition, in high parallel magnetic fields, $B \gtrsim 1.5$ T, a strong broad-band non-Josephson microwave emission which was not expected for conventional junctions was observed in the stack of intrinsic Josephson junctions of BSCCO. This was explained with respect to Cherenkov radiation taking place when the vortex velocity exceeds the lowest velocity of an electromagnetic wave in the stack [66].

3.4 Behavior of Vortices in Stacks of Josephson Junctions

In order to understand better the static and dynamic behavior of vortices in stacks of intrinsic Josephson junctions in layer-parallel magnetic fields, we here show some results of numerical simulations for layered stacks of Josephson junctions, based on (14). Figure 16 shows static solutions for 7-fold stacks of short ((a) $\ell = 1$) and long ((b) $\ell = 10$) junctions, where the junction length is normalized by λ_J. In this figure, the horizontal scale is the special

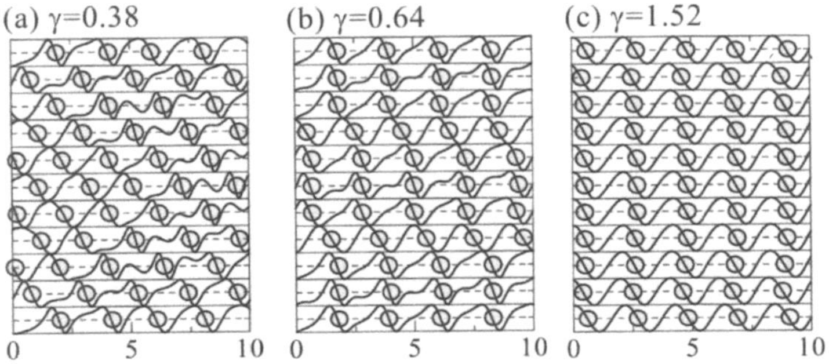

Fig. 17. Snapshots of the vortex flow states for different γ in an 11-fold stack with $\ell = 10$: (**a**) $\gamma = 0.38$, (**b**) $\gamma = 0.64$, (**c**) $\gamma = 1.52$

coordinate inside the stack. The circles correspond to the vortex and the curves account for the supercurrent distribution. The magnetic field h is normalized by $h_0 = \Phi_0/\ell\lambda_J(t+d)$. For short stacked junctions it is found that the supercurrent flowing in the different junctions due to a magnetic field are in phase. On the other hand, for long stacked junctions vortices start to penetrate the middle junction and then the outer junctions in the stack with increasing h. At higher h it is seen that the vortices tend to form a triangular lattice.

As mentioned above, a stack of N identical junctions has N different velocities of elctromagnetic waves, and they also characterize the vortex flow in the stack [62,63]. Figure 17 shows one of the simulated snapshots of vortex flow for different normalized bias current γ in the 11-fold stack in relatively high h. One can see that at lower γ vortices penetrate all the stacked junctions and flow with a triangular lattice, which corresponds to the lowest velocity mode. With increasing γ the vortex lattices changes from triangular to rectangular corresponding to the highest velocity mode, i.e. from out of phase to in phase modes, accompanied by an increase in the velocity of vortex flow. Recently, the in phase vortex flow state has been found to exist stably in a wide region of the I-V characteristics, by large scale numerical simulation for the intrinsic Josephson junctions with realistic dimensions in a magnetic field ($> 1\,\mathrm{T}$) [67].

3.5 Summary

Strongly anisotropic layered high-T_c superconductors such as BSCCO represent natural stacks of atomic scale intrinsic Josephson junctions. When a magnetic field is applied parallel to the layers of the stacks, the intrinsic Josephson junctions with dimensions close to or smaller than λ_J show the Fraunhofer-type magnetic field dependence of I_c, but for those with dimensions much larger than λ_J it is considerably disturbed by Josephson vortices (fluxons) entering the junctions at $B > B_{c1}$. In this case the vortices can be induced between the superconducting layers of the junctions and strong inductive coupling between them can occur. Under the influence of a bias current a collective motion of the vortices, which can excite the Fiske resonance, occurs with a high velocity of $\sim 10^6$ m/s in the stacked junctions, being different from that in a single Josephson junction. This may hopefully lead to new types of oscillators, detectors and transistors operating in the range of ~ 10 GHz– ~ 1 THz.

4 Optical Control of Vortices

Excitation and observation of ultrafast phenomena in the solid state are an essential interest in the field of condensed matter physics. As one of the attractive fields, one can study the ultrafast phenomena in quantized system by means of fs optical excitation. Superconductivity can provide an excellent platform to open a new field such as transient quantum physics. The coherent states quantizing macroscopically are based on the minimum free energy of the ensemble system. The fs laser illumination of the system could bring out the strong local perturbations, which would propagate in both the time and spatial domains, and result in the generation of new quantized states.

In a series of recent reports, the authors proposed and studied the optical generation and control of magnetic flux quanta (MFQs) in high-T_c superconductors (HTSs) [68,69,70,71,72,73,74,75,76,77,78,79]. When parts of superconductive thin film loops or strip-lines carrying transport supercurrent are illuminated by the fs laser, the magnetic flux is generated and controllable in them with the laser power, bias current, and other excitation parameters. The generation is itself regarded as a new ultrafast phenomenon breaking the conventional rule of superconductivity. The origin of the optical process can be understood as an ultrafast partial supercurrent modulation without phase transition: optical pair breaking, hot-quasiparticle scattering, and recombination to pairs [80,81].

In this subchapter optical vortex generation in the superconductors as a consequence of its ultrafast optical response is reviewed. Up to the present, the following notable performances have been achieved; MFQs are generated by even a single shot of the fs pulse [76,77] and the fs optical beam profile is transferred into the 2D vortex distribution in the superconductive strip-lines [78,79]. Prior to the above results, an idea for optical vortex generation in superconductive thin films is proposed in the follwing section.

4.1 Idea for Optical Vortex Generation

An intensive study of the ultrafast optical response of high-T_c superconductors has been carried out by a number of researchers [82,83,84]. In most of the reports, the change of the optical reflectance and/or transmitance in the

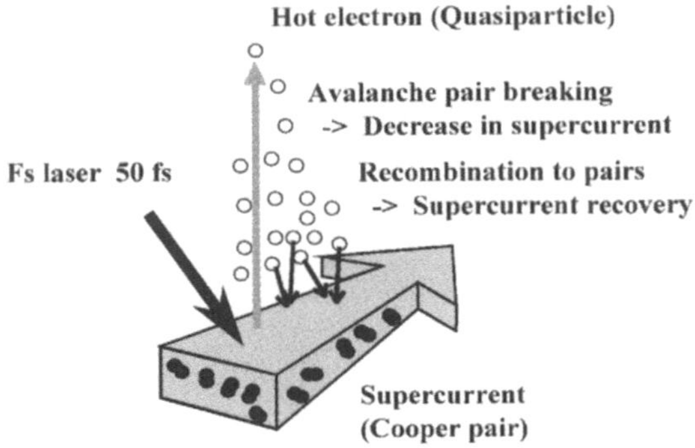

Fig. 18. Schematic illustration to explain the carrier dynamics in optically excited superconductors

time domain just after the photon injection is observed to determine the carrier relaxation. However, little has been done to study the nonequilibrium state in the optically excited HTSs carrying transport current [85,86,87]. Let us first consider the effect of the fs optical excitation on the moving carriers (Fig. 18). When the fs optical pulses illuminate the superconductors, the arrival photons break the Cooper pairs which scatter with the other carriers simultaneously. The scattered particles, hot quasiparticles, cannot carry the supercurrent anymore, and transform themselves into pair breakers. Thus the avalanche process reduces the great number of the pair density. In the generally accepted picture of optically excited superconductors [85,86,87], the leftover pairs are accelerated and the total current is kept constant as given by

$$\frac{\partial J_{\mathrm{s}}}{\partial t} = ev_{\mathrm{s}}\frac{\partial n_{\mathrm{s}}}{\partial t} + en_{\mathrm{s}}\frac{\partial v_{\mathrm{s}}}{\partial t} = 0, \tag{16}$$

where J_{s} is the supercurrent density, n_{s} is the density of the Cooper pairs, and v_{s} is their drift velocity. However, the compensation is not a instant process but requires a certain time. Thus the supercurrent in the optically excited area decreases so that an ultrashort electromagnetic wave is emitted into free space in proportion to the time derivative of the supercurrent. We discovered such emission from $YBa_2Cu_3O_{7-\delta}$ (YBCO) thin films, which is referred to as THz radiation because of its ultra-wide band frequency component up to the THz region. The radiated THz energy is considered as a part of the inductively stored energy in the illuminated area. Note that although the strip-line is connected to the current source, it is completely isolated from the source becasue the system response time is limited by many parameters such as the capacitance, resistance and inductance of both the system and devices in addition to pure carrier transient behavior.

Suppose that the superconductive thin film strip-line is carrying transport current as shown in the upper panel of Fig. 19. The supercurrent dis-

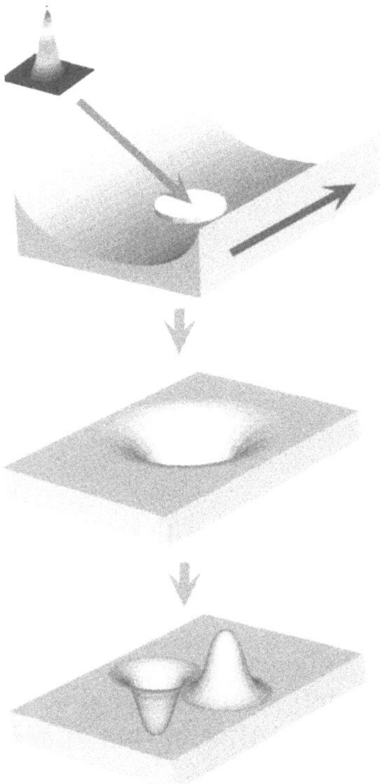

Fig. 19. Idea for generating vortices using ultrafast optical supercurrent modulation. The supercurrent flows in the thin film strip-line following the *arrow* in the upper panel. The lower panel describes the birth of the magnetic field in the strip-line with the optical excitation

tributes in the whole area accross the strip-line even in a Meissner state as expected from critical state theory for thin films and confirmed by many researchers [88,89,90]. When the small area is excited with the fs laser, the arrival photons break the pairs which reduces the supercurrent in the illuminated area. Since the current is reduced in proportion to the local photon number, the reduced supercurrent distribution reflects the optical beam profile as illustrated in the middle panel. Once this scenario is accepted, it means that inside the stripline there exists magnetic flux induced by the surroundung supercurrent; the leftover pairs at the optically excited area cannot block its penetration.

In other words, the initial conditions are changed to accept the existence of the magnetic field (the lower panel). Then the system undergoes relaxation into the equilibrium state of superconductivity. During the relaxation, the magnetic flux quantizes into the vorticies, which are expected to be trapped at the pinning center. Finally, vortex and anti-vortex bundle pairs can be generated and the distribution of the vorticies may reflect the laser beam profile. This is a basic idea for "optical vortex generation in superconductors".

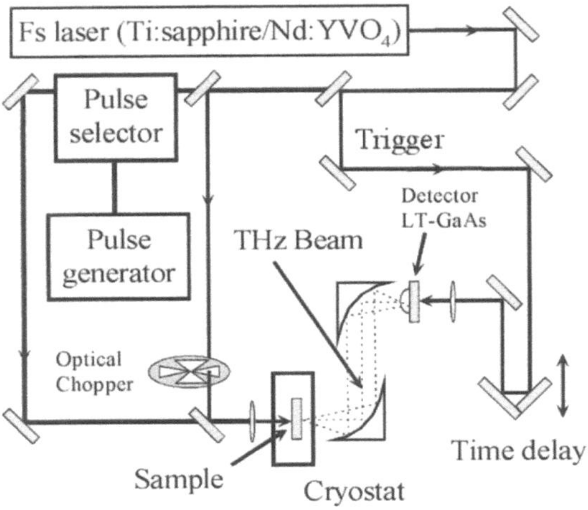

Fig. 20. Experimental setup for exciting the cryogenic samples with the selected pulses and visualizing the supercurrent distributions

4.2 Experimental Setup

100 nm thick YBCO thin films are patterened into several types of strip-line structures to study the optical vortex generation. After cooling to 17 K supercurrent is applied to the sample. The area referred to as a "laser spot" is then illuminated by fs laser pulses at various laser powers and bias currents. The spot size is controlled by changing the distance between the sample and an objective lens. Figure 20 shows the experimental setup to generate the vortices with the fs laser pulses and to visualize the supercurrent distributions by THz radiation imaging (Sect. 4.5). A pulse selector with a Bragg-angle accousto-optic modulator is employed to control the number of optical pulses, which is triggered by a pulse generator. After removal of the bias current, the vortex distributions are examined by observing persistent supercurrent distributions. Only the x (or z)-component of the supercurrent distribution is observable in the present study.

4.3 Optical Vortex Generation with Optical Pulses

Vortex generation effect in the YBCO thin film strip is confirmed by illumination using a conventional fs laser at a repetition rate of 82 MHz. The 114 μm wide strip-line, half of which has an ordered array of antidots, is prepared as shown in Fig. 21. The antidots are introduced to enhance the pinning force for the generated vortices. The positions A, B and C correspond to the area excited at an I_B of 200 mA and a P_P of 30 mW.

Figure 22 shows two-dimensional images of the x-component of the supercurrent distribution in the remnant state. After laser illumination at position A, the clockwise and counterclockwise supercurrents were found to

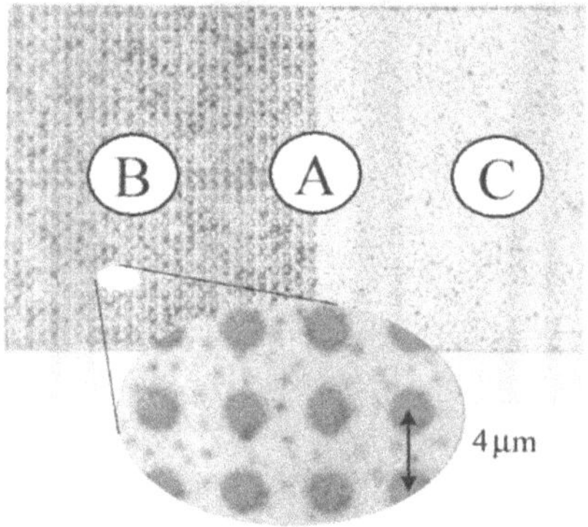

Fig. 21. Fabricated YBCO thin film strip-line, half of which has ordered arrays of the antidots. The positions A, B, and C correspond to the irradiation spots

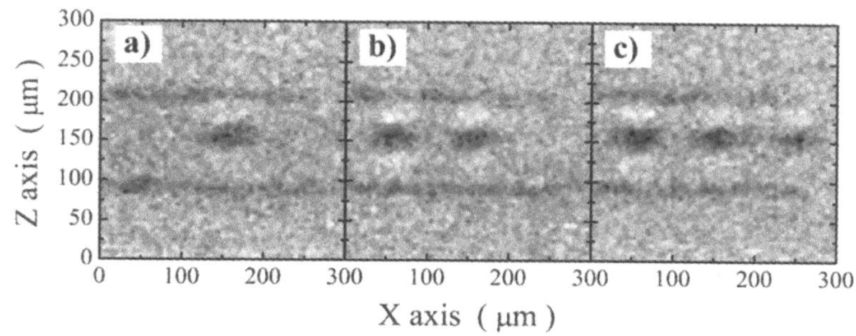

Fig. 22. Supercurrent distributions after vortex-generation. ©1993/2001 IEEE

circulate in the upper and lower part of the excited area, respectively, which indicates that a bundle of vortices with different polarity was generated near the excited area. The generation was also confirmed when positions B and C were excited. Although the structure of the excited areas differ from each other, persistent supercurrent distributions are clearly observed at all the positions. The poor contrast of the image for the position C in Fig. 22c originates from the low THz emission effciency of the strip-line without the antidots.

The magnetic field distribution is calculated by solving Biot–Savart's law. The edge current effect due to the trapped vortices was eliminated by subtracting the supercurrent distribution along the z-axis in the unilluminated area from that at the laser spots. Figure 23 shows the calculated distributions at positions A, B, and C along the z-axis. Within the present experimental resolution, they distribute indentically. The distance from the maximum positions for the positive magnetic field to the minimum for the negative field is estimated to be about 28 μm, which is of the order of the laser beam profile. This suggests that the vortices are generated by the supercurrent modulation and frozen at the generated areas regardless of the existence of the antidots. An abnormal temperature dependence of the THz radiation from the flux

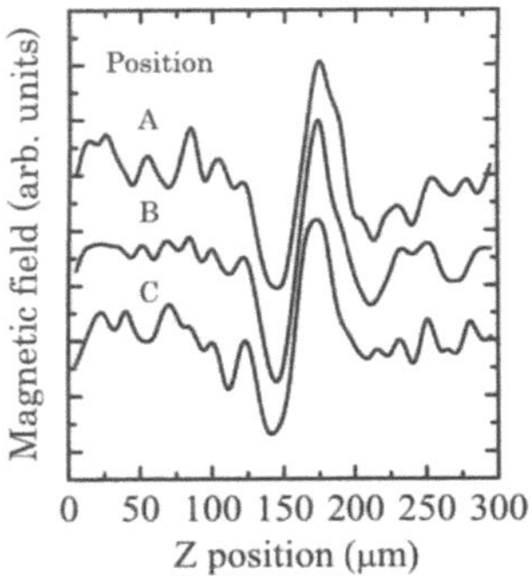

Fig. 23. Magnetic field distributions at the positions A, B, and C along the z-axis. The effect of the trapped vortices at the strip-line edges was eliminated

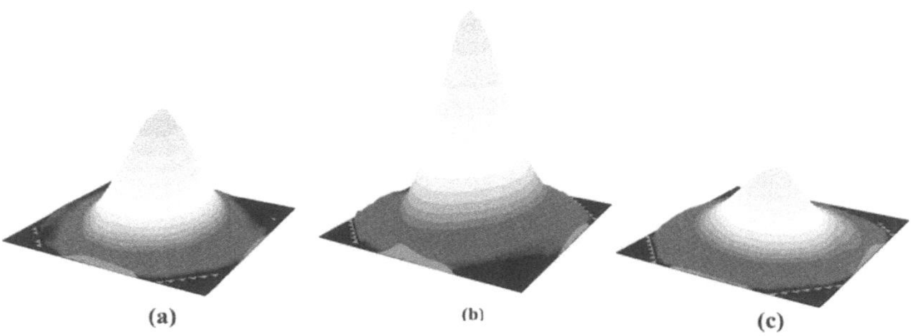

Fig. 24. Supercurrent distribution images after vortex generation. Each bundle was generated by optical pulses with different beam diameters

trapped YBCO films has been observed so far [91]. Thus, one can expect different effects of the antidots on the temperature dependence.

Optical vortex generation with different laser beam diameters is then examined. The optical beam profile was controlled by changing the distance between the sample and an objective lens. The diameter and the laser power were varied between 7 and 30 μm. Some of the measured beam profiles are depicted in Fig. 24. Then the sample with the array of antidots was illumated at $P_{\rm P}$ of 80 mW for 10 seconds to generate the vortices. Figure 25 shows images of the supercurrent distributions in the remnant state. The vortex bundles at the left, middle and right postions were generated at beam diameters (FWHM) of 18 μm, 13 μm, and 38 μm. We observe images determined by the convolution between the laser beam profile and the real supercurrent distribution. The numerically obtained magnetic flux density using Biot–Savart's law. Taking the convolution into account suggests that the observed images reflect the real laser beam profile to some extent. Namely the fs laser beam profile can be transferred into the vortex distribution. Since the THz radiation imaging utilizes the scanning the laser beam, it takes a bit longer time to visualize. One can expect to be able to build a fast imaging system utilizing the MO imaging technique with a CCD camera.

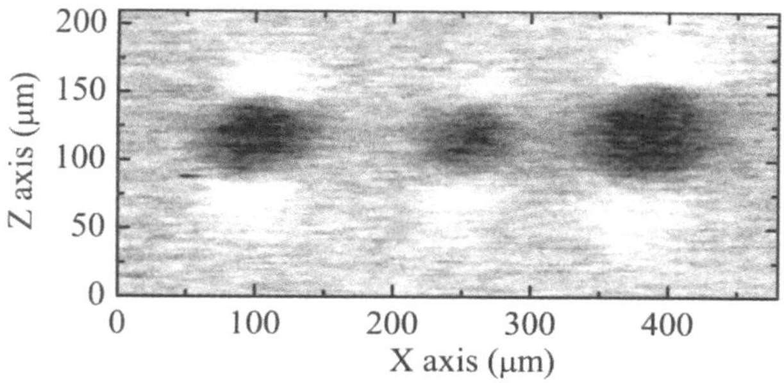

Fig. 25. Supercurrent distribution images after vortex generation. Each bundle was generated by optical pulses with different beam diameters

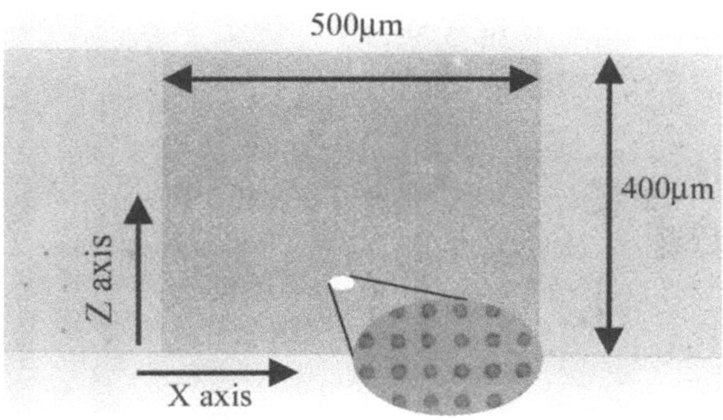

Fig. 26. Future prospect of the present research

4.4 Single Shot Pulse Operation

Optical vortex generation with a single shot optical pulse is then demonstrated. A 100 nm thick YBCO thin film was patterned into 400 μm wide strips with an ordered array of antidots with a diameter of 1 μm at an interval of 4 μm as shown in Fig. 26. The experiments were done at an I_B between 200 and 800 mA, a P_P between 0.3 and 2 nJ for a single pulse, and 17 K. The injection photon density under the above conditions is roughly estimated to be between $5 \times 10^{18}\,cm^{-3}$ and $1 \times 10^{20}\,cm^{-3}$. The supercurrent distributes over the whole strip while biasing the transport current to the sample as depicted in Fig. 27. A slight asymmetric distribution is due to misalignment of the experimental setup. After removal of the bias current without the vortex generation procedures, no supercurrent was observable except near the edge of the sample [92,93,94].

Figure 28 shows the supercurrent distribution images (x-component) turning the optical vortex generation procedures. After illumination at the three spots at an I_B of 600 mA, redistribution of the supercurrent while the strip carrys the transport current is clearly observed. A similar vortex-bandle is generated even by a single shot fs pulse. This is what we expected based on

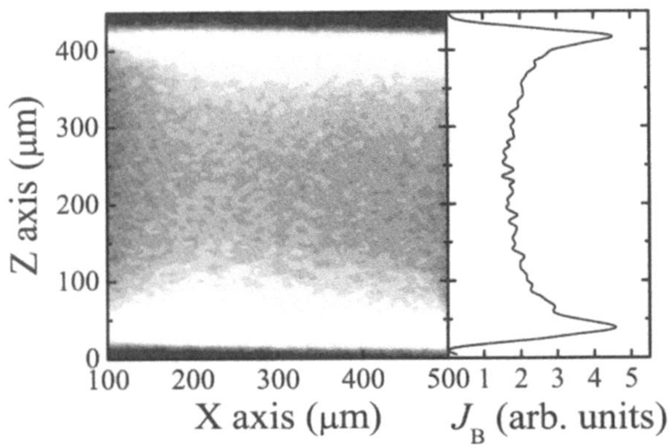

Fig. 27. Supercurrent distribution image at $I_B = 600\,\text{mA}$

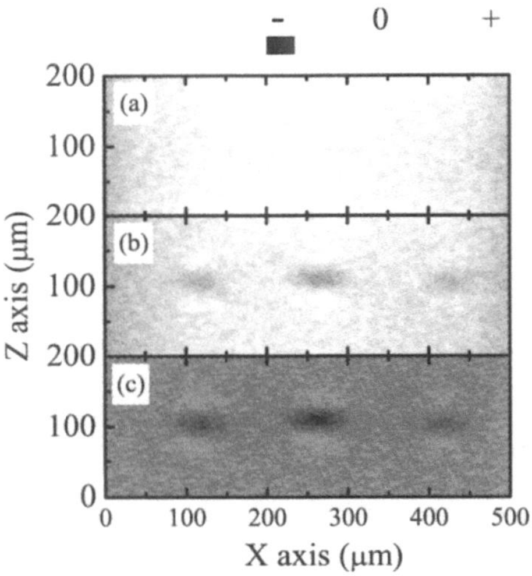

Fig. 28. Supercurrent distributions (**a**) at $I_B = 600\,\text{mA}$, (**b**) after optical excitation, and (**c**) after removal of the transport current

the proposed idea. One can eliminate the vortices by inducing the flux flow with a large bias current, increasesing the temeprature above T_c, or illuminating them with the fs laser at a large power.

Generation with two different laser beam diameters was then examined. The supercurrent distribution images with x- and z-components clearly indicate that the supercurrent circulates around the excited area, and the distribution area increased on increasing the laser beam diameter as depicted in Fig. 29. The vortex distribution was estimated by taking the imaging resolusion into account, which agrees with the laser mean profile. This suggets

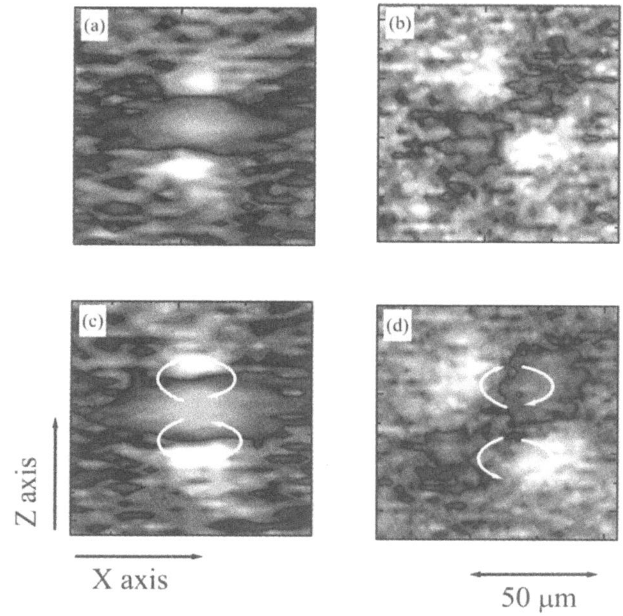

Fig. 29. Enlarged images of the supercurrent distribution near the vortex bundle dot. (**a**) represents the x-component of the supercurrent; (**b**) the z-component

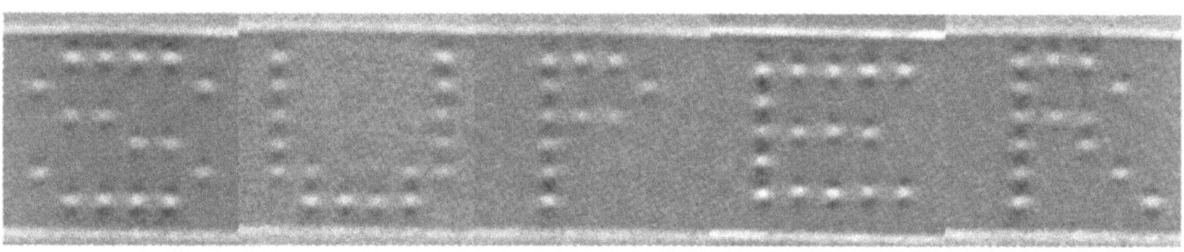

Fig. 30. Vortex letters generated by optical-vortex generation

that a single shot fs optical two-dimensional signal can be transferred into the vortex distributions.

Finally, we demonstrate generating vortex letters with the fs optical pulse. The images for 5 letters "SUPER" are shown in Fig. 30. Each pixcel is written by a single shot of fs pulse. The minimum pixcel size is about 60 μm in diameter.

4.5 Summary

Optical vortex generation with a femtosecond laser was reviewed. First of all, we proposed a new research field to realize ultrafast modulation of the superconductive quantum state with femtosecond optical excitations and mentioned that it could possibly break the conventional rule of superconductivity.

Vortex and anti-vortex bundle pairs were created inside the superconductive thin films without their penetration from the outside of the supercondutors. It was also found that the generated vortex distributions reflect the fs optical beam profile.

Acknowledgements

We are grateful to V. B. Geshkenbein, T. Hanaguri, N. Ichioka, Y. Kato, T. Kita, N. B. Kopnin, M. Ogata, M. Takigawa, A. Tanaka, M. Hangyo and H. Murakami for their helpful discussions. We thank K. Kakuyanagi, T. Nagaoka, K. Nozaki, and T. Fukui for their technical assistance.

References

1. P. G. de Gennes: *Superconductivity of Metal and Alloys* (Addison-Wesley, Reading 1989)
2. D. I. Khomskii, A. Freimuth: Phys. Rev. Lett. **75**, 1384 (1995)
3. G. Blatter, M. Feigelman, V. Geshkenbein, A. Larkin, A. van Otterlo: Phys. Rev. Lett. **77**, 566 (1996)
4. T. Mishonov: LANL preprint cond-mat/0004286.
5. N. Hayashi, M. Ichioka, K. Machida: J. Phys. Soc. Jpn. **67**, 3368 (1999)
6. J. Koláćek, P. Lapavský, E. H. Brandt: Phys. Rev. Lett. **86**, 312 (2001)
7. A. van Otterlo, M. Feigelman, V. Geshkenbein, G. Blatter: Phys Rev. Lett. **75**, 3736 (1995)
8. M. V. Feigelman, V. Geshkenbein, A. I. Larkin, V. M. Vinokur: JETP Lett. **62**, 834 (1995)
9. Y. Kato: J. Phys. Soc. Jpn. **68**, 3798 (1999)
10. K. Kumagai, K. Nozaki, Y. Matsuda: Phys. Rev. B **63**, 144502 (2001)
11. Y. Matsuda, K. Kumagai: *Vortices in Unconventional Superconductors and Superfluid-Microscopic Structure and Dynamics*, R. P. Huebener, N. Schopohl, G. E. Volovik (Eds.): Springer Ser. Solid-State Sci. **132**, 283–298 (Springer, Berlin, Heidelberg 2002)
12. F. Gygi, M. Schlüter: Phys. Rev. B **43**, 7609 (1991)
13. N. Schopohl, N. K. Maki: Phys. Rev. B **52**, 490 (1995)
    M. Ichioka, N. Hayashi, N. Enomoto, K. Machida: Phys. Rev. B **53**, 15316 (1996)
14. P. I. Soininen, C. Kallin, A. J. Berlinsky: Phys. Rev. B **50**, 13883 (1994)
    Y. Ren, J. H. Xu, C. S. Ting: Phys. Rev. Lett. **74**, 3680 (1995)
    A. J. Berlinsky, A. L. Fetter, M. Franz, C. Kallin, P. I. Soininen: Phys. Rev. Lett. **75**, 2200 (1995)
15. Y. Wang, A. H. MacDonald: Phys. Rev. B **52**, R3876 (1995)
16. Y. Morita, M. Kohmoto, K. Maki: Phys. Rev. Lett. **78**, 4841 (1997)
17. D. P. Arovas, A. J. Berlinsky, C. Kallin, S. C. Zhang: Phys. Rev. Lett. **79**, 2871 (1997)
18. A. Himeda, M. Ogata, Y. Tanaka, S. Kashiwaya: J. Phys. Soc. Jpn. **66**, 3367 (1997)

19. M. Franz, Z. Tesanovic: Phys. Rev. Lett. **80**, 4763 (1998)
20. K. Yasui, T. Kita: Phys. Rev. Lett. **83**, 4168 (1999)
21. B.M. Andersen, H. Bruus, P. Hedegard: Phys. Rev. B **61**, 6298 (2000)
22. I. Maggio-Aprile, Ch. Renner, A. Erb, E. Walker, Ø. Fischer: Phys. Rev. Lett. **75**, 2754 (1995)
23. S.H. Pan, E.W. Hudson, A.K. Gupta, K.W. Ng, H. Eisaki, S. Uchida, J.C. Davis: Phys. Rev. Lett. **85**, 1536 (2000)
24. M. Eschrig, J.A. Sauls, D. Rainer: Phys. Rev. B **60**, 10447 (1999)
25. S.J. Hagen, A.W. Smith, M. Rajeswari, J.L. Peng, Z.Y. Li, R.L. Greene, S.N. Mao, S. Bhattacharya, Q. Li, C.J. Lobb: Phys. Rev. B **47**, 1064 (1993) and references therein
26. T. Nagaoka, Y. Matsuda, H. Obara, A. Sawa, T. Terashima, I. Chong, M. Takano, M. Suzuki: Phys. Rev. Lett. **80**, 3594 (1998) and references therein
27. J. Bardeen, M.J. Stephen: Phys. Rev. **140**, A1197 (1965)
P. Noziéres, W.F. Vinen: Philos. Mag. **14**, 667 (1966)
28. K. Schwarz: Phys. Rev. B **42**, 2051 (1990)
Y. Ohta, W. Koshibae, S. Maekawa: J. Phys. Soc. Jpn. **61**, 2198 (1992)
K. Hanzawa: J. Phys. Soc. Jpn, **62**, 3302 (1993)
29. G. Zheng, Y. Kitaoka, K. Ishida, K. Asayama: J. Phys. Soc. Jpn. **64**, 2524 (1995)
30. H. Yasuoka: *Spectroscopy of Mott Insulator and Correlated Metals*, A. Fujimori, Y. Tokura (Eds.), Springer Ser. Solid-State Sci. **119**, 213 (Springer, Berlin, Heidelberg 1995)
31. A.A. Gippius, E.V. Antipov, W. Hoffmann, K. Luders: Physica C **276**, 57 (1997)
32. G.C. Cater: *Metallic Shifts in NMR* (Pergamon, New York 1977)
33. H. Ikuta, N. Hirota, Y. Nakayama, K. Kishio, K. Kitazawa: Phys. Rev. Lett. **70**, 2166 (1993)
34. Y. Matsuda, S. Komiyama, T. Terashima, K. Shimura, Y. Bando: Phys. Rev. Lett. **69**, 3228 (1992)
35. Z.D. Wang, J.M. Dong, C.S. Ting: Phys. Rev. Lett. **72**, 3875 (1994)
36. P. Ao: J. Phys. C **10**, L677 (1998)
37. A.T. Dorsey: Phys. Rev. B **46**, 8376 (1992)
38. N.B. Kopnin, B.I. Ivlev, V.A. Kalatsky: J. Low Temp. Phys. **90**, 1 (1993)
39. A.G. Aronov, S. Hikami, A.I. Larkin: Phys. Rev. B **51**, 3880 (1995)
40. A. Ino, T. Mizokawa, A. Fujimori, K. Tamasaku, H. Eisaki, S. Uchida, T. Kimura, T. Sasagawa, K. Kishio: Phys. Rev. Lett. **79**, 2101 (1997); We take μ as the chemical potential of the electron
41. J. Shoenes, E. Kaldis, J. Karpinski: Phys. Rev. B **48**, 16869 (1993)
42. V.B. Geshkenbein, L.B. Ioffe, A.I. Larkin: Phys. Rev. B **55**, 3173 (1997)
43. N.B. Kopnin, A.V. Lopatin: Phys. Rev. B **51**, 15291 (1995)
44. C.P. Poole, Jr., H.A. Farach, R.J. Creswick: *Superconductivity* (Academic, San Diego 1995)
45. M.J. Naughton, R.C. Yu, P.K. Davies, J.E. Fischer, R.V. Chamberlin, Z.Z. Wang, T.W. Jing, N.P. Ong, P.M. Chaikin: Phys. Rev. B **38**, 9280 (1988)
46. H. Maeda, K. Togano (Ed.): *Bismuth-Based High-Temperature Superconductors* (Marcel Dekker, New York 1996)
47. R. Kleiner, F. Steimeyer, G. Kunkel, P. Müller: Physica C **185-189**, 2617 (1991)
R. Kleiner, F. Steimeyer, G. Kunkel, P. Müller: Phys. Rev. Lett. **68**, 2394 (1992)

48. G. Oya, N. Aoyama, S. Kishida, H. Tokutaka: Physica C **185-189**, 2453 (1991) G. Oya, N. Aoyama, A. Irie, S. Kishida, H. Tokutaka: Jpn. J. Appl. Phys. **31**, L829 (1992)
49. P. Müller: Intrinsic Josephson effects in layered superconductors, Festkörperprobleme/Adv. Solid-State Phys. **34**, R. Helbig (Ed.) (Vieweg, Braunschweig 1994) pp. 1–33
50. M. Tachiki, T. Yamashita (Ed.): *Intrinsic Josephson Effects and THz Plasma Oscillations in High-T_c Superconductors*, Proc. 1st RIEC Int'l Symposium, Physica C **293** (Elsevier, Amsterdam 1977)
51. G. Oya, A. Irie: Intrinsic Josephson effects and vortex dynamic properties in layered high-temperature superconductors, Recent Res. Develop. Appl. Phys. **2**, S. G. Pandalai (Ed.) (Transworld Research Network, Trivandrum 1999) pp. 429–452
52. A. A. Yurgens: Intrinsic Josephson junctions: Recent developments, Supercond. Sci. Technol. **13**, R85–R100 (2000)
53. M. Tachiki, T. Yamashita (Ed.): *Intrinsic Josephson Effects and Plasma Oscillation in High-T_c Superconductors*, Proc. 2nd JST Int'l Symposium, Physica C **362** (Elsevier, Amsterdam 2001)
54. T. Kobayashi, M. Tonouchi (Ed.): Extended Abstracts 8th Int'l Superconductive Electronics Conference (ISEC'01), Osaka, Japan, June 2001
55. J. R. Cooper, L. Forró, B. Keszei: Nature (London) **343**, 444 (1990)
56. S. Sakai, P. Bodin, N. F. Pedersen: J. Appl. Phys. **73**, 2411 (1993)
57. A. Irie, T. Mimura, M. Okano, G. Oya: Supercond. Sci. Technol. **14**, 1097 (2001)
58. L. N. Bulaevskii, J. R. Clem, L. I. Glazman: Phys. Rev. B **46**, 350 (1992)
59. A. Irie, S. Heim, S. Schromm, M. Mosle, T. Nachtrab, M. Godo, R. Kleiner, P. Müller, G. Oya: Phys. Rev. B **62**, 6681 (2000)
60. Yu. I. Latyshev, J. E. Nevelskaya, P. Monceau: Phys. Rev. Lett. **77**, 932 (1996)
61. A. Irie, S. Kaneko, G. Oya: Int. J. Mod. Phys. B **13**, 2678 (1999)
62. R. Kleiner: Phys. Rev. B **50**, 6919 (1994)
63. S. Sakai, A. V. Ustinov, H. Kohlstedt, P. Petraglia, N. F. Pedersen: Phys. Rev. B **50**, 12905 (1994)
64. G. Hechtfischer, R. Kleiner, K. Schlenga, W. Walkenhorst, P. Müller: Phys. Rev. B **55**, 14638 (1997)
65. A. Irie, Y. Hirai, G. Oya: Appl. Phys. Lett. **72**, 2159 (1998)
66. G. Hechtfischer, R. Kleiner, A. V. Ustinov, P. Müller: Phys. Rev. Lett. **79**, 1365 (1997)
67. M. Machida, T. Koyama, A. Tanaka, M. Tachiki: Physica C **330**, 85 (2000)
68. M. Tonouchi, N. Wada, M. Hangyo, M. Tani, K. Sakai: Appl. Phys. Lett. **71**, 2364 (1997)
69. T. Kiwa, M. Yamashita, O. Morikawa, T. Nakamura, M. Tonouchi, M. Hangyo: *Advances in Superconductivity XI*, N. Koshizuka, S. Tajima (Eds.) (Springer, Tokyo 1999) pp. 219
70. H. Saijo, M. Yamashita, T. Kiwa, T. Kondo, O. Morikawa, M. Tonouchi, M. Hangyo: *Advances in Superconductivity XI*, N. Koshizuka, S. Tajima (Eds.) (Springer, Tokyo 1999) pp. 199
71. K. Shikita, M. Yamashita, T. Kiwa, O. Morikawa, M. Tonouchi, M. Hangyo: *Advances in Superconductivity XI*, N. Koshizuka, S. Tajima (Eds.) (Springer, Tokyo 1999) pp. 215
72. M. Tonouchi: Jpn. J. Appl. Phys. **40**, L542 (2001)

73. M. Tonouchi, S. Shikii, M. Yamashita, K. Shikita, M. Hangyo: IEEE Trans. Appl. Supercond. **9**, 4467 (1999)
74. K. Shikita, T. Kiwa, M. Tonouchi, M. Hangyo: *Advances in Superconductivity XI*, T. Yamashita, K. Tanabe (Eds.) (Springer, Tokyo 2000) pp. 230
75. K. Shikita, M. Yamashita, M. Tonouchi, M. Hangyo: *Advances in Superconductivity XI*, T. Yamashita, K. Tanabe (Eds.) (Springer, Tokyo 2000) pp. 233
76. M. Tonouchi, K. Shikita, M. Morimoto, M. Hangyo: IEEE Trans. Appl. Supercond. **11**, 3939 (2001)
77. M. Tonouchi, K. Shikita: Physica C **367**, 37 (2002)
78. T. Fukui, A. Moto, H. Murakami, M. Tonouchi: Physica C **357–360**, 454 (2001)
79. T. Fukui, H. Murakami, M. Tonouchi: IEICE Trans. Electron. **E85-C**, 818 (2002)
80. M. Tonouchi, M. Tani, Z. Wang, K. Sakai, S. Tomozawa, M. Hangyo, Y. Murakami, S. Nakashima: Jpn. J. Appl. Phys. **35**, 2624 (1996)
81. M. Tonouchi, M. Tani, Z. Wang, K. Sakai, N. Wada, M. Hangyo: Jpn. J. Appl. Phys. **35** (1996) L1578
82. S. G. Han, Z. V. Vardey, K. S. Wong, O. G. Symko: Phys. Rev. Lett. **65**, 2708 (1990)
83. C. J. Stevens, D. Smith, C. Chen, J. F. Ryan, B. Podobnik, D. Mihalovic, G. A. Wagner, J. E. Evetts: Phys. Rev. Lett. **87**, 2212 (1997)
84. T. Mertelj, J. Demsar, B. Podobnik, I. Poberaj, D. Mihalovic: Phys. Rev. B **55**, 6061 (1999)
85. N. Bluzer: Phys. Rev. B **44**, 10222 (1991)
86. F. A. Hegmann, J. S. Preston: Phys. Rev. B **48**, 16023 (1993)
87. C. Jaekel, H. G. Roskos, H. Kurz: Phys. Rev. B **54**, R6889 (1996)
88. E. Zeldov, J. R. Clem, M. McElfresh, M. Darwin: Phys. Rev. B **49**, 9802 (1994)
89. J. C. Culbertson, H. S. Newman, C. Wilker: J. Appl. Phys. **84**, 2768 (1998)
90. S. Shikii, T. Kondo, M. Yamashita, M. Tonouchi, M. Hangyo: Appl. Phys. Lett. **74**, 1317 (1999)
91. S. Shikii, N. Wada, M. Tonouchi, M. Hangyo, M. Tani, K. Sakai: *Advances in Superconductivity X*, K. Osamura, I. Hirabayashi (Eds.) (Springer, Tokyo 1998) pp. 1231
92. M. Tonouchi, M. Yamashita, M. Hangyo: J. Appl. Phys. **87**, 7366 (2000)
93. M. Tonouchi, A. Moto, M. Yamashita, M. Hangyo: IEEE Trans. Appl. Supercond. **11**, 3230 (2001)
94. A. Moto, M. Hangyo, M. Tonouchi: IEICE Trans. Electron. **E84-C**, 67 (2001)

High-T_c SQUIDs

K. Enpuku[1], S. Kuriki[2], and S. Tanaka[3]

[1] Graduate School of Information Science and Electrical Engineering, Kyushu University, Higashi-ku, Hakozaki 6-10-1, Fukuoka 812-8581, Japan
enpuku@ed.kyushu-u.ac.jp
[2] Research Institute for Electronic Science, Hokkaido University, N12-W6, Kita-ku, Sapporo 060-0812, Japan
sk@es.hokudai.ac.jp
[3] Ecological Engineering, Toyohashi University of Technology, Hibarigaoka, Tempaku-cho, Toyohashi, Aichi 441-8580, Japan
tanakas@eco.tut.ac.jp

Abstract. This chapter reviews the development of magnetometers based on high-T_c superconducting quantum interference devices (HTSC-SQUID) operated in liquid nitrogen. A HTSC-SQUID is usually made of a grain-boundary Josephson junction, and the SQUID inductance is directly connected to a pickup coil with a typical area of 10 mm × 10 mm. This type of magnetometer is called a direct-coupled magnetometer, and a field sensitivity around 50–100 fT/Hz$^{1/2}$ has been obtained at $T = 77$ K in the white noise region. In order to obtain such high performance SQUIDs, optimization of the SQUID parameters has been done, including the effects of the thermal noise at $T = 77$ K and the large dielectric constant of $SrTiO_3$ substrate. The main difficulty of the HTSC-SQUID was the strong $1/f$ noise. The $1/f$ noise arises both from the critical-current fluctuation of the junction and from the motion of vortices trapped in the superconducting thin film. The former noise has been solved by using readout electronics utilizing the so-called bias-reversal scheme. The latter noise has been avoided by optimizing the geometry of the pickup coil as well as by using the so-called flux dam. As a result, the $1/f$ noise of HTSC-SQUIDs has been much improved.

1 Introduction

Many studies have so far been made for the development of highly sensitive SQUID magnetometers operating at 77 K. These studies include the optimization of the SQUID parameters, investigation of the origin of the $1/f$ noise, design of the pickup coil to avoid the flux trapping, and invention of the so-called flux dam. The read-out electronics that can suppress the $1/f$ noise has also been developed. In the following, we discuss the details of these improvements. In Sect. 2 the fundamental properties of the dc SQUID magnetometer are discussed. First, transport and noise properties of the bicrystal junction are clarified both analytically and experimentally. Next, optimization of the SQUID parameters, such as the junction critical current I_c, the junction resistance R_s and the SQUID inductance L_s, is shown in order to obtain high performance SQUIDs. Properties of the high performance SQUIDs

T. Kobayashi et al. (Eds.): Vortex Electronics and SQUIDs,
Topics Appl. Phys. **91**, 141–184 (2003)

utilizing the bicrystal junction with a misorientation angle of 30° are also shown. Finally, design of the pickup coil, which can realize a large effective area and avoid flux trapping, is shown. Properties of the flux dam that is inserted in the pickup coil are also analyzed. It is shown that the flux dam needs a long time for the completion of the flux entry since the thermal activation process dominates the flux entry. In order to solve this long waiting time a so-called switch is proposed and its usefulness is demonstrated.

In Sect. 3 the readout electronics of HTSC-SQUIDs are discussed. Fluctuations of the critical current of grain boundary junctions of HTSC films is much larger than the SIS junctions of Nb-based LTSC. Because of these fluctuations, strong $1/f$ noise appears in the FLL output of SQUIDs. This noise is effectively reduced in the output of the bias-reversal readout. The operation mechanisms of different bias-reversal schemes with and without flux modulation are discussed.

In Sect. 4 the low-frequency noise due to the thermal fluctuations of vortices existing in the superconducting film is discussed. It is shown that incorporation of slots and holes into the superconducting film reduces trapping of vortices when the device is cooled in a static magnetic field. The slot structure is also effective in suppressing the long-distance motion of vortices along wide grain boundaries that are formed in the pickup coil at a site where the superconducting film crosses over the bicrystal line of the substrate. The vortex penetration can be reduced by use of a flux dam, suppressing the shielding current induced by a field change. The Effects of narrowing the width of the superconducting film of the flux dam by incorporation of holes are also examined.

In Sect. 5 we discuss a direct-coupled SQUID with a high μ-metal needle for a SQUID microscope. The magnetic field is collected at the tip of the needle and is guided to the pickup coil of the magnetometer. Using this configuration we can improve the spatial resolution of the microscope.

2 HTSC Junctions for SQUIDs

Many types of HTSC junctions have been developed for device application, as shown in Fig. 1. Among them, step edge junctions and bicrystal junctions are usually used for the SQUID application. In these junctions an insulating barrier that is naturally formed at the grain boundary is used for making the Josephson junction. Therefore, these junctions are also called grain-boundary junctions. In the following we discuss the transport and noise properties of the bicrystal junction.

2.1 Transport Properties

In Fig. 2a a simple model of the bicrystal junction is shown [1,2]. It is assumed that an insulating barrier is formed at the grain boundary. The junction is

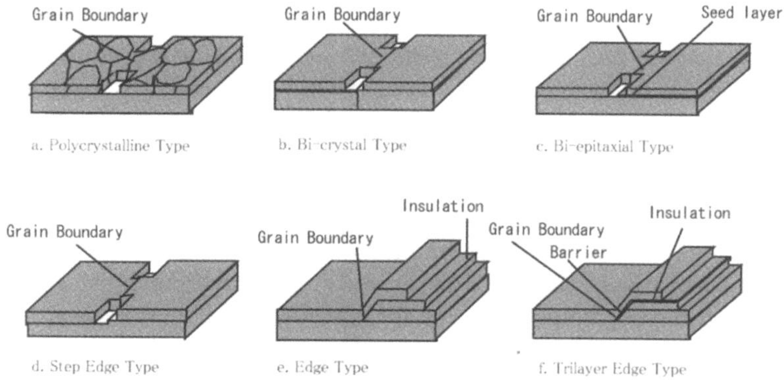

Fig. 1. Types of HTSC Josephson junctions

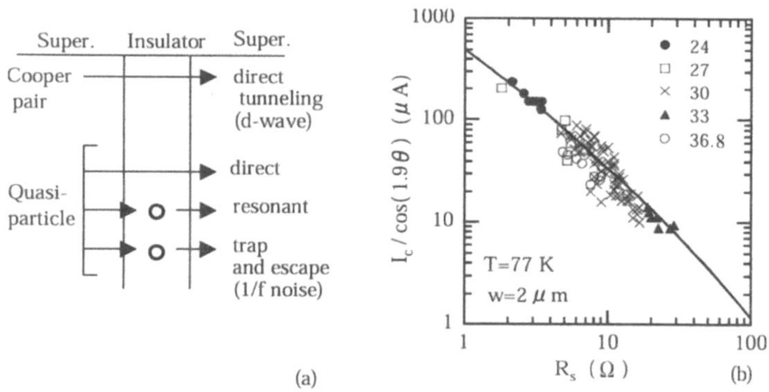

Fig. 2. (**a**) Transport model of the bicrystal junction. *Circles* show the localized states in the barrier. (**b**) Relationship between the normalized current $I_c/\cos(1.9\theta)$ and the misorientation angle θ. The *solid line* is calculated from (3)

made by the structure of superconductor/insulator/superconductor (SIS) and the current flows by a tunneling mechanism. The thickness of the barrier d is changed by both the misorientation angle θ of the bicrystal substrate and the fabrication conditions. In the insulating barrier there exist localized states due to imperfections in the barrier, which are schematically shown by circles.

It is suggested that a quasiparticle current flows with both direct tunneling and resonant tunneling via localized states in the barrier. In this case the conductivity of the junction σ is given by [1,2]

$$\sigma = \sigma_{\mathrm{do}} \exp(-2ad) + \sigma_{\mathrm{ro}} \exp(-ad)\,, \tag{1}$$

where $1/a$ is a decay length and σ_{do} and σ_{ro} are parameters representing the degree of direct and resonant tunneling, respectively.

Superconducting current flows only with direct tunneling. It is pointed out, however, that the d-wave symmetry of the superconducting order parameter affects the tunneling. When a $d_{x^2-y^2}$ wave symmetry is assumed the amplitude of the order parameter, i.e., the energy gap Δ of the superconducting electrode, becomes dependent on the wave vector k and is given by $\Delta_k = \Delta_0 \cos(2\varphi)$, where φ is the angle measured from the a-axis [3]. When we use a bicrystal junction with misorientation angle θ the a-axis becomes different between the two electrodes by the angle θ. This means that overlap of the superconducting order parameter between the two electrodes changes with the angle θ; the overlap becomes smaller with increasing of θ. Since the critical current density j_c of the junction depends on the overlap of the order parameter, j_c becomes dependent on the misorientation angle θ even when the thickness of the insulating barrier d is constant. The dependence of j_c on θ has been theoretically studied in [3]. Taking account of this effect, we can obtain an expression for j_c as

$$j_c(\theta, T, d) = j_{co}(\theta, T)\exp(-2ad) \approx K(T)\cos(1.9\theta)\exp(-2ad) \tag{2}$$

at high temperatures.

A theoretical expression for $j_{co}(\theta, T)$ was given by (2) [3], and the θ-dependence can be calculated numerically. The last expression in (2) can be obtained by approximating the numerical result of $j_{co}(\theta, T)$ with a cosine function; the approximation can be shown to become accurate at high temperatures. As shown in (2), reduction of j_c due to the d-wave symmetry of the order parameter becomes more significant when the angle θ becomes larger.

From (1) and (2), a relationship between $I_c = j_c A_j$ and $R_s = 1/(A_j \sigma)$ is given by

$$\frac{1}{R_s} = G_{dt} + G_{rt} = \frac{C_1 I_c}{\cos(1.9\theta)} + C_2\sqrt{\frac{I_c}{\cos(1.9\theta)}}, \tag{3}$$

where $C_1 = \sigma_{do}/K(T)$, $C_2 = \sigma_{ro}(A_j/K)^{1/2}$ and A_j is a junction area. Note that the first term G_{dt} and the second term G_{rt} represent the conductance due to the direct and resonant tunneling of the quasiparticle, respectively.

In Fig. 2b experimental results are compared with (3). Here the vertical axis represents the value of $I_c/\cos(1.9\theta)$. Symbols show the experimental results obtained from $YBa_2Cu_3O_{7-x}$ bicrystal junctions, which are fabricated on $SrTiO_3$ bicrystal substrate with angles of $24° < \theta < 36.8°$ [4]. The thickness of the film and junction width are nominally 150 nm and 2 μm, respectively. The solid line is calculated from (3) with parameters $C_1 = 1650$ S/A and $C_2 = 7.5$ S/A$^{1/2}$. As shown, the line calculated from (3) is in good agreement with the experimental results for all misorientation angles. It should be noted that the effect of the d-wave symmetry of the order parameter must be included in order to obtain good agreement. If this effect is neglected, deviation between experiment and analysis becomes large, especially for the case of $\theta = 36.8°$.

2.2 Noise Rounding and Excess Current

When we estimate the junction critical current I_c and resistance R_s, we have to take into account noise rounding of the *I-V* curve and the so-called excess current. Since the noise parameter $\Gamma = 2\pi k_B T/I_c\Phi_0$ is not small at $T = 77\,K$, the critical current is suppressed by the thermal noise, i.e., the experimentally observed critical current is smaller than the *true* critical current. Moreover, there exists the so-called excess current I_{ex} which does not behave as the Josephson current. In this case, the junction current I_J can be expressed as $I_J = I_c \sin\phi + I_{ex}$. Since the excess current does not contribute to the SQUID operation, it should be much smaller than the Josephson current.

In Fig. 3a the experimental *I-V* curve is compared with the numerical simulation including the thermal noise and the excess current. The junction is a 30° bicrystal junction with a width of 4 µm. The solid line is the experimental *I-V* curve and the symbols show the simulation ones [5]. As shown, good agreement is obtained between the experiment and simulation. From the comparison we can estimate the junction parameters as $I_c = 56\,\mu A$, $R_s =$ 4.5 Omega and $I_{ex} = 10\,\mu A$ at $T = 77\,K$. We note that the excess current becomes large when the critical current I_c becomes large. For example, the 24° bicrystal junction has much larger excess current than the case of the 30° junction.

In Fig. 3b a simulation result is shown on the suppression of the critical current due to the thermal noise. The horizontal axis shows the experimentally observable maximum zero voltage current I_{max}, while the vertical axis shows the *true* critical current I_c of the junction. As shown, the value of I_{max} becomes small compared to I_c at $T = 77\,K$. For example, we cannot observe the zero voltage current in the case of $I_c < 7\,\mu A$.

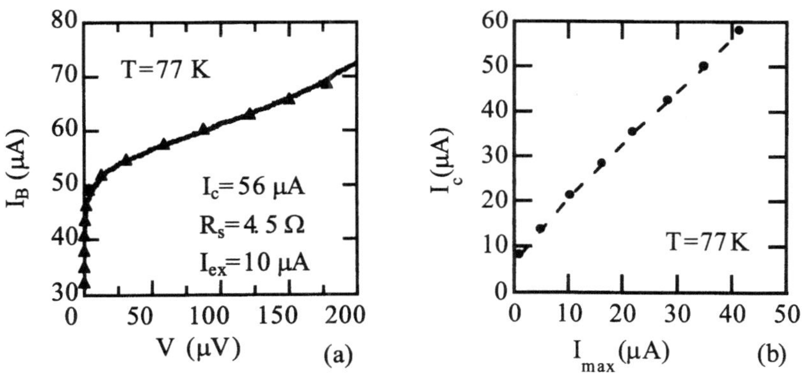

Fig. 3. (**a**) *I-V* curve of the bicrystal junction. (**b**) Relationship between the maximum zero voltage current I_{max} and the critical current I_c

2.3 $1/f$ Noise

We now discuss the $1/f$ noise of the bicrystal junction. In the insulating barriere there exists another type of localized state that can capture the quasiparticle, as shown in Fig. 2a. The quasiparticle is trapped in these states for a certain time and escapes from the states due to thermal activation. This capture and emission of the quasiparticle in the localized states causes charging and discharging of the states and results in a change of the barrier height for the tunneling of other electrons (Cooper pairs and quasiparticles). This barrier height fluctuation causes low frequency fluctuations of the conductivity σ and the critical current density j_c of the junction and results in the $1/f$ noise of the junction [2,6].

From (2), fluctuations of the critical current density δj_c due to barrier height fluctuations are given by

$$\delta j_c/j_c = -2da(\delta a/a) = -\ln\left[K\cos(1.9\theta)/j_c\right](\delta a/a), \tag{4}$$

where the barrier height fluctuation is represented by the fluctuation of the decay length $1/a$.

Fluctuations of the critical current δI_c due to the fluctuations in δj_c can be obtained as follows. We assume that the barrier height locally changes in the area of A_e when the quasiparticle is trapped in a single localized state. The total number of fluctuation sites is given by $N = nA_j d$ when the volume density of the fluctuation site is n. Therefore, the fluctuation of the critical current is given by $\delta I_c = \sum_{i=1}^{N} A_{e,i}\delta j_{c,i}$, where $A_{e,i}$ and $\delta j_{c,i}$ are the values for the ith fluctuation site. When we neglect the correlation between the fluctuations at each site and assume that $A_{e,i}$ and $\delta j_{c,i}$ are uniform at every site, the power spectral density of the critical current fluctuation S_{I_c} is given by

$$S_{I_c} = NA_e^2(2ad)^2 j_c^2 S_a/a^2 = \alpha^2 I_c^2\left[\ln(K\cos(1.9\theta)/j_c)\right]^3/A_j\,, \tag{5}$$

where $\alpha^2 = (nA_e^2/2a)(S_a/a^2)$ and S_a is the power spectral density of the fluctuations of a.

In Fig. 4 experimental results for critical current fluctuations are compared with (5) [7]. Here vertical and horizontal axes represent the normalized fluctuation $A_j^{1/2}S_{I_c}^{1/2}/I_c$ at $f = 1\,\mathrm{Hz}$ and the critical current density $j_c/\cos(1.9\theta)$, respectively. Triangles and rectangles show the results obtained for the case of $\theta = 30°$ and $\theta = 24°$, respectively. As shown, the normalized fluctuations become dependent on the critical current density $j_c/\cos(1.9\theta)$. The solid line in the figure is calculated from (5) with parameters $\alpha = 14.6\times10^{-12}$ and $K = 5.6\times10^5\ \mathrm{A/cm^2}$. As shown, the analytical result is in good agreement with the experiment.

As shown in Fig. 4, the normalized fluctuation $A_j^{1/2}S_{I_c}^{1/2}/I_c$ increases with decreasing of $j_c/\cos(1.9\theta)$. The increase becomes considerable when the current density is small. This result means that the fluctuation of the critical current $S_{I_c}^{1/2}$ does not decrease in proportional to I_c, but remains high

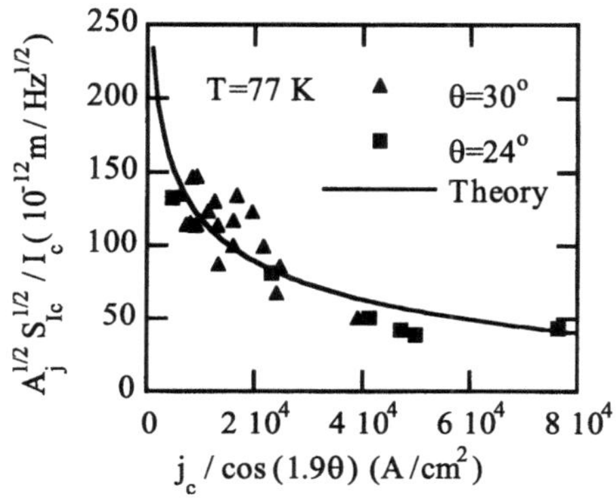

Fig. 4. Comparison between analysis and experiment of the critical-current fluctuation at $f = 1\,\text{Hz}$ and $T = 77\,\text{K}$

when I_c becomes small. When the current density becomes large, on the other hand, the normalized fluctuation approaches the value of $A_j^{1/2} S_{I_c}^{1/2} / I_c = 45 \times 10^{-12}\ \text{m}/\,\text{Hz}^{1/2}$. For a junction area of $A_j = 0.3\,\mu\text{m}^2$ we obtain the relation $S_{I_c}^{1/2}/I_c = 8.2 \times 10^{-5}\ /\text{Hz}^{1/2}$ at $f = 1\,\text{Hz}$. This value is consistent with the previous report [8].

Making a similar analysis, we can obtain an expression for the fluctuations of the resistance $S_R^{1/2}$. An empirical relation of $S_R^{1/2} = 3.2 \times 10^{-5} R_s^{1/2}$ at $f = 1\,\text{Hz}$, which was obtained experimentally in [9] can be explained reasonably well with the analysis.

2.4 Dependence of SQUID Performance on Junction Parameters

Since the performance of the SQUID depends on the junction parameters, i.e., the junction critical current I_c and the resistance R_s, it is important to optimize these parameters in order to realize a highly sensitive magnetometer. In Fig. 5a a simulation result is shown on the voltage modulation depth ΔV of the SQUID for different values of I_c and R_s. In the simulation we use a SQUID inductance of $L_s = 100\,\text{pH}$ and the operation temperature $T = 77\,\text{K}$. As shown, we can obtain large voltage modulation for large values of $I_c R_s$. Furthermore, the value of ΔV becomes a maximum at $I_c = 10\,\mu\text{A}$ for a fixed value of $I_c R_s$. The value $I_c = 10\,\mu\text{A}$ corresponds to the well-known condition of $\beta = 2 L_s I_c / \Phi_0 = 1$ for the case of $L_s = 100\,\text{pH}$.

In Fig. 5b a simulation result is shown for the flux noise $S_\Phi^{1/2}$. As shown, the flux noise becomes smaller for larger values of $I_c R_s$. The flux noise becomes a minimum at $I_c = 10\,\mu\text{A}$ for a fixed value of $I_c R_s$. Therefore it is important to use the junction with large $I_c R_s$ keeping the condition of $\beta = 1$ in order to obtain a high performance SQUID. Since typical values of the

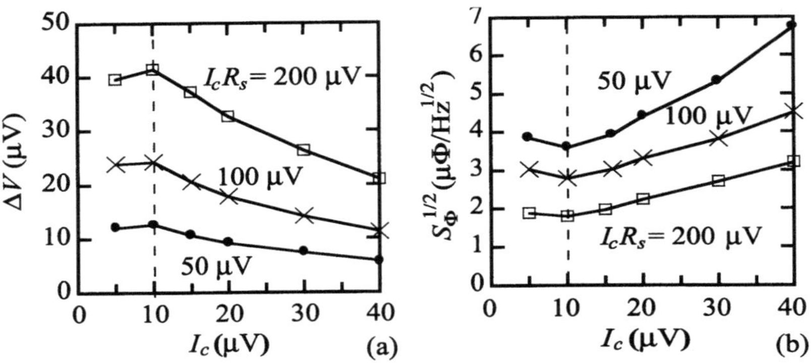

Fig. 5. Dependence of the SQUID performance on junction parameters. (**a**) Simulation results on the voltage modulation depth ΔV. (**b**) Flux noise spectrum S_Φ. Other parameters are $L_s = 100\,\text{pH}$ and $T = 77\,\text{K}$

SQUID inductance range from $L_s = 30\,\text{pH}$ to $L_s = 100\,\text{pH}$, the critical current of the junction should be from $I_c = 30\,\mu\text{A}$ to $I_c = 10\,\mu\text{A}$. Then the junction resistance should be as large as possible. For example, a junction resistance from $R_s = 6\ \Omega$ to $R_s = 10\ \Omega$ is desired.

From the numerical simulations shown in Fig. 5a and Fig. 5b we can obtain expressions for the modulation voltage ΔV and the flux noise S_Φ as [10,11]

$$\Delta V = \frac{4}{\pi}\frac{I_c R_s}{1+\beta} \exp\left(-3.5\pi^2 \delta\Phi_n^2/\Phi_0^2\right), \tag{6}$$

$$S_\Phi \approx \frac{1.5 k_B T L_s^2}{R_s}\left[4 + \left(1+\frac{1}{\beta}\right)^2 \exp\left(7\pi^2\delta\Phi_n^2/\Phi_0^2\right)\right], \tag{7}$$

with

$$\delta\Phi_n = \sqrt{k_B T L_s}\,. \tag{8}$$

It must be noted that the $\delta\Phi_n$ given in (8) shows the total noise flux existing in the inductance L_s. This total noise flux is caused by the thermal noise and becomes large for large values of L_s and T. For example, we obtain $\delta\Phi_n = 0.157\Phi_0$ for $L_s = 100\,\text{pH}$ and $T = 77\,\text{K}$. As shown in (6) and (7), the values of ΔV and S_Φ depend exponentially on the noise flux $\delta\Phi_n$. These exponential terms are peculiar to the high-T_c SQUID operating at $T = 77\,\text{K}$. Due to these terms, ΔV and S_Φ depend on the inductance L_s more strongly than in the case of the low T_c SQUID at $T = 4.2\,\text{K}$. As a result, the value of L_s is limited to $L_s < 150\,\text{pH}$ at $T = 77\,\text{K}$.

In order to realize a junction that satisfies the above-mentioned requirements several types of junction have been developed. Among them the bicrystal junction is typically used for high-T_c SQUIDs. In this case the values of I_c

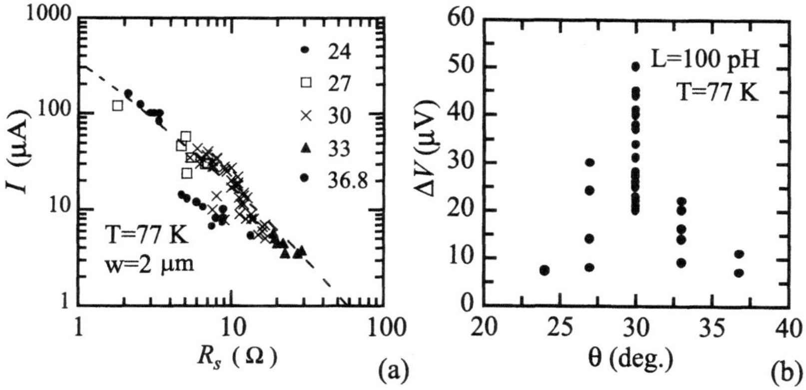

Fig. 6. (**a**) Relationship between I_c and R_s for the bicrystal junction at $T = 77\,\text{K}$. Experimental results are obtained from bicrystal junctions with $24° < \theta < 36.8°$. The *broken line* is to guide the eye. (**b**) The voltage modulation depth ΔV of the 100 pH SQUID with different misorientation angles

and R_s can be changed by changing the misorientation angle θ, as mentioned before. In Fig. 6a the relationship between I_c and R_s is shown when different misorientation angles are used [4]. As shown in Fig. 6a, we obtain typical values of $I_c = 100\,\mu\text{A}$ and $R_s = 2.5\ \Omega$ for the case of $\theta = 24°$. Although the value of $I_c R_s = 250\,\mu\text{V}$ is very large, the critical current is too high and the junction resistance is small for application to the SQUID. If we can make a junction with a width less than 1 μm we can use this junction for the SQUID. On the other hand, in the case of $\theta = 36.8°$ we obtain typical values of $I_c = 10\,\mu\text{A}$ and $R_s = 7\Omega$. In this case the value of $I_c R_s = 70\,\mu\text{V}$ is relatively small.

When we use the junction with $\theta = 30°$ we obtain the typical values of $I_c = 20\,\mu\text{A}$ and $R_s = 8\,\Omega$. In this case we can satisfy the requirement for the high performance SQUID, i.e., we can satisfy both conditions of $\beta \approx 1$ and large $I_c R_s$. Therefore, the junction with $\theta = 30°$ is suitable for the high performance SQUID. In Fig. 6b the voltage modulation depth ΔV of the SQUID is compared when junctions with different misorientation angles θ are used [4]. In the experiment the SQUID inductance is chosen as $L_s = 100\,\text{pH}$ and the operation temperature is $T = 77\,\text{K}$. As expected, we can obtain large values of ΔV when the junction with $\theta = 30°$ is used.

We obtain values ranging from $\Delta V = 20\,\mu\text{V}$ to $\Delta V = 50\,\mu\text{V}$ for $L_s = 100\,\text{pH}$ when we use the 30° bicrystal junction. It has been shown that these experimental values are 60 %–100 % of the theoretical values [4]. The experimental values of the flux noise are typically $S_\Phi^{1/2} = 10\text{–}15\,\mu\Phi_0/\text{Hz}^{1/2}$ for the case of $L_s = 100\,\text{pH}$. These values are about 2–3 times higher than the theoretical predictions calculated from (7). The reason for this deviation is not clear at present. When these SQUIDs are coupled to a pickup coil with an area of 9 mm × 9 mm we can realize highly sensitive magnetometers with 50 fT/Hz$^{1/2}$ field resolution [12].

It should be noted that the situation $\beta > 1$ often occurs due to the difficulty of controlling the value of I_c, though the value of ΔV is maximized when the condition of $\beta = 1$ is satisfied. In this case, the value of ΔV becomes very small compared to the value of $I_c R_s$. This problem can be solved when the SQUID inductance is shunted by a resistance, i.e., the so-called resistively shunted inductance is used [10]. In this case, the value of ΔV becomes almost independent of β and a large value of ΔV can be obtained even in the case of $\beta \gg 1$. The usefulness of this method has been demonstrated [13,14].

2.5 SQUID Inductance

Since the SQUID performance strongly depends on the inductance L_s as shown in (6) and (7), it is important to accurately evaluate the value of L_s. For the SQUID inductance the geometry shown in Fig. 7a is usually used. The inductance is made of a slit washer where the length of the slit, the width of the slit, the width of an electrode and the thickness of the electrode are denoted l_{sl}, s, w, and t, respectively. For this geometry the inductance L'_{sl} and parasitic capacitance C'_{sl} of the slit per unit length are given by [15,16]

$$L'_{sl} = L'_{m,sl} + L'_{k,sl}\,, \tag{9}$$

$$L'_{m,sl} = \mu_0 K(k)/K(k')\,, \tag{10}$$

$$L'_{k,sl} = \frac{2\mu_0\lambda_L^2}{twk'^2K^2(k')} \left[\frac{w}{s}\ln\left(\frac{4ws}{t(w+s)}\right) + \frac{w}{2w+s}\ln\left(\frac{4w(2w+s)}{t(w+s)}\right)\right], \tag{11}$$

$$C'_{sl} = \frac{\epsilon_r + 1}{2c^2 L'_{m,sl}}\,, \tag{12}$$

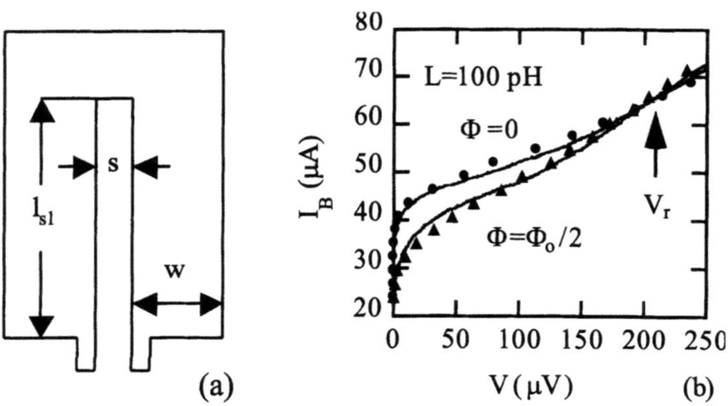

Fig. 7. (**a**) Geometry of the SQUID inductance. (**b**) Comparison between simulation and experiment of the *I*-*V* curve of the SQUID. The voltage V_r corresponds to the $5\lambda|/4$ strip-line resonance occuring in the SQUID inductance

where $L'_{m,sl}$ is the magnetic inductance, $L'_{k,sl}$ is the kinetic inductance, μ_0 is the permeability, $K(k)$ is the complete elliptic integral of the first kind with a modulus $k = s/(s+w)$, $k' = (1-k^2)^{1/2}$, λ_L is the penetration depth of the film, ϵ_r is the dielectric constant of the substrate and c is the velocity of light. For typical values of $s = 4\,\mu m$, $w = 15\,\mu m$, $t = 0.15\,\mu m$, $\lambda_L = 0.3\,\mu m$, and $\epsilon_r = 1930$ for $SrTiO_3$ substrate, we obtain $L'_{m,sl} = 0.56\,pH/\mu m$, $L'_{k,sl} = 0.15\,pH/\mu m$ and $C'_{sl} = 0.02\ pF/\mu m$. The SQUID inductance is given by $L_s = L'_{sl}l_{sl} + L_{pr}$ where L_{pr} is the parasitic inductance around the electrode of the junction. The value of L_{pr} is typically 15–25 pH.

Although the SQUID inductance can be designed by using (10) and (11), we note that the parasitic capacitance C'_{sl} of the slit becomes large due to the large dielectric constant of the $SrTiO_3$ substrate. This parasitic capacitance distributes along the slit of the washer. Therefore, the SQUID performance is affected by the parasitic capacitance, as shown below. First, it should be noted that microwave current is produced from the junction due to the ac Josephson effect. This microwave current flows along the slit of the washer. For example, if the operation voltage is $V = 40\,\mu V$, the frequency of the microwave current becomes $f = V/\Phi_0 = 20\,GHz$. In this case the wavelength of the electromagnetic wave becomes $\lambda = (L'_{sl}C'_{sl})^{-1/2}/f = 419\,\mu m$ for $L'_{m,sl} = 0.56\,pH/\mu m$, $L'_{k,sl} = 0.15\,pH/\mu m$ and $C'_{sl} = 0.02\ pF/\mu m$. Since the length of the slit of the washer is typically $l_{sl} = 100\,\mu m$, the slit length cannot be negligibly small compared to the wavelength. This means that the SQUID inductance cannot be expressed by a simple lumped element, but behaves as a distributed line for the microwave current caused by the ac Josephson effect. If the slit length l_{sl} becomes equal to a quarter wavelength ($\lambda/4$) a so-called standing wave occurs in the slit, i.e., so-called strip line resonance occurs.

Therefore we have to treat the SQUID inductance as a distributed resonator. In Fig. 7b the effect of the strip line resonance on the I-V curve of the SQUID is shown [16]. The solid lines show the experimental results for $\Phi = 0$ and $\Phi = \Phi_0/2$ when the slit length of the SQUID washer is $l_{sl} = 120\,\mu m$. Other parameters are $s = 4\,\mu m$ and $w = 30\,\mu m$ which give the SQUID inductance $L_s = 100\,pH$. As shown in Fig. 7b, a resonant current step is observed. We can see crossover of the I-V curve for $\Phi = 0$ and $\Phi = \Phi_0/2$ at $V_r = 206\,\mu V$.

The symbols show simulation results including the effect of the strip line resonance. As shown, the experimental I-V curves agree well with the numerical simulation. It can be shown that the frequency corresponsing to the quarter wavelength resonance ($l_{sl} = \lambda/4$) becomes $f_r = 20\,GHz$ in the present case. Therefore, the crossover voltage $V_r = 206\,\mu V$ corresponds to the $5\lambda/4$ strip-line resonance. When the length of the slit l_{sl} becomes small we can observe a much clearer resonant step corresponding to the $3\lambda/4$ strip-line resonance [16].

Finally we note that the strip-line resonance also occurs when the SQUID inductance is coupled to the pickup coil. In this case the standing wave occurs in the pickup coil or the input coil and causes distortion of the V-Φ curve of the SQUID [17,18,19]. The effect of the resonance becomes more significant in the case of flux-transformer coupling than in the case of direct coupling.

2.6 Pickup Coil and Coupling Circuit

When we make a magnetometer we have to couple the SQUID to the pickup coil. Several types of coupling scheme have been developed, i.e., direct coupling [20], flux-transformer coupling [17] and multiloop coupling [21]. In high-T_c SQUIDs the direct-coupling scheme is usually used due to its easiness of fabrication and high reliability. In this case the SQUID inductance L_s is directly connected to the pickup coil and the SQUID detects the current I_{cir} flowing in the pickup loop. The magnetic flux Φ_s, which interlinks the SQUID inductance, is given by

$$\Phi_s = I_{cir} L_c = \frac{B A_p L_c}{L_p + L_c} = B A_{eff} , \tag{13}$$

where A_p and L_p are the pickup area and the inductance of the pickup coil, respectively, and L_c is the coupling part of the SQUID inductance. The value A_{eff} is the effective area of the magnetometer. For the geometry of the pickup coil shown in Fig. 8a we obtain $A_p = a(a - 2w)$. The inductance is roughly given by [22]

$$L_p = (2/\pi)\mu_0 (a - w) \ln\left(\frac{a - w}{w} + 0.5\right) , \tag{14}$$

where the effect of the kinetic inductance is neglected. Since $L_p \gg L_c$ in the usual case, we have to maximize the value of A_p/L_p in order to obtain a large effective area A_{eff}. This requirement is satisfied when the condition of $w = a/3$ is satisfied [20]. Therefore, we have to use a wide thin film for the pickup coil. For example, a pickup coil with $a = 9\,\mathrm{mm}$ and $w = 3\,\mathrm{mm}$ gives $A_p = 27\,\mathrm{mm}^2$ and $L_p = 4.4$ nH. When a SQUID inductance of $L_c = 80\,\mathrm{pH}$ is coupled to the pickup coil the effective area of the magnetometer becomes $A_{eff} = 0.48\,\mathrm{mm}^2$. If the flux noise of the SQUID is $S_\Phi^{1/2} = 10\mu\Phi_0/\,\mathrm{Hz}^{1/2}$ we can expect the field noise of the magnetometer to be $S_B^{1/2} = 44\,\mathrm{fT}/\,\mathrm{Hz}^{1/2}$.

When a wide thin film is used for the pickup coil, however, the problem of flux trapping in the film becomes serious. When the magnetometer is cooled in the presence of a residual magnetic field the field is easily trapped in the film during the cooling process and the trapped flux causes strong $1/f$ noise. In order to avoid this problem the residual field must be below a threshold field $B_{th} = \pi\Phi_0/4w^2$.

The above equation means that we have to use a narrow thin film in order to avoid the flux trapping during field cooling. For example, if the magnetometer is cooled under the earth magnetic field of about 50 μT, the width w must be smaller than 6 μm in order to avoid flux trapping. This requirement of the narrow thin film contradicts the requirement of the wide thin film for the large effective area. In order to realize a large effective area and avoid flux trapping, the pickup coils shown in Fig. 8b and Fig. 8c have been developed [23,24,25,26]. In Fig. 8b the wide thin film is divided into parallel loop of narrow thin film while the wide film is changed to a mesh structure in Fig. 8c. The effective areas of these pickup coils are almost the same as that of the wide pickup coil shown in Fig. 8a, while the flux trapping during the field cooling in the earth magnetic field can be avoided.

It should be mentioned that the pickup coils shown in Fig. 8b and Fig. 8c cannot solve the flux trapping when a large field change occurs after the magnetometer is cooled. When the field B is applied after cooling, the circulating current $I_{\mathrm{cir}} = BA_{\mathrm{p}}/L_{\mathrm{p}}$ flows in the pickup coil. The values of I_{cir} are almost the same as that of the case shown in Fig. 8a. For example, the circulating current becomes $I_{\mathrm{cir}} = 6\,\mathrm{mA}/\mu\mathrm{T}$ for the case of $a = 9\,\mathrm{mm}$ and $w = 3\,\mathrm{mm}$. This current flows in the SQUID inductance. Since the SQUID inductance L_{s} is made of a narrow thin film, e.g. 10 μm width, the current density and the self-field become very high at the SQUID inductance. For example, the current $I_{\mathrm{cir}} = 6\,\mathrm{mA}/\mu\mathrm{T}$ gives the current density $j = 4 \times 10^5\ \mathrm{A/cm^2}/\mu\mathrm{T}$ if the SQUID inductance is made of a thin film with a width of 10 μm and a thickness of 0.15 μm. As a result flux trapping will easily occur in the SQUID inductance due to the large circulating current.

In order to solve the flux trapping that is caused by the large circulating current due to the large field change, a so-called flux dam is used [27]. For example, a flux dam is inserted in each loop of the pickup coil as shown in Fig. 8b. The flux dam is made of a Josephson junction with a critical current

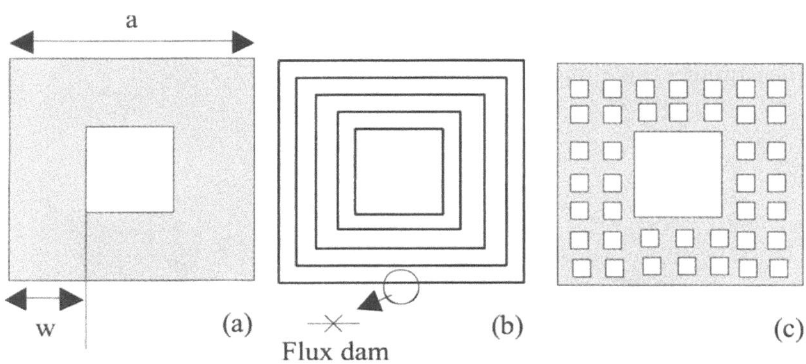

Fig. 8. Pickup coil of the directly coupled SQUID magnetometer. (**a**) Pickup coil made of wide thin film. (**b**) Pickup coil made of a parallel loop of narrow thin film with addition of the flux dam. (**c**) Pickup coil made of the mesh structure

I_c and limits the circulating current in the pickup coil to less than I_c. As a result, $1/f$ noise due to the large circulating current can be prevented.

It should be mentioned that it takes a long time for the flux dam to completely close the pickup coil. It has been shown that the transition from the open state (voltage state) to the closed state (zero voltage state) is strongly affected by the thermal noise, i.e., the transition is dominated by the thermal process [28]. Due to the thermal process, it takes a long time for completion of the transition. This problem will be discussed in the next section.

2.7 Thermal Activation in the Flux Dam

We study the behavior of flux entry through the flux dam. For this purpose we consider the simple case where one flux dam is inserted in the pickup coil, as shown schematically in Fig. 9a. When the external field B_{ex} is applied to the pickup coil, a circulating current $I_{cir} = A_p (B_{ex} - B_{in})/L_p$ flows in the coil. If the value of I_{cir} exceeds the critical current I_c of the flux dam the external field enters the pickup coil where B_{in} is the field inside the pickup coil. Then the difference between B_{ex} and B_{in} becomes small and I_{cir} decreases. When the value of I_{cir} becomes lower than I_c the flux dam begins to close.

However, it should be noted that the circulating current is near I_c at this time, i.e., the operation point of the flux dam is at point "A" in the *I-V* curve of the flux dam, as shown in Fig. 9b. In this case thermal noise easily causes flux entry through the flux dam. This thermally activated flux entry continues for a long time, i.e., it takes a long time for the completion of this thermal process. Since the circulating current I_{cir} decreases if flux entry occurs, the value of I_{cir} becomes small when the thermally activated flux entry proceedes sufficiently. In this case the operation point of the flux dam moves from point "A" to point "B" in Fig. 9b. Then the system becomes stable, i.e., we can neglect thermally activated flux entry during the measurement time. Therefore we need a long waiting time for the stabilization of the system before measurement, e.g. more than 5 min.

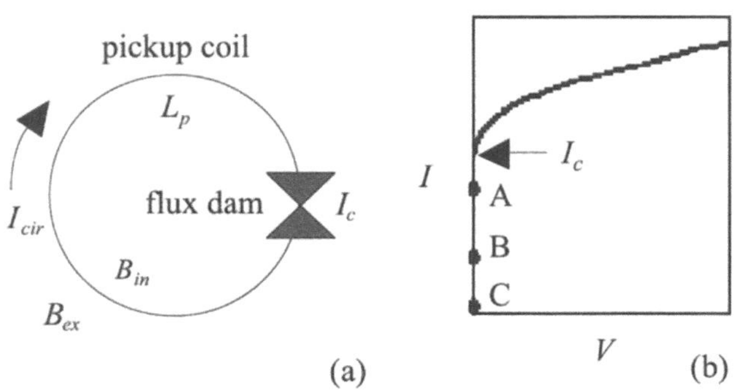

Fig. 9. (**a**) Schematic figure of the pickup coil with the flux dam. (**b**) *I-V* curve of the flux dam

In the following we analytically discuss the thermally activated flux entry in terms of the circulating current I_{cir}. Since the present situation shown in Fig. 9a is similar to an rf SQUID, we can discuss the thermally activated flux entry on the basis of the rf SQUID [29,30].

Using the phase difference ϕ across the flux dam we can obtain the potential energy U of the system as $U(\phi) = (I_c\Phi_0/2\pi)\left[1 - \cos\phi + (\phi - \phi_{ex})^2/2\beta_L\right]$, where $\beta_L = 2\pi L_p I_c/\Phi_0$ and $\phi_{ex} = 2\pi\Phi_{ex}/\Phi_0 = 2\pi A_p B_{ex}/\Phi_0$.

In Fig. 10a the potential U is shown as a function of ϕ. As shown, the potential has local minima and maxima. The difference between them gives the potential barrier ΔU. The local minimum is a stable point in the absence of thermal noise. When thermal noise exists, however, the phase ϕ moves around the stable point. If the thermal noise is large enough to get over the barrier ΔU, the operation point moves to the next local minimum. This means that one flux quantum enters into the pickup coil due to the thermal noise.

In Fig. 10b triangles show the simulation results for ΔU [28]. The values of ΔU are shown as a function of the circulating current I_{cir} for different values of I_c. As shown, the potential barrier ΔU becomes a maximum at $I_{cir} = 0$ and decreases with increasing of I_{cir}. In the present case where $\beta_L \gg 1$ the dependence of ΔU on I_{cir} shown in Fig. 10b can be well approximated by the analytical expression for ΔU that was obtained for the case of a single junction. If we substitute the circulating current I_{cir} for the bias current of the single junction, we obtain the analytical expression as

$$\Delta U(i_{cir}) = \frac{I_c\Phi_0}{\pi}\left[\sqrt{1 - i_{cir}^2} + i_{cir}\left(\arcsin i_{cir} - \frac{\pi}{2}\right)\right], \qquad (15)$$

where $i_{cir} = I_{cir}/I_c$ is the normalized circulating current. The solid lines in Fig. 10b are calculated from (15). As shown, good agreement is obtained between the simulation and (15).

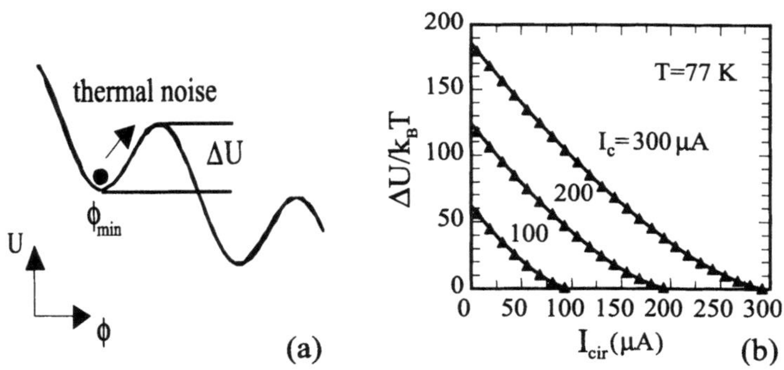

Fig. 10. (**a**) Potential energy U of the system shown in Fig. 10a. (**b**) Dependence of the potential barrier ΔU on the circulating current I_{cir}. *Triangles* are simulation results and the *solid line* shows the analytical result calculated from (15)

The lifetime τ, during which the phase ϕ remains at the local minimum, is given by [29,30]

$$\frac{1}{\tau} = f_0 \left(1 - i_{\mathrm{cir}}^2\right)^{\frac{1}{4}} \exp\left(-\frac{\Delta U}{k_{\mathrm{B}}T}\right), \tag{16}$$

where $f_0 = 10^{11}$ Hz is the characteristic frequency.

Using the potential barrier ΔU and the lifetime τ, we can discuss the thermally activated flux entry. When the operation point of the flux dam is point "A" in Fig. 9b, i.e., the value of I_{cir} is close to I_{c}, the potential barrier ΔU is small and the lifetime is short. In this case, therefore, thermal noise easily causes flux entry into the pickup coil. When the operation point of the flux dam moves from point "A" to point "B" in Fig. 9b, the potential barrier ΔU becomes large as shown in Fig. 10b. As a result, we have a long lifetime τ, and thermal activation can be neglected during the measurement. For example, if we want to realize the lifetime $\tau > 100$ s, the condition of $\Delta U/k_{\mathrm{B}}T > 30$ must be satisfied as can be seen from (16). As shown in Fig. 10b, this condition can be satisfied when the value of I_{cir} is less than the threshold value of 120 μA for the case of $I_{\mathrm{c}} = 200$ μA.

We now discuss the decay process of the circulating current due to the thermal activation. Using the lifetime τ, we can describe the decay of the circulating current I_{cir} as [31]

$$\frac{dI_{\mathrm{cir}}}{dt} = -\frac{(\Phi_0/L_p)}{\tau} \tag{17}$$

$$\approx -0.85 f_0 \frac{\Phi_0}{L_{\mathrm{p}}} \exp\left[\frac{2}{\Gamma}\left(0.77\frac{I_{\mathrm{cir}}}{I_{\mathrm{c}}} - 0.7\right)\right] \text{for } 0.55 < \frac{I_{\mathrm{cir}}}{I_{\mathrm{c}}} < 0.85\,, \tag{18}$$

where $\Gamma = 2\pi k_{\mathrm{B}}T/I_{\mathrm{c}}\Phi_0$ is the noise parameter. The last equation is obtained by approximating the relationship between ΔU and I_{cir} that is shown in Fig. 10b with the linear relation of $\Delta U(i_{\mathrm{cir}}) = (I_{\mathrm{c}}\Phi_0/\pi)(0.7 - 0.77 i_{\mathrm{cir}})$ in the region of $0.55 < i_{\mathrm{cir}} < 0.85$. We note that thermal activation in this region becomes important in practical cases.

The solution of (18) is

$$\frac{I_{\mathrm{cir}}(t)}{I_{\mathrm{c}}} = 0.9 - 0.65\Gamma \ln\left(\frac{0.65 f_0 \Phi_0^2}{\pi k_{\mathrm{B}} T L_{\mathrm{p}}} t\right). \tag{19}$$

Equation (19) means that the decay of the circulating current due to the thermal activation proceeds logarithmically with time. Due to the logarithmic decay of I_{cir}, we have to wait a long time in order to decrease the circulating current sufficiently. For example, if we want to realize the lifetime $\tau > 100$ s, the value of I_{cir} must be less than 120 μA for the case of $I_{\mathrm{c}} = 200$ μA, as mentioned before. In order to realize the condition $I_{\mathrm{cir}} < 120$ μA we need a waiting time of 500 s, which can be calculated from (19). This long waiting time is problem in practical applications.

2.8 Switch for Opening and Closing the Pickup Coil

In order to solve the problem of the long waiting time it is necessary to develop a method that can forcibly move the operation point of the flux dam from point "A" to point "B" in Fig. 9b, instead of the thermally activated process. A method that can rapidly decrease the circulating current I_{cir} much below I_c of the flux dam is necessary. If this method is developed, the thermally activated flux entry will be prevented without the long waiting time. For this purpose a switch instead of the flux dam has been proposed [32].

In Fig. 11a a magnetometer involving the switch is schematically shown. The outer size of the pickup coil is 3 mm by 6 mm and the pickup coil consists of three parallel loops with a width of 120 μm and a spacing of 80 μm. The switch shown in Fig. 11b or Fig. 11c is inserted in the pickup coil. In Fig. 11b the switch consists of 4 bicrystal junctions where the width and critical current of each junction are 10 μm and $I_{cj} = 90\,\mu\text{A}$, respectively. As shown in Fig. 11b, terminals P and Q are connected to the pickup coil. The 4 junctions can be biased with a current I_s from the terminals R and S. When the current I_s exceeds $2I_{cj}$, the 4 junctions become a voltage state. In this case, the circulating current I_{cir} becomes zero. Therefore, we can forcibly move the operation point from the point "A" to the point "C" in Fig. 9b by applying the switch current I_s. Note that the switch current I_s does not cause any current flow in the pickup coil if the parameters of the 4 junction are equal to each other. Therefore we can use the 4 junctions as an electrical switch that can control the opening and closing of the pickup coil. Since we use bicrystal junctions, the pattern shown in Fig. 11b is used for the 4 junctions, where the broken line indicates the grain boundary (G. B.) of the substrate.

In Fig. 11c another type of switch is shown where two junctions are used to form the switch. The two junctions are current-biased through the resistance

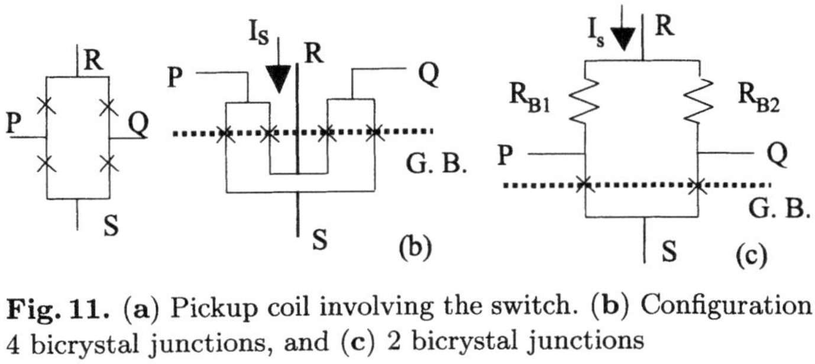

Fig. 11. (a) Pickup coil involving the switch. (b) Configuration of the switch with 4 bicrystal junctions, and (c) 2 bicrystal junctions

R_{B1} and R_{B2}. When the parameters of the two junctions are different from each other, the values of R_{B1} and R_{B2} are adjusted so that no current flow occurs in the pickup coil due to the switch current I_s.

Using the switch shown in Fig. 11b or Fig. 11c we can decrease the circulating current I_{cir} in a short time. Therefore we can avoid the thermally activated flux entry in this case. In the following we compare the experimental results between the case of the flux dam and the switch. First, we study the case of the flux dam. Note that the switch works as a conventional flux dam when no switch current I_s is supplied. In Fig. 12a the output of the FLL is shown after the field is increased from $B_{ex} = 0$ to $B_{ex} = 6\,\mu T$ for a time of 5 s. The result is shown in the section "S1" in the figure. Here, the output of the FLL is bandpass filtered from $f = 0.5\,Hz$ to $f = 10\,Hz$. As shown, many voltage spikes are observed in this case. Each voltage spike is caused by the thermally activated flux entry from the flux dam. The time interval between the voltage spikes represents the time interval for each flux entry. The voltage scale is $10\,mV/div$, corresponding to a flux scale of $2.9\,m\Phi_0/div$.

After the measurement of the section "S1", the FLL is opened for $T_w = 15\,s$. After this waiting time, the FLL is closed again, and the output of the FLL is recorded. The result is shown in the section "S2" in Fig. 12a. As shown, many voltage spikes are observed in this case, too. This means that the waiting time of $T_w = 15\,s$ is not enough to complete the thermally activated flux entry.

In Fig. 12b an experimental result is shown when the 4-junction switch is used. In the experiment the pickup loop is opened by applying the switch current $I_s = 220\,\mu A$ to the 4-junction switch. Then the external field is

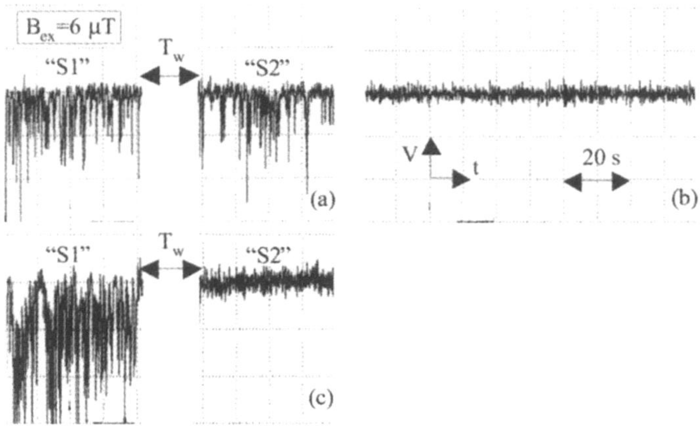

Fig. 12. Output of the FLL after the field is increased from $B_{ex} = 0$ to $B_{ex} = 6\,\mu T$. (**a**) In the case when the switch is not used, i.e., in the case of the conventional flux dam. (**b**) In the case when the switch is opened during the increase of B_{ex}, and then the switch is closed. (**c**) In the case when the field is increased without using the switch ("S1"). Then, the switch is opened and closed ("S2")

increased from $B_{ex} = 0$ to $B_{ex} = 6\,\mu T$. After the field becomes $B_{ex} = 6\,\mu T$, the switch current I_s is reduced to $I_s = 0$ for a time of 5 s and the FLL circuit is closed. The output of the FLL is shown in Fig. 12b. As shown, the output of the FLL is smooth and no voltage spikes are observed in this case. This result can be compared to that shown in Fig. 12a where many voltage spikes due to the flux entry are observed. The difference between the two results is due to the different values of I_{cir}. In the case of Fig. 12a, the operation point is the point "A" in Fig. 9b, while the operation point is the point "C" in the case of Fig. 9b.

In Fig. 12c we show another experimental result showing the effect of the switch. In this case the external field is increased from $B_{ex} = 0$ to $B_{ex} = 6\,\mu T$ while no switch current I_s is supplied. In this case the operation point is the point "A" in Fig. 9b and the output of the FLL shows many voltage spikes due to the thermally activated flux entry, as mentioned before. The result is shown in the section "S1" in Fig. 12c. Then the FLL circuit is opened and the switch current I_s is supplied. In this case, the pickup loop is opened and the circulating current becomes zero, i.e., the operation point moves from the point "A" to the point "C" in Fig. 9b. After, the switch current is reduced to zero the FLL circuit is closed and the output of the FLL is recorded. The result is shown in section "S2" in Fig. 12c. As shown, the voltage spikes disappear and a smooth output is observed.

The results shown in Fig. 12b and Fig. 12c demonstrate the usefulness of the switch. We can control the opening and closing of the pickup coil with the switch, and therefore we can forcibly reduce the circulating current to zero. In this case we can solve the problem of the thermally activated flux entry.

3 SQUID Control System

Readout electronics for HTSC-SQUIDs is an important factor for realizing its high performance. This section reviews some of the readout techniques especially designed for HTSC-SQUIDs; an additional positive feedback circuit, the bais-reversal method, a flux moduation FLL circuit, and so on.

3.1 Input Equivalent Noise

The output voltage of HTSC-SQUIDs is reduced from that of LTSC SQUIDs by the thermal flux noise effect [33], which originates in the current noise of the normal resistance R_n coupled to the SQUID inductance L_S. Using the thermal noise reduced voltage swing (output voltage) ΔV, the flux-to-voltage transfer function of HTSC-SQUIDs is given by $\partial V/\partial \Phi \approx \pi \Delta V/\Phi_0$. A typical value of ΔV is 10–20 μV for a SQUID made of YBCO films, assuming $L_S \approx$ 70–100 pH, but can be as high as 50 μV for good quality junctions. The thermal noise of R_n of two junctions generates a voltage noise

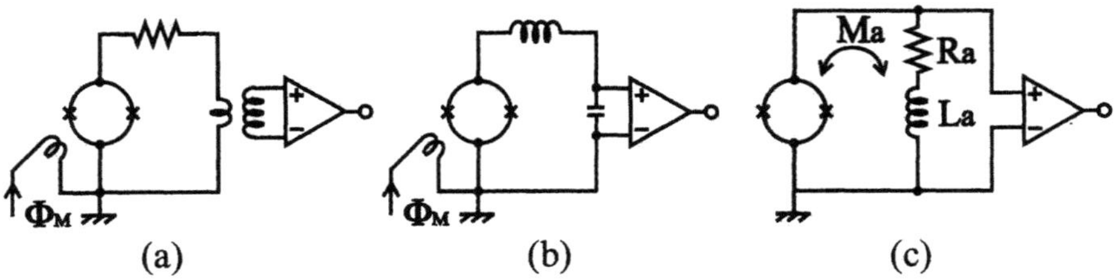

Fig. 13. Different methods to amplify the SQUID output of (**a**) a transformer, (**b**) a tank circuit, and (**c**) additional positive feedback of flux

power of $S_v = 2k_BTR_D^2/R_n$ at the SQUID output and a flux noise power of $S_I = 2k_BTL_S/R_n$ in the SQUID inductance, where k_B is Boltzman constant and R_D is the dynamic resistance at the operating point in the current-voltage characteristic. Then, the input equivalent flux noise power of the SQUID is

$$S_\Phi = \frac{S_v + (V_n/A)^2}{\partial V/(\partial \Phi)^2 + S_I L_S^2}, \tag{20}$$

where V_n is the voltage noise of the preamplifier in the readout electronics and A is the amplification factor of a circuit element that is inserted between the SQUID and the amplifier.

The magnitude of S_V in the first term of (20) is estimated to be $S_V^{1/2} \approx 2 \times 10^{-10}$ V/Hz$^{1/2}$ assuming typical values of $R_n \approx 5\,\Omega$ and $R_D \approx 10\,\Omega$ of the grain boundary junction and $T = 77$ K. On the other hand, low-noise operational amplifiers have V_n of the order of 1×10^{-9} V/Hz$^{1/2}$. Although the V_n can be lowered by about a factor of two with low-noise FETs added to the input, the amplifier noise is still larger than the intrinsic noise S_V of the SQUID. The role of the amplification element is to raise the output of the SQUID without thermal noise, reducing the contribution of the amplifier noise effectively by the amplification factor A. A so-called impedance matching transformer (Fig. 13a) is often used, while a tank circuit (b) can also be used to amplify a frequency component of flux modulation. These methods are applied in the readout electronics that operate in the conventional flux modulation scheme. Additional positive feedback (APF) (c) has been developed in the direct readout scheme in which flux modulation is omitted to simplify the electronics [34]. In the APF circuit a small change of the flux in the SQUID produces an additional flux change, in a manner of positive feedback, through the magnetic coupling M_a between an inductance L_a and the SQUID. This flux feedback increases $\partial V/\partial \Phi$, instead of A, in (20), and reduces the contribution of V_n to the flux noise of the SQUID.

3.2 Bias-Reversal Schemes

Grain boundary junctions of HTSC thin films have high low-frequency noise [8,35], which is much larger than the noise in LTHC junctions of Nb/AlO$_x$/Nb.

The low frequency noise is observed as a strong $1/f$ component in the power spectrum of the voltage noise. Such $1/f$ noise is known to originate in the fluctuation of junction parameters such as the critical current I_c and normal resistance R_n. In YBCO junctions the magnitude of the critical current fluctuation $\delta I_c/I_c$ is of the order of 10^{-4}. It is the major source of low frequency noise in HTSC-SQUIDs.

When the critical current of two junctions fluctuates in time, its out-of-phase change yields a circulating current in the SQUID loop and leads to an effective flux change. The direction of this flux change, which is referred to as differential-mode flux fluctuation (differential-$\delta\Phi$), is opposite for positive and negative bias currents. On the other hand, the direction of the flux change due to external (signal) fields or movements of trapped flux is independent of the direction of the bias current. This flux change is referred to as common mode flux fluctuation (common-$\delta\Phi$). The bias current reversal method is used in HTSC-SQUID control systems to cancel the $1/f$ noise due to the critical current fluctuation. The principle of the bias-reversal scheme is based on the distinct effects of the common $\delta\Phi$ and differential $\delta\Phi$ on the output voltage of the SQUID.

In the conventional FLL (Flux-Locked-Loop) readout a modulation flux of amplitude of $\pm\Phi_0/4$ is applied. The flux modulation is a standard technique to raise the signal frequency to a high (10^1–10^2 kHz) modulation frequency and amplify the signal in ac mode without the influence of low frequency drift. The amplified voltage is demodulated, low-pass filtered and fed back to the SQUID as a flux (see Fig. 16). On the other hand, the bias current I_B is alternated with time in the bias-reversal scheme. Bias flux Φ_B, which is applied in synchrony to I_B, and a modulation flux Φ_M are employed, or not, depending on the method. Various methods of bias-reversal schemes in FLL operation have been reported [36,37,38,39,40,41,42,43]. They can be classified into three regimes of $f_M \gg f_B$, $f_M = f_B$, and $f_M \ll f_B$ by the relation between the flux modulation frequency f_M and the bias current reversal frequency f_B. There is another regime of $f_M = 0$, i.e., with no flux modulation. This is the most simplified version proposed by *Drung* et al. [40,43] in which the output of the SQUID is directly read out.

Figure 14 illustrates schematic diagrams of bias-reversal methods for the direct readout (left column) and the flux modulation readout (right column) of $f_M \gg f_B$. Here Fig. 14a shows the waveform of the variables I_B (and Φ_B) and Φ_M (and the voltage signal V_{DM} for demodulation) and output voltages V_C and V_D for a flux change of common $\delta\Phi$ and differential-$\delta\Phi$, respectively. The shaded waveform indicates the output voltage after demodulation. The time course of the change of operating point is shown in Fig. 14 on the voltage-flux (V-Φ) characteristics for positive and negative I_B. Open and solid circles indicate the operating points without and with input flux, respectively. In order to represent the flux input of common $\delta\Phi$ for positive and negative I_B, the operating points are shifted in a positive Φ direction from the

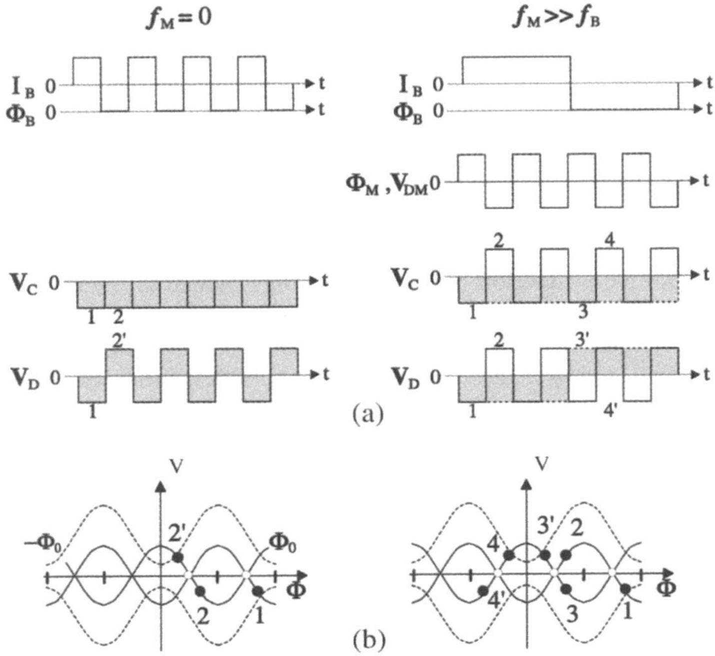

Fig. 14. Schematic diagrams of bias-reversal methods for the direct readout (*left column*) and the flux modulation readout (*right column*) of $f_M \gg f_B$. (**a**) Waveform of variables and (**b**) time course of operating point on the V-Φ characteristics for positive and negative I_B. (See the text for explanation of symbols)

points without flux input. The flux input of differential $\delta\Phi$ is represented by a positive and negative shift of the operating point for common $\delta\Phi$ and differential $\delta\Phi$, respectively. An offset voltage that is generated with I_B reversal is cancelled with an appropriate subtraction circuit for proper operation.

In the direct readout scheme [43] shown in the left column of Fig. 14, Φ_B is changed between 0 and $\Phi_0/2$ while I_B is alternated between $\pm I_B$ in synchrony. The operating point is switched between 1 and 2 on the V-Φ curve when a common $\delta\Phi$ is present, while it is switched between 1 and 2′ with a differential $\delta\Phi$. The modulated voltage includes a finite component, i.e., nonzero output, for the common $\delta\Phi$ but becomes zero for differential $\delta\Phi$ after low-pass filtering. Thus, the voltage noise that is caused by the out of phase fluctuation of the critical current is cancelled in the output, provided that the fluctuation frequency is lower than the bias-reversal frequency. On the other hand, in phase fluctuation of the critical current shifts the voltage along the V axis of V-Φ characteristics in opposite directions for positive and negative I_B, which results in zero averaged output voltage.

Koch et al. [36] reported a bias-reversal readout in the regime of $f_M \gg f_B$, where Φ_B is alternated between 0 and $\Phi_0/2$ in synchrony to I_B. The right

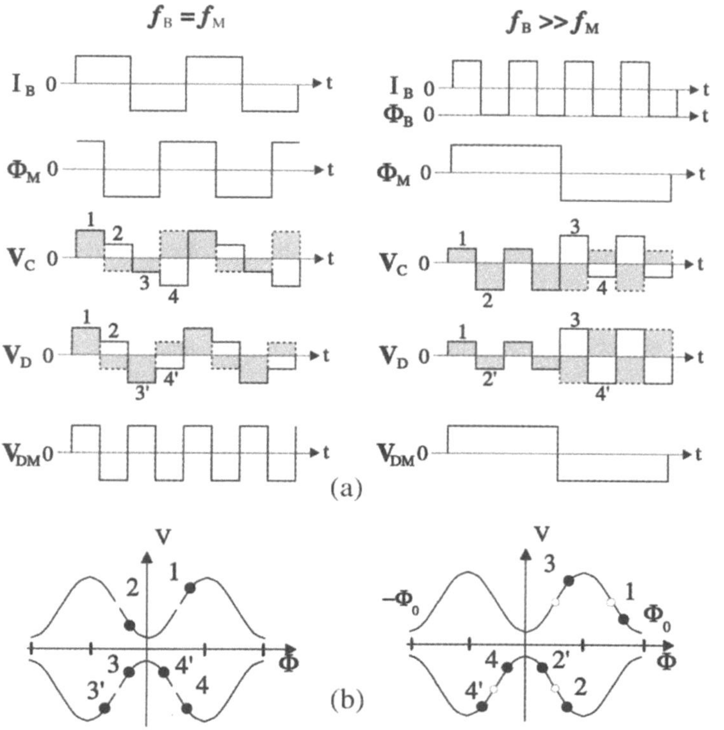

Fig. 15. Schematic diagrams of bias-reversal readout with flux modulation of $f_B = f_M$ (*left column*) and $f_B \gg f_M$ (*right column*). (**a**) Waveform of variables and (**b**) time course of operation point on the V-Φ characteristics for positive and negative I_B. (See the text for explanation of symbols)

column of Fig. 14 shows the waveform of different signals and the time course of the operating point. As Φ_M is alternated between $\pm\Phi_0/4$, the operation point switches between 1 and 2 on the V-Φ curve during the period of positive I_B for common $\delta\Phi$ (and differential $\delta\Phi$), and switches between 3 and 4 (or 3′ and 4′) for common $\delta\Phi$ (or differential $\delta\Phi$) after I_B is reversed. Demodulation is done by multiplying the output voltage V_C (or V_D) for common $\delta\Phi$ (or differential $\delta\Phi$) with a demodulation signal V_{DM}. This process results in a finite readout for the common $\delta\Phi$ and zero output for the differential $\delta\Phi$ after low-pass filtering. A modification of this method into a simpler scheme has been reported [39] in which application of bias flux Φ_B is omitted but the same output is obtained by using a slightly different reference signal for demodulation.

Figure 15 illustrates schematic diagrams of the bias-reversal methods for two flux modulation readouts of $f_M = f_B$ (left column) and $f_B \gg f_M$ (right column). There are some slightly different versions that belong to the same regime of $f_M = f_B$ [37,38,41]. Here we describe the one reported for a magnetometer in biomagnetic use [41]. As can be seen in the left column of Fig. 15, the bias current and flux modulation are alternated with a phase shift of $\delta\Phi$ between them. There is no application of bias flux. In the time course, the operating point proceeds through 1-2-3-4 as the I_B and Φ_M are alternated with an input flux of common $\delta\Phi$. With a differential $\delta\Phi$, the operating point proceeds through 1-2-3′-4′ in a way that is different during the period of negative I_B. Because of this difference, the resulting output of V_D for the differential $\delta\Phi$ has zero time average. This method has an obvious advantage of higher frequency response than other flux modulation regimes of $f_B \gg f_M$ and $f_B \ll f_M$, in which the maximum frequency response is limited by the lower frequency of either flux modulation or bias reversal.

The method proposed by *Kuriki* and *Suzuki* [42] for the readout of high-T_c SQUIDs is in the last regime of $f_M \ll f_B$, which is shown in the right column of Fig. 15. The operating point alternates between 1 and 2 (or 3 and 4) during the half period of $\Phi_M = \Phi_0/4$ (or $\Phi_M = -\Phi_0/4$) of flux modulation for common $\delta\Phi$, while between 1 and 2′ (or 3 and 4′) during the half period of $\Phi_M = \Phi_0/4$ (or $\Phi_M = -\Phi_0/4$) for differential $\delta\Phi$. Using a demodulation signal of the same waveform as Φ_M, the modulated voltage has nonzero and zero averaged value for common $\delta\Phi$ and differential $\delta\Phi$, respectively. It is found that synchronization is not necessary between I_B/Φ_B and Φ_M in this scheme, so far as $f_B \gg f_M$. Therefore, a high frequency bias-reversal/flux-bias component can be separated in the electronics from the other FLL circuit that includes the flux modulation. This is an obvious advantage in multichannel SQUID control systems. A free-running common square wave of a frequency of f_B can be used as the bias-reversal/flux-bias signal in all the channels, which may simplify the whole electronics and reduce high frequency interference among channels.

3.3 Readout Electronics

An example of the FLL circuit that may be used for controlling multichannel HTSC-SQUIDs is illustrated in Fig. 16. Here the bias-reversal scheme belongs to $f_B \gg f_M$ (right column of Fig. 15) and slightly modified from the version developed for YBCO SQUIDs [44]. Here, a high frequency ($f_B \sim 200$ kHz) generator, which supplies I_B and Φ_B signals to all the channels, is separated from the individual-channel boards. The currents that give appropriate amplitudes of I_B and Φ_B are controlled with simple resistors in a way to minimize the phase shift and difference between them. The flux modulation ($f_M \sim 30$ kHz) and the associated FLL operation, i.e., demodulation, integration and flux-feedback, are closed in individual channels. A bias resistor (RB), adopted from *Drung*'s method [43], is used to reduce the voltage jump generated by the I_B reversal. Any type of amplification element of the SQUID output, shown in Fig. 13, may be used between the SQUID and a low noise differential amplifier, to reduce effectively the amplifier noise. The bandwidth

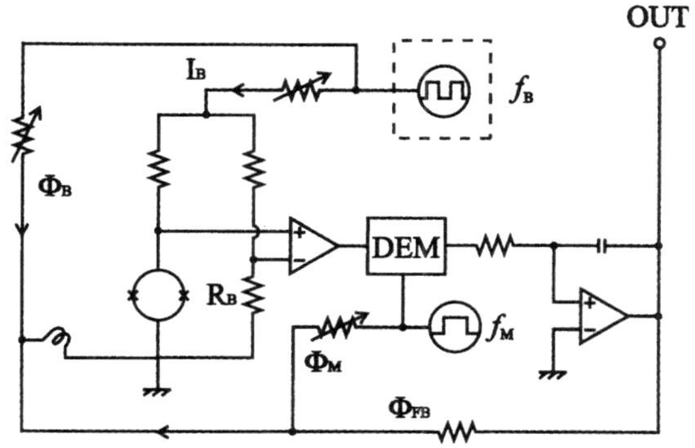

Fig. 16. Schematic drawing of a flux modulation FLL circuit, where DEM is a demodulator

required for the differential amplifier is given by the frequency components of the flux-modulated signal, but not the bias-reversal signal.

An advantage of flux modulation FLL readout is the stability of operation against ambient temperature change compared with direct readout. There is little drift of the FLL output, which may be serious in the measurement of low frequency signals. Thus a temperature controller (air conditioner) of the environment is not necessary and the measurement can be started without warming time of electronics. On the other hand, in high frequency measurements the direct readout FLL without flux modulation would be suitable. Even bias-reversal can be omitted if the signal frequencies are higher than the frequency of I_c fluctuations. A wide bandwidth of MHz can readily be obtained with a simple circuit in this scheme.

Based on recent progress in digital technology, different digital systems are incorporated in the readout electronics. Controlling the parameters such as I_B and Φ_B is automated by the aid of a digital processor or a personal computer [45,46]. The feedback voltage is digitized with an analog-to-digital converter (A/D) and the subsequent FLL is controlled in a fully digital manner. Signals exceeding Φ_0 can be measured by counting the Φ_0 steps without breaking the FLL. Some readout electronics implemented with such digital controlling/processing systems are commercially available [47].

4 Noise Reduction in HTSC-SQUIDs

The sensitivity of SQUIDs in measuring external signals is determined by different noises that are input into and generated within the SQUID. Various shielding or canceling techniques of magnetic fields can be used to reduce the external noise. The external noise also includes the contribution from readout electronics, which is mainly given by the preamplifier noise of the control system, as we have seen in the previous section.

4.1 Low Frequency Noises in HTSC-SQUIDs

The noise generated in the SQUID determines the ultimate sensitivity of the SQUID. In the frequency domain, the intrinsic noise of HTSC-SQUIDs has a white noise and a low frequency ($1/f$) component. The main source of the white noise is the Johnson noise of the resistive component (normal resistance) of the SQUID. Important formulas that describe the flux noise in terms of the parameters of SQUIDs, i.e., R_n, L_s, etc., are described later. Based on such equations one can optimize the parameters to obtain the lowest white noise of the SQUID. On the other hand, the low frequency noise has two main sources of critical current fluctuation and thermal motion of flux vortices in the SQUID. Here, the noise originating in the critical current fluctuation is removed from the output of the readout electronics, as described in the previous section, by the use of the bias-reversal method. We describe hereafter the behavior of the flux vortices trapped/penetrating into HTS-SQUIDs and the method to reduce the low frequency noise that is caused by those vortices.

When a SQUID is cooled down in a static magnetic field through T_c flux vortices are pinned at defects in the HTS film, giving rise to excess low frequency noise. Such a situation is serious, especially in unshielded operation of the SQUID. One way to eliminate this noise is to fabricate the device with narrow linewidths so that vortices are not trapped in the superconducting film below a threshold field. This effect was first confirmed in SQUIDs composed of a small-area washer (pickup coil) [23]. Recent studies of the field effect and the reduction of low frequency noise in HTS-SQUIDs are focused on magnetometers that have a large area pickup coil. Here, measuring low frequency signals with sensitive magnetometers is one of the important applications of the HTS-SQUID, such as biological examination and non-destructive evaluation of materials. In magnetocardiography (MCG), electrophysiological functions of the heart muscle are examined by the use of conventional Nb-based SQUIDs. For the development of easy-handling and compact MCG systems, low noise HTS magnetometers that can be operated in unshielded or weakly shielded space are required.

4.2 Direct-Coupled Magnetometers

Figure 17 shows the geometry of representative types of direct-coupled magnetometers that have been developed, with their equivalent circuits. These devices utilize grain boundary Josephson junctions formed on a bicrystal line (broken line in the figure) of $SrTiO_3$ substrate. A large-washer geometry with a SQUID fabricated within the washer film [20,48] is the most conventional magnetometer, but we describe here a somewhat different version, as shown in Fig. 17a, that has a SQUID in the central hole of the washer [49]. This device has flux dams in the superconducting lines that connect the pickup coil and the SQUID. Details of the flux dam are described later. The double-hole

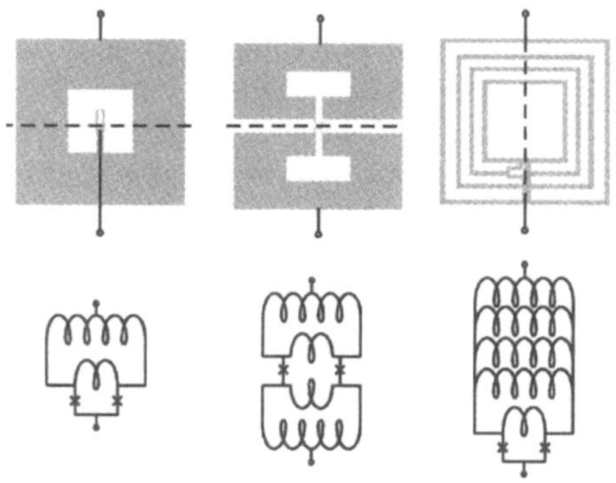

Fig. 17. Geometry of different types of magnetometer. (**a**) Washer type, (**b**) double-hole type and (**c**) multi-loop magnetometers

magnetometer shown in Fig. 17b has no part of superconducting film crossing over the bicrystal line of the substrate, except two Josephson junctions, and thus no flux dams [50]. The multi-loop magnetometer in Fig. 17 is composed of parallel pickup loops where each loop has a flux dam at the crossover with the bicrystal line [24,25].

4.3 Effects of Slots and Holes on the Reduction of Low frequency Noise

The superconducting film of the washer type magnetometer in Fig. 17a has slots of 5 μm × 45 μm on the whole area of the washer and holes of 5 μm × 5 μm on the connecting line. Thus, the whole film consists of 5 μm wide strips. Such fine structure of the superconducting film may suppress flux trapping in magnetic fields. It is expected from Clem's equation ($B_{\text{th}} = \pi\Phi/4w^2$, where w is the linewidth) that the threshold field B_{th} for the flux trapping is about 65 μT. After this device was cooled in different static magnetic fields, the measured low frequency field noise did not show appreciable change up to 30μT, above which an abrupt increase of the noise was observed [51]. This threshold field is apparently lower than the above prediction, which might be due to the field effect on the flux dams of this device (described later). In a similar washer type magnetometer, composed of holes and narrow lines of superconducting film, the threshold field of the increase of low frequency noise is reported to be 35 μT [52], which is close to the value expected (42 μT) from the width (6.2 μm) of the strips. It is also reported that the steep edges, in addition to the fine structure, of the pickup loop are important to obtain a low frequency noise in field-cooled magnetometers [53].

As described in Sect. 5 in the Chapter by Kuriki et al. in this volume) the direct flux detection method has shown that long-distance motion of flux vortices is initiated along a wide (3 mm) grain boundary junction, which mimics the bicrystal grain boundary in the washer type magnetometer (Fig. 17a), at a field change of 1.4 μT [54]. A steep increase in low frequency noise was observed in this washer type device when a small field change of 2 μT was applied. It is believed that long-distance vortex motion is responsible for the increase of this low frequency noise. The grain boundary junction serves as a channel for vortex motion far below the threshold field of the flux penetration into the superconducting film [55]. The direct flux detection method has also proved that the long-distance vortex motion can be well suppressed by making many slots along the wide grain boundary [54]. Here these slots constitute a parallel array of SQUIDs. Each SQUID satisfies a condition of $\beta_L \gg 1$ to keep stably a single flux quantum, where $\beta_L (= 2LI_c/\Phi_0)$ is the inductance parameter. After forming an array of slots (right side of Fig. 18) with the same size as the slots on the washer, along the 3 mm wide grain boundary, the magnetometer did not show a substantial increase in low frequency noise for a field change of 10μT (Fig. 18). Therefore, the slot structure has the effect of suppressing not only flux trapping in the superconducting film, but also flux movement through the grain boundary junction.

The double-hole magnetometer of Fig. 17b has the same slot and hole structures as the device of Fig. 17a and is composed of 5 μm wide strips of superconducting film. Figure 19 shows the low frequency noise, represented by the noise at 1.5 Hz in the measured spectrum, after the device is cooled in various static magnetic fields [26]. The low frequency noise shows clear onset of increase at a threshold field of about 100 μT, which is much larger than the theoretical prediction of 65 μT for 5 mm wide strips. This threshold field far exceeds the earth magnetic field (about 50 μT). This means that the device can be cooled in an environmental field with no magnetic shielding. It is considered that the high threshold field obtained in this double-hole type

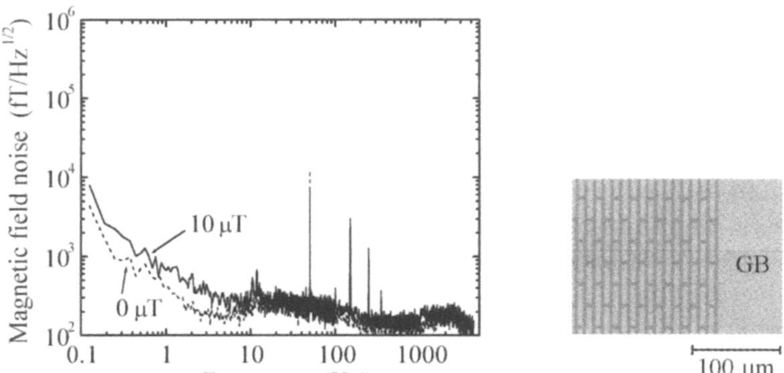

Fig. 18. Field noise spectra of the washer type magnetometer, which has slots along wide grain boundaries (*right side*) after application of a static field

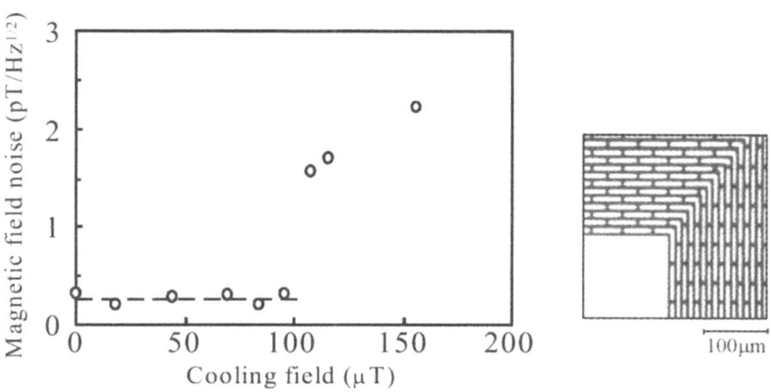

Fig. 19. Magnetic field noise at 1.5 Hz of the double-hole magnetometer after cooling in various static fields. Slot patterns at the corner of the pickup coil are shown on the *right-hand side*

magnetometer is because the device does not include any grain boundary junctions in the superconducting coil, except the Josephson junctions.

High performance has been reported for the multi-loop magnetometers of Fig. 17c after cooling in or application of static magnetic fields. The low frequency noise does not increase substantially for a cooling field of 80μT [24]. This device has 16 loops of 50 μm wide parallel lines, where each loop has a flux dam crossing over a bicrystal line. A similar type of parallel loop magnetometer shows stable low noise operation when a field of 60μT is applied and maintained during measurement [25]. All the loops of this device have flux dams made of SNS junctions, which are opened by the shielding current at a field of 0.056 μT. It is suggested that the SNS junctions rapidly damp the shielding current when the junction becomes normal and recover in a short time to the superconducting state.

4.4 Flux Penetration

When the direct-coupled magnetometer is moved from one place to another the background field may change. Such a field change in the pickup coil induces a shielding current, which flows through the connecting line and the inductance of the SQUID. The shielding current exerts a Lorentz force on the flux at the edge of the superconducting film. According to the model by *Sun* et al. [56], massive penetration of flux vortices occurs when the Lorentz force exceeds the surface barrier of the edge and the pinning force of the superconducting film. This threshold current I_p is given by

$$I_p = (J_c d w) \left[2.46 \left(\frac{\Phi_0 / w^2}{H^*} \right)^{1/4} \right]^2 , \tag{21}$$

where J_c is the critical current density of the superconducting film, d is the thickness, w is the width and $H^* = 6.28 \times 10^{-7} J_c d$ in MKS units. Flux penetration in the direct-coupled magnetometer may be observed as an increase in the low frequency noise due to thermal fluctuation of the penetrating vortices. Using the values of J_c, d, and w for 5 μm wide strips of the double-hole magnetometer of Fig. 17b, the threshold current was calculated to be $I_p = 0.9 J_c dw$ $(= I_c)$. Therefore substantial flux penetration may occur at a shielding current slightly lower than the critical current.

Figure 20 shows the low frequency field noise at 1 Hz of the double-hole magnetometer for different applied fields. Here the noise measurement was made after zero-field cooling, when a field was applied and maintained (in-field noise) and switched off (off-field noise). It is found that the off-field noise increases steeply above about 4 μT and becomes close to the in-field noise at about 9 μT, suggesting that nearly all the penetrating flux vortices remain within the film. In the waveform of the output of the SQUID a very rapid shift in one direction was observed around 9 μT. It is inferred that massive penetration of vortices occurred across a superconducting strip, resulting in the entry of fluxes into the pickup loop. The site of the flux penetration may be in the connecting line near the SQUID (see the right side of Fig. 20), at which the shielding current is most concentrated. The threshold field change that induces the shielding current of I_p is estimated to be 8.3 μT, using the inductance and the area of the pickup. This threshold field is compatible with the field of the massive flux penetration.

The drift of the SQUID output was observed, first at a slow rate, above the applied field of 4 μT, which coincides with the onset field of the increase of the off-field noise. This correspondence indicates initiation of flux penetration into the superconducting film. The exact mechanism that causes the

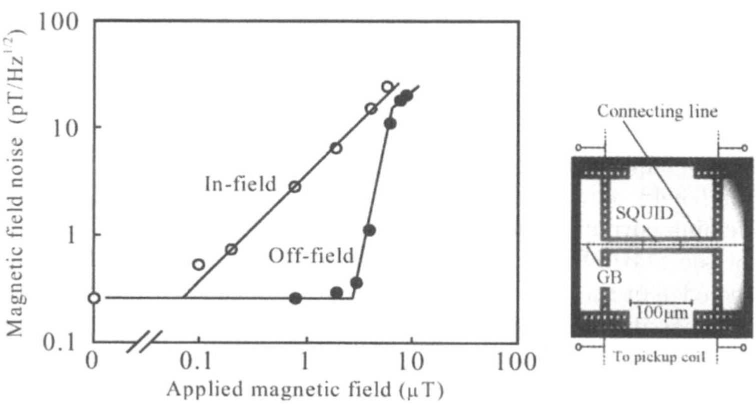

Fig. 20. Magnetic field noises at 1 Hz with maintained applied field (in-field) and after turning off the field (off-field) in the double-hole magnetometer. The SQUID and connecting lines are indicated on the *right side*

reversible in-field noise below 4 μT is not clear, but a tentative explanation is as follows [55]. Assuming that the penetration of fluxes at low applied fields takes place preferentially at the edge of the pickup loop, the penetrating flux vortices proceed inward and drop into the outermost slots successively. At the moment of flux drop, the current distribution around the flux line changes significantly and the magnetic field generated by the vortex current does as well. This large field change may cause low frequency noise in the SQUID. Since the flux drop stops when the applied magnetic field is turned off, the low frequency noise is seemingly reversible. In superconducting strip lines separated by slits, such transport current induced flux motion is described as "quasistatic flux penetration" in [57].

The outer edge of the pickup coils of the magnetometer was shaped by lithography and etching, removing the outermost part of the device [55]. The effect of this edge sharpening was remarkable. The in-field noise was reduced by nearly one order of magnetite but the off-field noise was nearly unchanged. This result supports the idea of flux penetration at the edge of the pickup coils as the source of reversible noise. The effect of steep edges of the SQUID and the pickup coil is also demonstrated in the measurement of magnetic hysteresis [58].

4.5 Behaviors of Flux Dams

The flux dam was introduced in direct-coupled magnetometers to allow fluxes to enter the pickup loop when the shielding current exceeds a critical value [27,59]. The flux dam reduces the field-change-induced noise by suppression of the shielding current. Most recently fabricated direct-coupled devices have flux dams [24,25], but the detailed characteristics of the flux dams have not been reported. We describe here experimental results on the flux dams examined in the washer type magnetometer, which has a SQUID and flux dams in the central hole of the pickup coil (Fig. 17a).

4.5.1 Grain Boundary Junction Flux Dam

Figure 21 shows an optical micrograph and a SEM photograph of the central part of the magnetometer. The SQUID is composed of 5 μm wide parallel strips with a 20 μm wide slit and has two connecting lines on both sides. The superconducting film of the connecting lines crosses a bicrystal line of the substrate and forms a flux dam on each side of the grain boundary junction. The bias current to the SQUID is fed through the two connecting lines symmetrically, while each line is connected to one end of the washer-type pickup coil. All the structures, except the SQUID and the vicinity of the flux dam, are made of 5 μm strips of superconducting film separated by 5 μm × 5 μm holes. The flux dam in Fig. 21 has a width of 20 μm, where the largest size of the superconducting film, as encircled, around the flux dam is 12 μm. Several

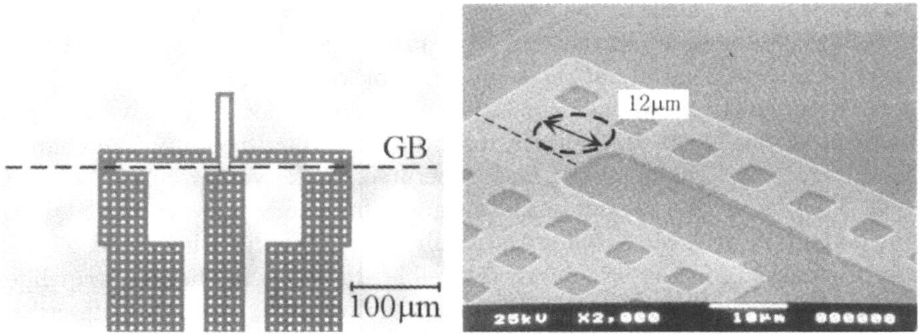

Fig. 21. Optical micrograph (*left*) showing a SQUID, two flux dams, and connecting lines, and SEM photograph (*right*) around a flux dam

devices were fabricated having different widths of the flux dam, ranging from 20–55 μm.

4.5.2 Critical Current of the Flux Dam

When the voltage-flux characteristic was measured while the flux applied to the SQUID was varied, collapse of the modulated voltage was observed above a threshold flux [49], as shown in Fig. 22a. The modulation disappeared during the flux sweep in one direction, but recovered immediately when the sweep direction was reversed. The same behavior was reproduced in alternating sweeps, where the maximum amplitude of the applied flux for continuous modulation corresponded to $25\Phi_0$ linking in the SQUID hole. From the width of the flux dam, the film thickness and the critical current density of the film, the critical current I_c of the flux dam on each side of the connecting line is estimated to be 0.4 mA, which corresponds to a field strength of 0.14 μT applied to the magnetometer. On the other hand, the observed sweep width of $25\Phi_0$ corresponds to a current of 0.5 mA flowing through the SQUID inductance, in reasonable agreement with the I_c of the flux dam. Therefore, the collapse of the modulated voltage is interpreted to reflect the opening of the flux dam. A trace of the estimated supercurrent through the flux dam during the flux sweeps is shown schematically in Fig. 22b. The immediate recovery of the flux dam on reversal of the bias flux suggests that the vortices moving through the grain boundary junction are easily swept out by the reversed Lorentz force. High mobility of vortices within grain boundaries is suggested from the direct observation by low temperature electron microscopy [60].

The critical current of the flux dam was measured as a function of applied static field. To do this, the maximum range of the flux that could be swept while the flux dam remained in the superconducting state was measured, as in Fig. 22, under various static fields that were applied after cooling the device in zero field. The obtained results showed a clear Fraunhofer-like dependence of the critical current on the applied field (Fig. 23a) [49]. It is noted that the

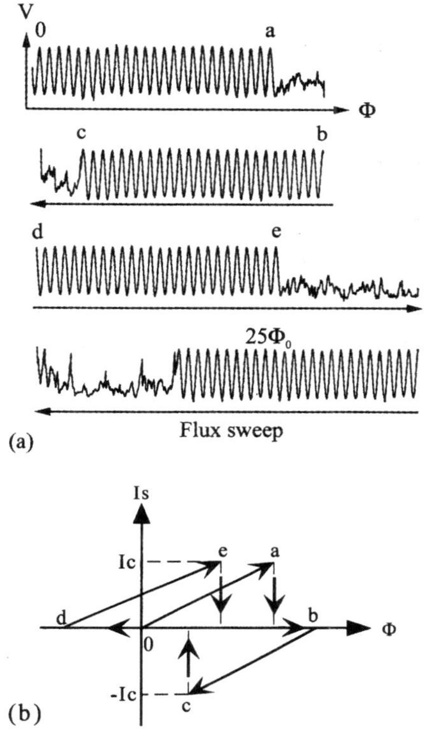

Fig. 22. (**a**) Voltage flux characteristics of a washer type magnetometer with flux dams observed when an applied flux was swept forward and backward, and (**b**) estimated supercurrent through the flux dam during the above process

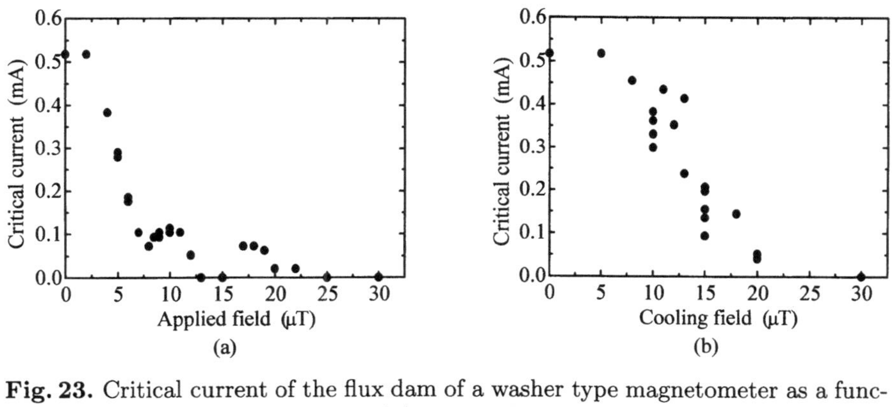

Fig. 23. Critical current of the flux dam of a washer type magnetometer as a function of (**a**) applied static field and (**b**) cooling field

flux dam of the device was opened for a field change of about 0.14 μT, while the field was applied up to 30μT.

Therefore, the observed suppression of the critical current of the flux dam is not the effect of shielding current generated in the pickup coil. It is inferred that the applied field around the flux dam is amplified by the supercurrent flowing near the edge of the connecting line and along the grain boundary junction of the flux dam. In the thin film limit of $d/\lambda \leq 1$, where λ is the penetration depth, this flux-focusing effect [61] predicts a critical field of

$$B_{\mathrm{th}} = \frac{1.84\Phi_0}{w^2}\,, \tag{22}$$

at which the I_c of the junction first becomes zero. Since the film thickness of the device λ (0.2 μm) is of the same order (0.2 mm for YBCO at 77 K), the above assumption should hold, and (22) gave $B_{\mathrm{th}} = 9.2\,\mu\mathrm{T}$ for a flux dam of $w = 20\,\mu\mathrm{m}$. This value of B_{th} coincides well with the result of Fig. 23a. The agreement may be fortuitous, however, if we consider that the current distribution within the connecting line would be much distorted from the theoretical current (Fig. 2a of [61]) because of the presence of many holes. In any case, the depression of the critical current of the flux dam would be a limiting factor for the operation of the magnetometer in a static field.

The critical current I_c of the flux dam of the same device was also measured after cooling the device in various static fields. The obtained I_c as a function of the cooling field, shown in Fig. 23b, was scattered, but the critical field of about 30μT for the disappearance of I_c is larger than the B_{th} obtained in (a). This result may be explained by the penetration of flux lines through the superconducting film of the connecting line during field cooling, which leads to reduction of the effective applied field. For the largest size of 12 μm (Fig. 21) of the superconducting film around the flux dam, the threshold field for flux trapping is estimated from Clem's relation to be $B_{\mathrm{th}} = 11\,\mu\mathrm{T}$. This field value is compatible with the cooling field around which the scatter of I_c becomes prominent. Further, a flux shift was observed in the voltage-flux characteristic when the device was cooled at 15 μT so that the modulated voltages, such as those in Fig. 22a, measured at different positive bias currents were shifted along the flux axis from those for negative bias currents. This kind of flux shift may be ascribed to the trapped flux vortices in the connecting line, possibly in the area around the flux dam, and their displacement with bias current. On the other hand, the voltage-flux characteristics measured at a cooling field of 10μT were completely symmetric without flux shift.

Flux dams are often used in parallel-multiloop magnetometers. Those flux dams inserted in the loop have the same width as a loop of $w =$ 50–100 μm [24,25]. In such wide junctions a very low value of the critical field for the suppression of I_c is expected from (22). However, normal operation of the multiloop devices is reported for cooling and changing fields of high as 50–60 μT [24,25]. The exact mechanism of such high field operation is not clear. One possibility is that penetration of abundant flux lines through the wide superconducting film reduces the field effectively applied and insures a finite critical current of the flux dam. Low noise operation was also observed at a cooling field of 10μT in a washer type magnetometer, which has wide

(w = 55 μT) flux dams (Fig. 18). Here the critical field for I_c suppression is estimated to be 1.2 μT while the threshold field for flux trapping is 0.52 μT.

Although such devices work seemingly stably with low noise, detailed inspection of the noise spectrum indicated a bump structure or peaks at low frequencies. A broad bump is observed between 10 and 100 Hz in the field noise of Fig. 18. In the noise spectrum of the parallel loop magnetometer many peaks can be seen in a frequency range centered at the peak of line frequency noise (Figs. 4 and 5 of [25]). It is inferred that these noise structures reflect a motion of vortices in the wide junction of the flux dam. After initiation by external disturbances, the vortices can easily move within the flux dam because of high mobility and produce a spectrum of low frequency noise. This kind of instability may be a drawback of the use of wide flux dams in some applications where low frequency signals are detected.

4.5.3 Parallel Junction Flux Dam

In order to suppress the high mobility vortex motion in flux dams we examined the effect of narrowing the width of grain boundary junctions of the flux dam. As shown in the right side of Fig. 24, two holes were formed along a wide (55 μm) grain boundary in the two connecting lines. The resulting flux dams have three narrow parallel junctions of w = 5–7 μm, where the ratio of w to the Josephson penetration depth is w/λ_J = 2.02–2.8. A hole and two junctions at both sides form a SQUID in which the inductance parameter was estimated to be β_L = 7.4–10.3. It is expected that the holes in the flux dam can hold a few flux quanta stably. The field noise spectra measured under different static applied fields, after the device was cooled in zero-field, is shown in Fig. 24 [62]. This noise should be compared with the spectra in Fig. 18, which were obtained for the device having 55 μm wide flux dams without holes. It is found that the bump structure at 10–100 Hz observed in the original flux dams disappeared in the parallel junction flux dams. This

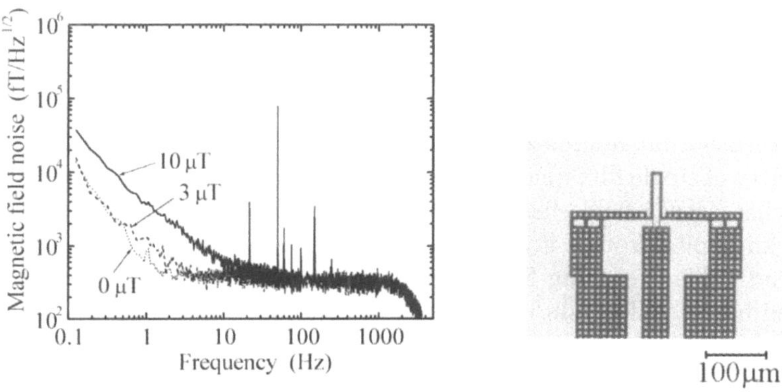

Fig. 24. Field noise spectra of the washer type magnetometer with narrow parallel junction flux dams (*right side*), after application of static fields

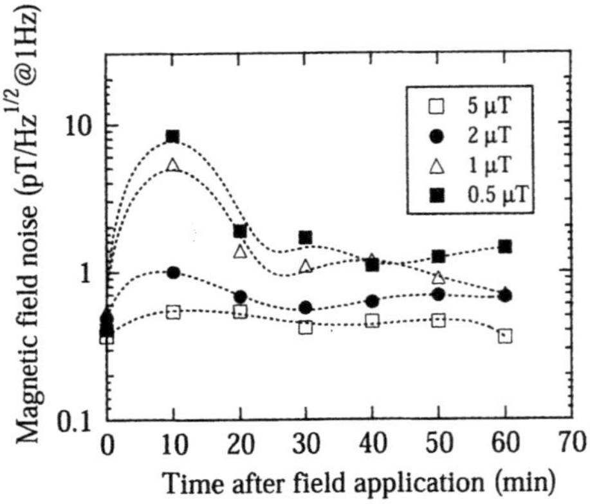

(a)

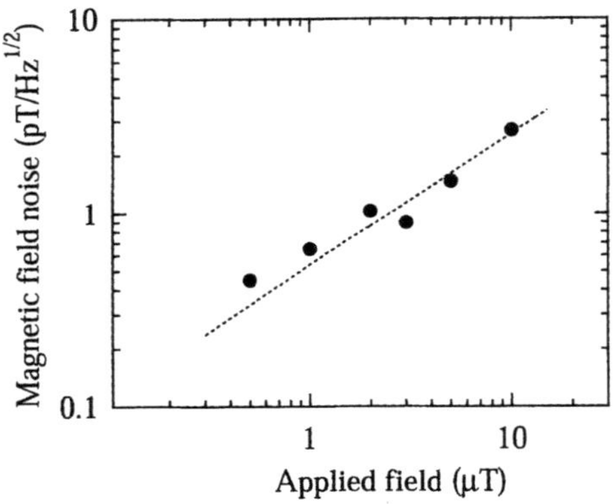

(b)

Fig. 25. (**a**) Dependence of low frequency noise on the time after various static fields are applied, and (**b**) residual low frequency noise at 1 Hz as a function of the applied field

may be the result of the suppression of free vortex motion along the wide grain boundary junction.

Figure 25a shows the dependence of the low frequency noise at 1 Hz on the time after different static fields were applied. This is the temporal change of the magnetic field noise induced by the opening of flux dams. During this change the FLL output fluctuated and voltage jumps that corresponded to successive entry of single flux quanta into the pickup coil were often observed. It is noted that even a field change of 1 μT brings about transfer of $10^4 \Phi_0$ into the pickup coil through flux dams. The relaxation time is longer for higher applied fields exceeding 30 min. A similar long relaxation time was also observed in the wide grain boundary flux dams without holes. Fig. 25b shows the residual low frequency noise at 1 Hz as a function of applied field. This noise is the values after the transient of flux dam opening is stabilized. It is found that the low frequency noise increases according to $B_{\mathrm{n}} \propto B_{\mathrm{appl}}^{0.5}$. Although the exact mechanism of this behavior is not clear, the increase of the low frequency noise may be due to thermally activated motion of flux quanta that are left in the holes of the flux dam. The thermal activation model of dc SQUIDs and the resulting low frequency noise have been described in the Chapter by Kuriki et al. in this volume).

The observed long relaxation time of flux dams is obviously a drawback in many applications of the magnetometers. The present and reported behaviors of different types of flux dams [25,28] suggest that high critical current junctions of RSJ-like characteristic are required for rapid damping of the shielding current and fast stabilization of FLL operation. Here, a high critical current with a thermal noise parameter of $\gamma \gg 100$ [63], where $\gamma = I_c h/2\pi k_B T$, h is Planck constant and e is the electron charge, leads to a sharp voltage onset at I_c. A single strip junction would be sufficient if its width is narrow ($w \leq 4\lambda$) but I_c satisfies the above condition.

5 Development of SQUIDs for Microscopes

In recent years SQUIDs have been used in a wide variety of applications due to their superior magnetic field sensitivity. In particular, SQUID microscopes have become a powerful tool for investigation of flux dynamics and other studies in physics [64,65,66,67,68,69]. For a low-T_c SQUID a small transfer coil was developed and connected to the SQUID [66,67]. However, for a high-T_c SQUID there is no technology to transfer a small magnetic field from a small area to the SQUID. Therefore, the high-T_c SQUID must be sufficiently small and the separation of the SQUID and the sample must be as small as possible. Some groups have proposed a high-T_c SQUID microscope using a high μ-metal tip or needle to solve the above-mentioned problems [67,68,69,70,71,72,73]. The advantage of the combination is that magnetization of the sample by the modulation coil of the SQUID can be avoided because the coil is far enough from the sample and it is couples not to the pickup coil directly but to the SQUID. One end of the flux guide penetrates through the pick up loop of the 77 K SQUID; the needle was held by the window with the other end sharpened and at room temperature.

Figure 26 shows a schematic drawing of a directly coupled dc SQUID with a flux needle. It was made of a 200 nm thick $YBa_2Cu_3O_{7-y}$ (YBCO) thin film on a 500 mm thick $SrTiO_3$ substrate by sputtering. The junctions utilized in the SQUID are a 30 degree bi-crystal type. The inductance of the SQUID and the pick up loop are 40 pH and 3 nH from calculations, respectively. The outer and inner diagonal dimensions of the pick up loop are ∅6.4 mm and ∅2.2 mm, respectively. The substrate at the center of the pick up loop was machined so that a 100–200 μm deep dimple was created for the needle space. It was not a through-hole but a dimple because the substrate was fragile.

A schematic cross-sectional view of the cryostat for the SQUID is shown in Fig. 27. Most of the parts of the cryostat are made of G-10 fiberglass and Delrin [68]. The cryostat contains a liquid N_2 copper reservoir having a volume of 0.8 l. The inside of the cryostat can be evacuated up to the order

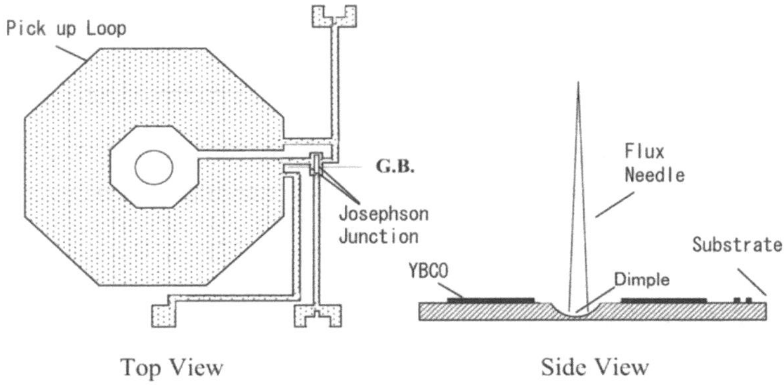

Fig. 26. Schematic drawing of a directly coupled dc SQUID

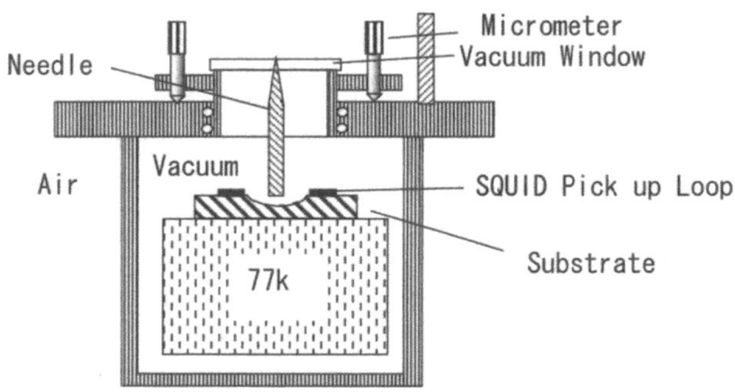

Fig. 27. Schematic cross-sectional view of the cryostat for a high-T_c SQUID

of 10^{-3} Pa by a vacuum pump and sealed off by an o-ring valve. The SQUID chip was attached to the top of a ∅12 mm sapphire rod with GE varnish, which was thermally anchored with the liquid N_2 reservoir. The SQUID was 50 mm away from the metallic reservoir. A 6 turn modulation coil and a heater were installed in the upper end of the sapphire rod. A copper wire step-up transformer was glued tightly to the top of the copper reservoir. The electrical contacts to the SQUID chip were made by applying conductive silver paint to the bonding pads and the side of the SQUID chip.

A needle made of a high m-metal (Fe–Ni base) was set at the center of the pick up loop. The length of the needle was 7 mm; its cross-section at the bottom was a 300 μm × 300 μm square shape. The top of the needle was filed so that it had a sharp edge. The shape of the top edge was not a circle but like an oval. The size of the oval shaped top edge was 50 μm × 10 μm from microscope observation. The needle penetrated a vacuum window through a hole. About 100 μm from the top of the needle was outside of the window. The hole was sealed with a silicone rubber glue after penetration. The distance between the needle and the pick-up loop was adjustable by turning three micrometers.

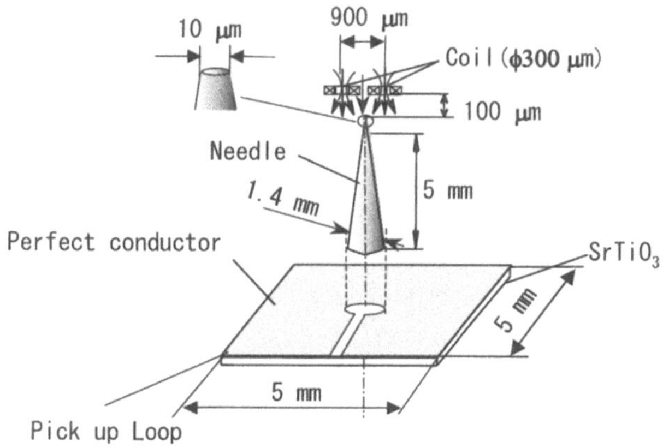

Fig. 28. Model of the simulation parameters

Here we show the computer simulated results for the SQUID with a needle. The three-dimensional FEED electro-dynamic simulation program Maxwell (supplied by Ansoft Japan) was used. We selected a perfect conductor as the material for the pick up loop instead of a superconductor because there was no superconducting material in the library. The permeability of the needle μ_r was selected as 60000. The details of the simulation parameters and dimensions are shown in Fig. 28. One set of small field coils with a separation of 900 μm, which generates a magnetic field of 10 mA/m, was positioned above the pick up loop. The relative position of the set of coils and the pick up loop was varied by steps of 150 μm in this simulation. The field calculation at the center of the pickup coil was performed at each position. Figure 29 shows the results of the simulation. In the case of Fig. 29a the needle penetrates through the loop. The bottom of the needle is 500 μm below the film in this case. In the case of Fig. 29b the needle is at the same level as the loop. In the case of Fig. 29c the needle is above the loop with a space of 500 μm. Two peaks, one at the positions of each field coil, were observed in all the cases. However, the field density in the case of Fig. 29a is the largest and one order of magnitude larger than that in the case of Fig. 29c. This suggests that the penetration of the needle through the pick-up loop is essential to obtaining a higher sensitivity.

Then we show the size effect of the top of the needle. In this simulation the space between the coil and the needle was set at 0.1 μm and the separation of the coils was 10 μm to obtain a good space resolution. The results are shown in Fig. 30. The peaks of the field with the needle of ∅1 μm are sharper than that with ∅10 μm. This result suggests that the needle with smaller top shows a better space resolution.

The performance of the direct-coupled SQUID magnetometer with a needle was investigated. A needle was positioned at the center of the pick up

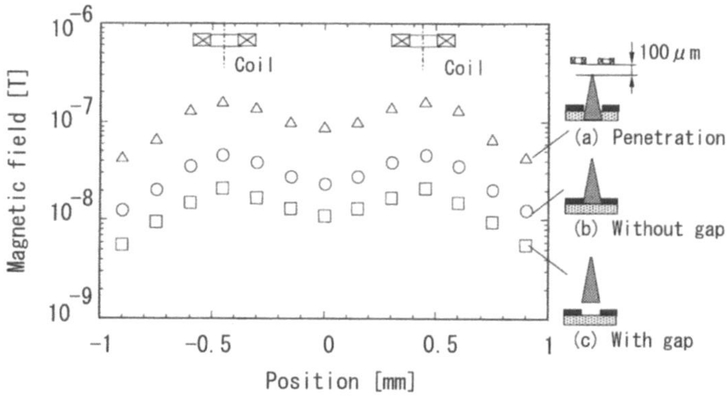

Fig. 29. Effect of the position of the needle in simulation

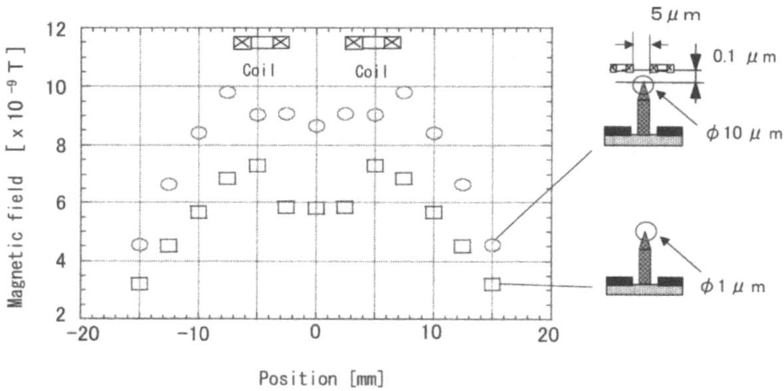

Fig. 30. Effect of size of the needle in simulation

loop. The needle was brought into contact once with the bottom of the dimple of the substrate by adjusting the micrometer and it was then lifted up a little bit above the bottom for thermal isolation.

The SQUID was driven by a flux-locked loop with a flux modulation frequency of 256 kHz. The output signal from the SQUID was amplified by double transformers, one at 77 K and the other at room temperature. The R_f/M_f is $0.5\,V/\phi_0$, where R_f is the feedback resistance and M_f is the mutual inductance between the SQUID magnetometer and the modulation coil. All of the measurements were performed in a magnetically shielded room, which had a shielding factor of -50 dB. A small one turn coil with a radius of 1 mm was prepared and put on the top end of the needle to measure the local effective area at the edge of the needle. A small coil rather than a large coil was used to avoid having the field from the coil couple directly to the pick up loop of the SQUID. A sinusoidal current of $1\,mA_{p\text{-}p}$ with a frequency of 100 Hz was used. Then we found that the effective area $A_{\text{eff-needle}}$ of the magnetometer was 540 µm^2. This value corresponds to a small SQUID with a washer size of about 15 µm × 15 µm [65]. The flux noise of the magnetometer was measured. The noise with needle was $80\mu\phi_0/\mathrm{Hz}^{1/2}$ at 100 Hz and $36\mu\phi_0/\mathrm{Hz}^{1/2}$ in the white noise region. By contrast, the flux noise $S_\phi^{1/2}(f)$ without needle was $28\mu\phi_0/\mathrm{Hz}^{1/2}$ at 100 Hz. Note that the noise at around 1 Hz is not reliable because of the small number of sampling points. Excess $1/f$ flux noise, which may be generated by the thermally activated hopping of vortices trapped during cooldown, was observed in the case with needle. However, since the optimal bias current was not changed in the measurements with and without the needle, the trapping was not significant. The higher noise level in the white noise region may be due to the operation temperature of the SQUID. The temperature at the top of the sapphire rod is about 2–3 K higher than the liquid nitrogen boiling point.

Laser printed outputs were used as a sample to evaluate the space resolution. Laser printer ink contains ferromagnetic particles and therefore it is easy to generate a bar line. The finest pattern we prepared was a line width of 100 µm and a spacing between lines of 200 µm. This pattern was limited by the resolution of the laser printer. The samples were moved in the direction that the needle crosses the bar lines. A schematic drawing of this scanning is shown in Fig. 31a [75]. The upper figure shows the top view and the lower shows the side view. The figure in the middle depicts the shape of the top edge of the needle. The scanning direction was along the longer axis of the oval top edge. The velocity of the scanning was 620 µm/s. The surface of the sample was brought into contact with the end of the needle during the scan. Figure 31b shows the output signal of the SQUID [74]. The signal

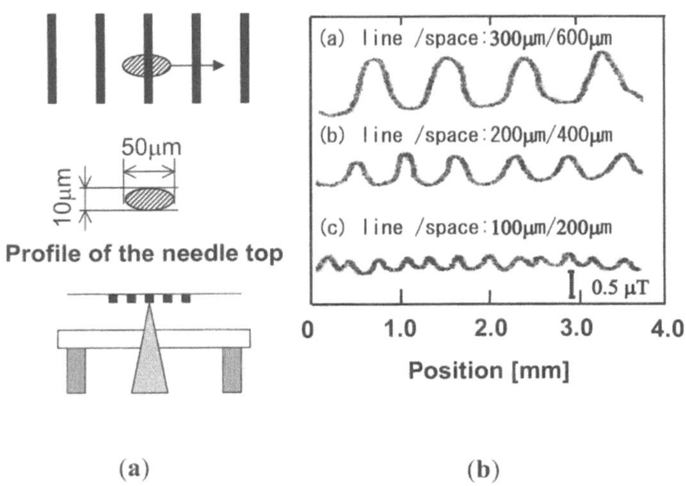

Fig. 31. Evaluation of the space resolution

was measured through a low pass filter of 5 kHz. All of the patterns were clearly observed. The peak to peak value of the signal that corresponds to the line was about 0.15 μT. Therefore this system has at least a resolution of line/space = 100 μm/200 μm.

References

1. R. Gross, L. Alff, A. Beck, O. M. Froehlich, D. Koelle, A. Marx: IEEE Trans. Appl. Supercond. **7**, 2929 (1997)
2. J. Halbritter: J. Appl. Phys. **58**, 1320 (1985)
3. Y. Tanaka, S. Kashiwaya: Phys. Rev. B **56**, 892 (1997)
4. K. Enpuku, T. Minotani, F. Shiraishi, A. Kandori, S. Kawakami: IEEE Trans. Appl. Supercond. **9**, 3109 (1999)
5. C. D. Tesche, J. Clarke: J. Low Temp. Phys. **29**, 301 (1977)
6. C. T. Rogers, R. A. Buhrman: IEEE Trans. Magn. **MAG-21**, 126 (1985)
7. K. Enpuku, T. Minotani: IEICE Trans. Electron. B**83**-C, 34 (2000)
8. M. Kawasaki, P. Chadhari, A. Guputa: Phys. Rev. Lett. **68**, 1065 (1992)
9. A. Marx, R. Gross: Appl. Phys. Lett. **70**, 120 (1997)
10. K. Enpuku, G. Tokita, T. Maruo, T. Minotani: J. Appl. Phys. **78**, 3498 (1995)
11. K. Enpuku: IEICE Trans. Electron. B**80**-C, 1240 (1997)
12. J. Beyer, D. Drung, F. Ludwig, T. Minotani, K. Enpuku: Appl. Phys. Lett. **72**, 203 (1998)
13. F. Kahlman, W. E. Booji, M. G. Blamire, P. F. McBrien, N. H. Peng, C. Jeynes, E. J. Romans, C. M. Pegrum, E. J. Tarte: IEEE Trans. Appl. Supercond. **11**, 916 (2001)
14. Y. Q. Shen, P. R. E. Petersen, T. Holst: IEEE Trans. Appl. Supercond. **11**, 920 (2001)
15. K. Yoshida, M. S. Hossain, T. Kisu, K. Enpuku, K. Yamafuji: Jpn. J. Appl. Phys. **31**, 3844 (1992)
16. K. Enpuku, T. Maruo, T. Minotani: J. Appl. Phys. **80**, 1207 (1996)
17. J. Ramos, A. Chwala, R. Ijsselsteijn, R. Stolz, V. Zakosarenko, V. Schultze, H. E. Hoenig, H. G. Meyer: IEEE Trans. Appl. Supercond. **9**, 3392 (1999)
18. K. Enpuku, R. Cantor, H. Koch: J. Appl. Phys. **71**, 2338 (1992)
19. T. Minotani, K. Enpuku, Y. Kuroki: J. Appl. Phys. **82**, 457 (1997)
20. L. P. Lee, J. Longo, V. Vinetskiy, R. Cantor: Appl. Phys. Lett. **66**, 1539 (1995)
21. F. Ludwig, E. Dantsker, R. Kleiner, D. Koelle, J. Clarke, S. Knappe, D. Drung, H. Koch, N. McN. Alford, T. W. Button: Appl. Phys. Lett. **66**, 1418 (1995)
22. F. W. Grover: *Inductance Calculations Working Formulas, Tables* (Dover, New York 1962)
23. E. Dantsker, S. Tanaka, J. Clarke: Appl. Phys. Lett. **70**, 2037 (1997)
24. F. Ludwig, D. Drung: Appl. Phys. Lett. **75**, 2821 (1999)
25. M. S. Dilorio, K. Y. Yang, S. Yoshizumi, S. G. Haupt, D. Haran, R. H. Koch, F. P. Milliken, J. R. Rozen, D. K. Lathrop, S. Kumar, H. S. Trammell: IEEE Trans. Appl. Supercond. **9**, 4428 (1999)
26. K. Yokosawa, S. Kuriki, S. Hirano, H. Oyama, D. Suzuki, K. Tsukada: J. Appl. Phys. **90**, 4049 (2001)
27. F. P. Milliken, S. L. Brown, R. H. Koch: Appl. Phys. Lett. **71**, 1857 (1997)

28. K. Enpuku, A. Nakahodo, M. Hotta, S. Hijiya, D. Tokimizu, D. Kuroda: Jpn. J. Appl. Phys. **40**, 4013 (2001)
29. J. Kurkijarvi: Phys. Rev. B **6**, 832 (1972)
30. L. D. Jackel, W. W. Webb, J. E. Lukens, S. S. Pei: Phys. Rev. B **9**, 115 (1974)
31. K. Enpuku, D. Tokimizu, D. Kuroda, S. Hijiya: IEICE Trans. Electron. B**85**-C, 681 (2002)
32. K. Enpuku, D. Tokimizu, D. Kuroda, S. Hijiya: Jpn. J. Appl. Phys. **40**, L869 (2001)
33. K. Enpuku, Y. Shimomura, T. Kisu: J. Appl. Phys. **73**, 7929 (1993)
34. D. Drung, R. Cantor, M. Peters, H. J. Scheer, H. Koch: Appl. Phys. Lett. **57**, 406 (1990)
35. R. H. Koch, W. Eidelloth, B. Oh, R. P. Robertazzi, S. A. Andrek, W. J. Gallagher: Appl. Phys. Lett. **60**, 507 (1992)
36. R. H. Koch, J. Clarke, W. M. Goubau, J. M. Martinis, C. M. Pegrum, D. J. Van Harlingen: J. Low Temp. Phys. **51**, 207 (1983)
37. V. Foglietti, W. J. Gallagher, R. H. Koch: IEEE Trans. Magn. **MAG-23**, 1150 (1987)
38. M. Gershenson, D. J. Fixsen: US Patent No. 4663590 (1987)
39. S. Kuriki, M. Matsuda, A. Matachi: J. Appl. Phys. **64**, 239 (1988)
40. D. Drung, E. Crocoll, R. Herwig, M. Neuhaus, W. Jutzi: IEEE Trans. Magn. **MAG-25**, 1034 (1989)
41. O. Dossel, B. David, M. Fuchs, W. H. Kullmann, K.-M. Ludeke: IEEE Trans. Magn. **MAG-27**, 2797 (1991)
42. S. Kuriki, D. Suzuki: Jpn. J. Appl. Phys. **32**, L1099 (1993)
43. D. Drung: Appl. Phys. Lett. **67**, 1474 (1995)
44. S. Kuriki, A. Hayashi, H. Oyama: Jpn. J. Apl. Phys. **38**, 4733 (1999)
45. T. Hirano, T. Nagaishi, H. Itozaki: Ext. Abst. 7th Int'l Superconductive Electronics Conf., Berkeley (June 1999) pp. 544–546
46. D. Drung: Physica C **368**, 134 (2002)
47. See, www.stl-gmbh.de; http://www.starcryo.com; http://www.tristantech.com
48. R. Cantor, L. P. Lee, M. Teepe, V. Vinetskiy, J. Longo: IEEE Trans. Appl. Supercond. **5**, 2927 (1995)
49. H. Oyama, S. Hirano, M. Matsuda, S. Kuriki: *Advances in Superconductivity*, T. Yamashita, K. Tanabe (Eds.) (Springer, Berlin, Heidelberg 2000) pp. 1027–1029
50. K. Yokosawa, E. Maruyama, Rashdi Sha Ahmad, H. Oyama, S. Kuriki, D. Suzuki, K. Tsukada: Jpn. J. Appl. Phys. **37**, 4362 (1998)
51. H. Oyama, S. Hirano, S. Kuriki: Ext. Abst. 8th Int'l. Superconductive Electronics Conference, Osaka (June 2001), pp. 327–328
52. S. Krey, H.-J. Barthelmess, M. Schilling: J. Appl. Phys. **86**, 6602 (1999)
53. H.-M. Cho, R. McDermott, B. Oh, K. A. Kouznetsov, A. Kittel, J. H. Miller Jr., J. Clarke: IEEE Trans. Appl. Supercond. **9**, 3294 (1999)
54. S. Hirano, H. Oyama, S. Kuriki, T. Morooka, S. Nakayama: Appl. Phys. Lett. **78**, 1715 (2001)
55. P. Selders, A. Castellanos, M. Vaupel, R. Wordenwerber: Appl. Supercond. **5**, 269 (1988)
56. J. Z. Sun, W. J. Gallagher, R. H. Koch: Phys. Rev. B. **50**, 664 (1994)
57. Y. Mawatari, J. R. Clem: Phys. Rev. Lett. **86**, 2870 (2001)
58. V. Schultze, N. Oukhanski, V. Zakosarenko, R. Ijsselsteijn, J. Ramos, A. Chwala, H.-G. Meyer: IEEE Trans. Appl. Supercond. **11**, 1 (2001)

59. R.H. Koch, J.Z. Sun, V. Foglietta, W.J. Gallagher: Appl. Phys. Lett. **67**, 709 (1995)
60. S. Keil, R. Straub, R.P. Huebener, D. Koelle, R. Gross, K. Barthel: IEEE Trans. Appl. Supercond. **9**, 2961 (1999)
61. P.A. Rosenthal, M.R. Beasley, K. Char, M.S. Colclough, G. Zaharchuk: Appl. Phys. Lett. **59**, 3482 (1991)
62. H. Oyama, S. Hirano, S. Kuriki, M. Matsuda: Supercond. Sci. Technol. **15**, 1 (2002)
63. V. Ambegaokar, B.I. Halperin: Phys. Rev. Lett. **22**, 1364 (1969)
64. R.C. Black, A. Mathai, F.C. Wellstood, E. Dantsker, A.H. Miklich, D.T. Nemeth, J.J. Kingston, J. Clarke: Appl. Phys. Lett. **62**, 2128 (1993)
65. T.S. Lee, E. Dantsker, J. Clarke: Rev. Sci. Instrum. **67**, 4208 (1996)
66. J.R. Kirtley, M.B. Ketchen, K.G. Stawiasz, J.Z. Sun, W.J. Gallagher, S.H. Blanton, S.J. Wind: Appl. Phys. Lett. **66**, 1138 (1995)
67. T. Morooka, S. Nakayama, A. Odawara, M. Ikeda, S. Tanaka, K. Chinone: IEEE Trans. Appl. Supercond. **9**, 3491 (1999)
68. S. Tanaka, O. Yamazaki, R. Shimizu, Y. Saito: Supercond. Sci. Technol. **12**, 809 (1999)
69. E.F. Fleet, S. Chatraphorn, F.C. Wellstood, L.A. Knauss: IEEE Trans. Appl. Supercond. **9**, 4103 (1999)
70. P. Pitzius, V. Dworak, U. Hartmann: 6th Int'l. Supercond. Electronics Conf. (ISEC'97) Ext. Abst. **3** (1997) p. 395
71. Y. Tavrin, M. Seigel: Applied Supercond. **1**, 719 (1998)
72. T. Nagaishi K. Minamimura, H. Itozaki: IEEE Trans. Appl. Supercond. **11**, 226 (2001)
73. S.A. Gudoshnikov, Y.V. Deryuzhkina, P.E. Rudenchik, Y.S. Sintnov, S.I. Bondarenko, A.A. Shablo, P.P. Pavlov, A.S. Kalabukhov, O.V. Snigirev, P. Seidel: IEEE Trans. Appl. Supercond. **11**, 219 (2001)
74. S. Tanaka, K. Matsuda, O. Yamazaki, M. Natsume, H. Ota, T. Mizoguchi: Jpn. J. Appl. Phys. **40**, L431 (2001)
75. S. Tanaka, K. Matsuda, O. Yamazaki, M. Natsume, H. Ota: Supercond. Sci. Technol. **15**, 146 (2002)

Applications of HTSC SQUIDs

H. Itozaki[1], K. Sakuta[2], T. Kobayashi[2], K. Enpuku[3], N. Kasai[4], Y. Fujinawa[4,5], H. Iitaka[4], K. Nikawa[6], and M. Hidaka[7]

[1] National Institute for Materials Science
1-2-1 Sengen, Tsukuba, Ibaraki 305-0047, Japan
itozaki.Hideo@nims.go.jp
[2] Graduate School of Engineering Science, Osaka University
1-3 Machikaneyama, Toyonaka, Osaka 560-8531, Japan
kobayashi@sup.ee.es.osaka-u.ac.jp
sakuta@ee.es.osaka-u.ac.jp
[3] Graduate School of Information Science and Electrical Engineering
Kyushu University, Higashi-ku, Hakozaki 6-10-1, Fukuoka 812-8581, Japan
enpuku@ed.kyushu-u.ac.jp
[4] Nanoelectronics Research Institute
National Institute of Advanced Industrial Science and Technology
1-1-4 Umezono, Tsukuba, Ibaraki 305-8568, Japan
kasai-naoko@aist.go.jp
[5] National Research Institute for Earth Science and Disaster Prevention
3-1 Tennodai, Tsukuba 305-0006, Japan
[6] Analysis Technology Development Division, NEC Electronics Corporation,
1753 Shimo-numabe, Nakahara-ku, Kawasaki
Kanagawa 211-8668, Japan
k.nikawa@necel.com
[7] Fundamental Research Laboratories, NEC Corporation
34 Miyukigaoka, Tsukuba, Ibaraki 305-8501, Japan
hidaka@bx.jp.nec.com

Abstract. Recently a high temperature superconductor (HTSC)-SQUID has been made with high magnetic field resolution and its application is expected to be practical for many fields. The HTSC-SQUID can work using liquid nitrogen which is handled much more easily than liquid helium, and thus a wide range of HTSC-SQUID applications is expected. This chapter describes recent HTSC-SQUIDs application researchs such as magnetocardiography, biological immunoassay, measurement of ultra low frequency to observe underground activities and a laser-SQUID microscope for LSI chip defects analysis.

1 Introduction

The HTSC-SQUID has been developed to have good performance as a high resolution magnetic sensor. In order to accelerate activities of SQUID application research, the SQUID chip and its electronics should be commercialized. CONDUCTUS in California sold "Mr. SQUID" that included a SQUID chip and its simple electronics. It is for education. Sumitomo Electric introduced "SQUID Kit" which has flux locked loop electronics. It is not only an

T. Kobayashi et al. (Eds.): Vortex Electronics and SQUIDs,
Topics Appl. Phys. **91**, 185–248 (2003)

educational kit but also a tool for SQUID application research. A direct read-out method and an additional positive feedback method were introduced to a new Sumitomo Electric kit "SEIQUID II". It can be controlled by a personal computer. In this way a high performance HTSC-SQUID can be obtained by SQUID application researchers who need not develop a SQUID chip and its electronics. This accelerates SQUID application research.

In Sect. 3 a magnetically shield-less HTSC SQUID magnetocardiograph named "Open-SQUID" system using an electronic gradiometer consisting of four SQUID magnetometers has been developed. A noise canceling process based on an adaptive filtering (ADF) a lgorithm and an active noise control system were also employed to reduce the noise component in the output signal and to improve the system stability. A smooth MCG waveform was thus obtained by the Open-SQUID system with supplemental procedures such as digital notch filtering and simple integration for only a 20 second interval. Automatic adjustment of filter coefficients by an ADF algorithm made the signal compensation between the sensor and the reference exact and avoided the necessity of complicated tuning. This adaptive noise canceling process is able to work in real time by virtue of a small amount of calculation. An active noise control (ANC) system was also applied to reduce the influence of the dc magnetic field fluctuations, which so far made a Flux Locked Loop (FLL) circuit out of range. A simple negative feedback loop from the reference SQUID to both the reference itself and the sensing SQUID through the feedback coils suppressed the magnetic fluctuations in the region of the SQUIDs, leading to stable operation of the Open-SQUID system even in the shield-less environment. Otherwise, the locking condition of FLL was completely missed for ten minutes or so.

In Sect. 4 a high T_c SQUID system is developed for application to biological immunoassay. In this application, a Fe_2O_3 nanoparticle is used as a magnetic marker in order to label the antibody and the binding reaction between an antigen, and its antibody is detected by measuring the magnetic field from the marker. Design and set up of the system is described. The minimum detectable amplitude of the magnetic flux is $0.6\,m\Phi_0$ and $0.12\,m\Phi_0$ for the measurement bandwidth from 0.2 Hz to 5 Hz when we use the magnetometer and the gradiometer, respectively. The system noise does not increase when a magnetic field of 1 mT is applied in parallel to the SQUID. An experiment to measure the antigen-antibody reaction shows that the sensitivity of the present system is 10 times better than that of the conventional method using an optical marker. Since the system performance strongly depends on the magnetic properties of the nanoparticle, it is important to develop a magnetic marker suitable for the present system, e.g., nanoparticles having the remanence. It will be easy to develop a system that is 100 times more sensitive than the conventional optical method.

Anomalous electromagnetic field variations in the ultra low frequency (ULF) band were measured as precursory phenomena of earthquakes or volcanic eruptions. In Sect. 5 a portable system with a single HTS-SQUID magnetometer and a small Dewar was constructed for monitoring the environmental magnetic field in the ULF range to study the generation mechanism of the ULF variations. The Dewar keeps liquid nitrogen about three weeks. The field test of the system was achieved at Mt. Bandai, an active volcano in Fukushima prefecture in Japan, from the end of October to the beginning of November 2000. The system worked for 36 hours safely. It shows a SQUID magnetometer operated by direct offset integrated technique can be used in the field with no magnetic shield.

In Sect. 6 a novel technique for electrical inspection and failure analysis that can detect open, high resistance and shorted interconnects without electrical contact is proposed and developed. The basic idea is detection of the magnetic field produced by OBIC (optical beam induced current) using a DC-SQUID (superconducting quantum interference device) magnetometer. Scanning laser-SQUID microscopy ("laser-SQUID" for short) was demonstrated with a spatial resolution of about 1.3 μm. Applications include IC defect detection before bond pad patterning and without pin selection on completed ICs. Localization areas range from whole die to a few square microns.

In Sect. 7 small scale HTSC digital applications are reviewed. The number of Josephson junctions in high-T_c superconductor (HTSC) digital circuits is much lower than that in Nb circuits. However, the small scale of HTSC circuits, has made them nearly practical because they are easy to be cooled. Samplers and analog-to-digital converters (ADC) whose operations are based on a single flux quantum (SFQ) have been developed by using HTSC materials. The time resolution of such sampler measurements is very fine. For example, the two prototype sampler measurement systems described in the article can measure waveforms of several gigahertz. The devices are cooled by single stage cryo-coolers. The ADC potentially has a large dynamic range and bandwidth and should be able to handle a high frequency input signal. The HTSC ADC may have application to software-defined radio for broadband mobile communications.

2 HTSC-SQUID for Commercial Use

2.1 SQUID Kit

A HTSC (High Temperature Superconductor)-SQUID has been developed with high sensitivity as a Nb-SQUID. A magnetic field resolution of less than $100\,\mathrm{fT/Hz^{1/2}}$ has been reported [1]. This indicates that the HTSC-SQUID is going to be used in practical applications in the near future [2]. Application research shown in Fig. 1 is active in the field of medical diagnosis [2], non-destructive evaluation [2,3], immunoassay [3,4,5], geological survey [7,8] and IC evaluation [9]. In the early stage of HTSC-SQUID development they

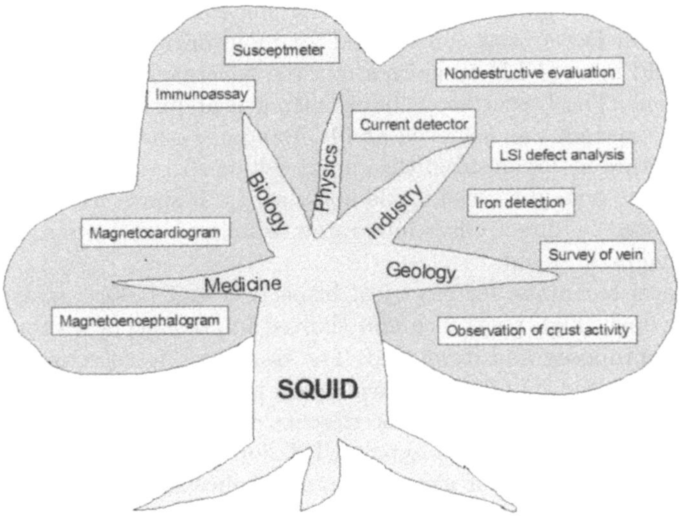

Fig. 1. Future prospects of SQUID applications

could be obtained by only researchers who could make them. Therefore application research of the HTSC-SQUID was very limited to them, even though its application field was so wide. The HTSC-SQUIDs should be supplied to application researchers by someone who could make them. In this situation, the small scale commercial product of HTSC-SQUIDs has begun to be sold shortly after the HTSC-SQUID chip was developed. This leading small product was a combination of a SQUID chip and minimal driving electronics. The most popular product was "Mr. SQUID" sold by CONDUCTUS in California. Since this was low priced and worked as a good SQUID demonstration kit for education, hundreds were sold in the world. However, it could not be used as a magnetic sensor for application researchers working on HTSC-SQUID because it did not have a FLL (flux locked loop) circuit which made the SQUID into a sensor with linear response to magnetic field. In Japan, Sumitomo Electric sells the "HTSC-SQUID Kit", which is shown in Fig. 2. This equipment cannot only demonstrate SQUID characteristics such as current vs. voltage and magnetic field vs. voltage shown in Fig. 3, but also works as a high sensitivity magnetic sensor with low noise FLL electronics. This equipment has been widely accepted by developers of SQUID applications. More than 70 units have been sold by the end of 2001. A SQUID chip of this kit is mounted on its chip carrier and is covered by a resin cap for protection from humidity (Fig. 4). This chip carrier has electric pins to connect to its SQUID holder easily. In the first step of HTSC-SQUID commercialization, it is important to make the SQUID simple in order to be used by beginners. The magnetic field sensitivity of this Sumitomo SQUID kit is about $1\,\mathrm{pT/Hz^{1/2}}$. This magnetic field resolution is not full performance of the HTSC-SQUID

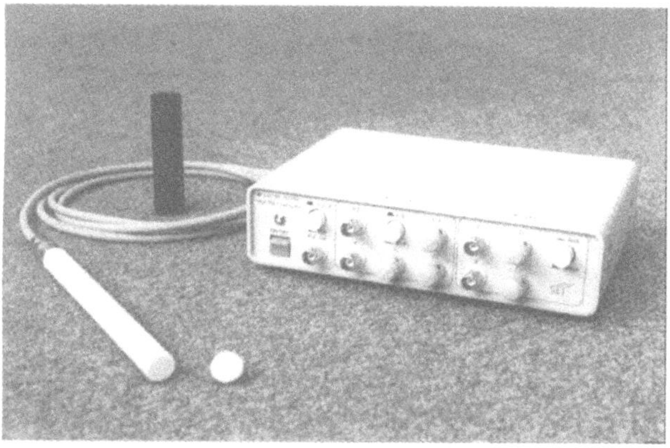

Fig. 2. "HTSC-SQUID Kit" by Sumitomo Electric Hightechs

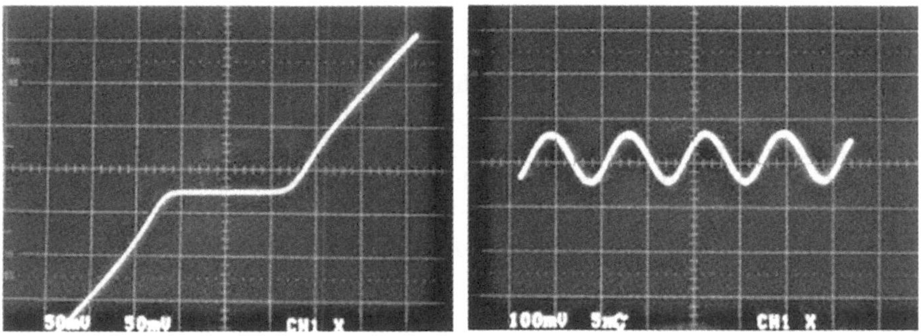

(a) (b)

Fig. 3. (**a**) Current vs voltage of HTSC-SQUID. (**b**) Magnetic field vs voltage of HTSC-SQUID

Fig. 4. HTSC-SQUID chip without protection cap (Sumitomo Electric Hightechs)

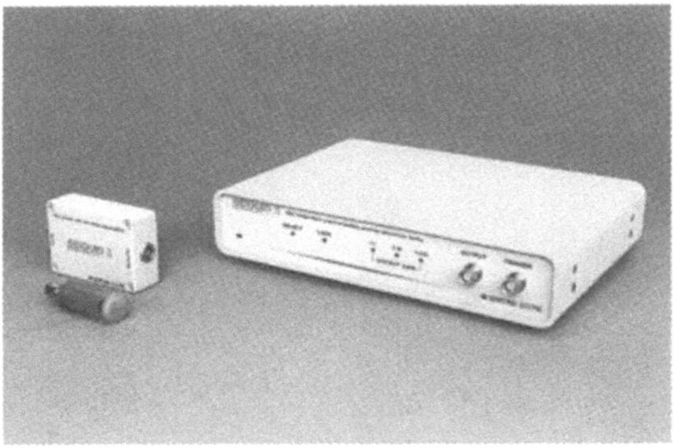

Fig. 5. "SEIQUID II" by Sumitomo Electric Hightechs

because it is difficult for beginners to handle a high sensitive SQUID with field resolution of less than $1\,\mathrm{pT/Hz^{1/2}}$. Resolution of $1\,\mathrm{pT/Hz^{1/2}}$ can be obtained by a small size SQUID such as 5 mm square. Its driving electronics uses the modulation of the magnetic field to reduce noise. It also uses a small transformer to increase the SQUID output voltage, and it has noise reduction. A low frequency noise cut filter, a high frequency noise cut filter and 50 Hz, 60 Hz notch filters are also included. These filters can reduce environmental noise and make signals clear. These functions are useful for SQUID beginners to start research and development of HTSC-SQUID applications.

2.2 Advanced SQUID Kit

While the SQUID kits were sold to beginners working with SQUIDs, a new requirement arose. A higher frequency response was required for detection of the screening current for non-destructive evaluation, detection of transition ground currents of geological surveys and so on. For these requirements, Sumitomo Electric begun to sell another advanced SQUID kit "SEIQUID II" shown in Fig. 5. This product has high frequency response with automatic tuning of SQUID operation.

2.2.1 High Frequency Response

FLL electronics with magnetic field modulation is generally used for SQUID. A schematic diagram of this electric circuit is shown in Fig. 6. Modulation signals are applied to the feedback coil. SQUID signals with modulation signals are only selected by a lock-in amplifier. This method increases the signal to noise ratio. However, it can not treat higher frequency signals than the modulation signal. In contrast, the FLL without modulation method shown in Fig. 7 can respond to high frequency [10]. This electronics is simpler than the modulation type because some electronics components such as the frequency generator and the lock-in amplifier can be omitted. It contributes to

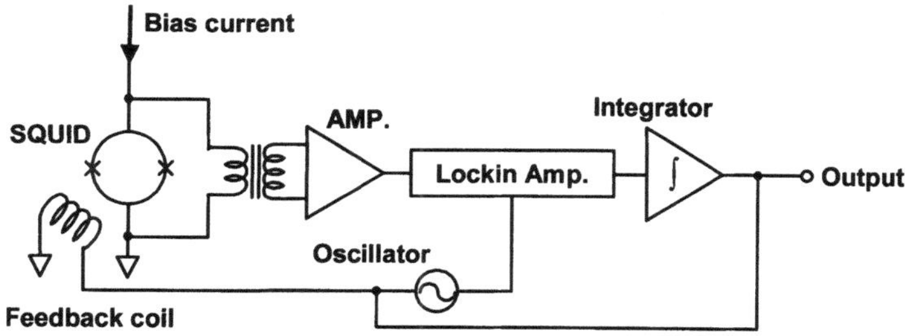

Fig. 6. Electric circuit of FLL with modulation

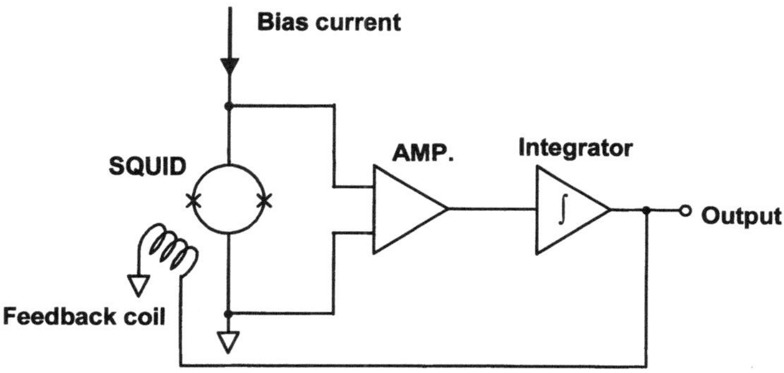

Fig. 7. Electric circuit of FLL without modulation

making the cost of a multi-channel system lower. This method can detect up to 300 kHz while the FLL with modulation can detect up to only 1 kHz. As this method does not use modulation, it cannot use a low noise transformer to amplify the SQUID output signal, but should use another technique. The SQUID signal can be increased by the APF (additional positive feedback) method [11]. A schematic diagram of this electronics is shown in Fig. 8. The SQUID output current adds additional magnetic field to the SQUID and it adds more SQUID current. This mechanism amplifies the SQUID signal. Figure 9 shows a SQUID chip with APF coil. Figure 10 shows the SQUID output voltage vs. magnetic flux. Its slope has been changed to be partly steeper. The steeper slope means that the transfer efficiency of magnetic field to SQUID voltage increases. These FLL without modulation and APF electronics are used by "SEIQUID II".

2.2.2 Automatic Tuning

It is necessary to tune the bias current and bias magnetic field to optimize the SQUID operation. Automation of this tuning process is kind to SQUID users. Figure 11 shows the schematic electronics diagram of "SEIQUID II" which can perform automatic tuning [12].

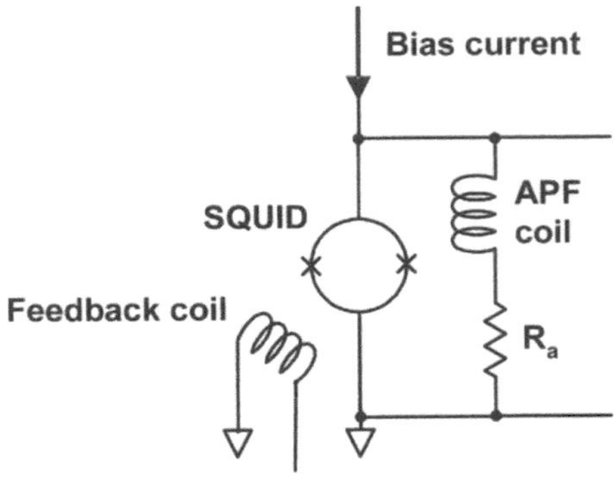

Fig. 8. Electric circuit of APF (Additional Positive Feedback)

Fig. 9. SQUID chip on chip carrier with APF coil

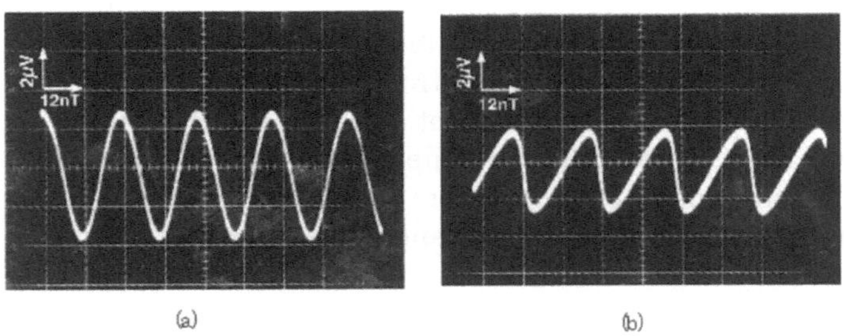

Fig. 10. (a) Voltage vs. magnetic field without APF. (b) Voltage vs. magnetic field with APF

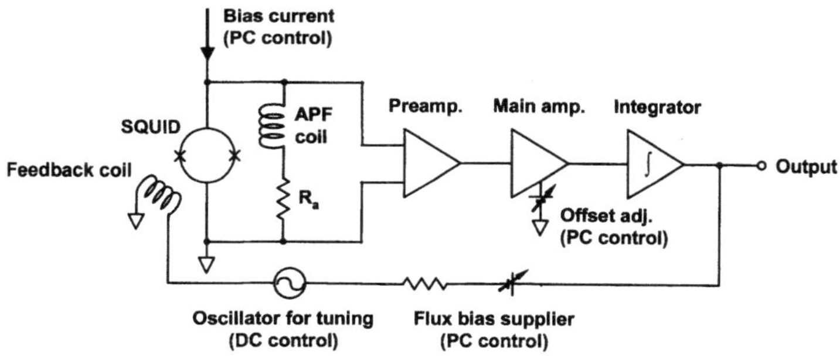

Fig. 11. Schematic diagram of FLL electronics in "SEIQUID II"

Automatic Tuning of Bias Current. The SQUID should operate with an optimized bias current which is slightly over the critical current because the optimized bias current gives maximum voltage response to the applied magnetic field. The following process makes it optimized using a personal computer. The SQUID output voltage is monitored when half of a magnetic flux ($\phi_0/2$) with frequency f is applied to the SQUID. Then the output voltage comes to have two frequency components, f and $2f$. The optimized point is when summation of both signals becomes a maximum during change of the bias current. Using this optimized bias current, the following bias magnetic flux optimization process is done.

Automatic Tuning of Bias Magnetic Flux. The SQUID has a non-linear response of magnetic flux shown in Fig. 10. The magnetic flux of the SQUID should be locked at the maximum slope of $dV/d\phi$. It is necessary to add a slight magnetic flux bias to the SQUID in order to get an optimally locked flux point. In this case also a half magnetic flux with frequency f is applied to the SQUID. Then the SQUID output voltage is monitored when the bias magnetic flux is swept with one flux. The optimized magnetic point is located at the maximum SQUID voltage when only a frequency f exists. In this case, not $2f$ but f should be selected and monitored because the $2f$ signal shows the minimum point.

After this optimizing process, we can use the SQUID properly. The process with personal computer helps the user of the SQUID a lot.

2.3 Commercialization of SQUID

The HTSC-SQUID has already been developed as a high sensitivity magnetic sensor. Application development has already been started in some fields. Now the small SQUID unit is sold to researchers and developers to make new SQUID applications. The HTSC-SQUID kit sold by Sumitomo Electric Hightechs has been used by the following researchers. Non-destructive evaluation is 30%. IC evaluation and geological survey are 20% each. MCG and immunoassay are 15% each. These applications are leading candidates for real applications. Although the HTSC-SQUID kit market is small now, the market of HTSC-SQUID applications will grow ten times in the near future. Production of the HTSC-SQUID is also on a small scale, it is necessary to scale up the engineering of its production, reliability and cost in order to be commercialized.

3 Challenge to Shield-Less HTSC-SQUID Magnetocardiography

The requirements for magnetocardiography (measurements of signals from human heart) are more easy than magnetoencephalography and as a result most biomagnetic measurements with HTSC-SQUIDs have focused on this application. Magnetocardiogram (MCG) measurements using SQUID magnetometers are usually performed in magnetically shielded rooms or tubes [13] to reduce the influence of electromagnetic disturbances from the environment. However, such shielding apparatus spoils the convenience of HTSC measurement systems. Moreover, these shields are expensive and immobile. Therefore, it is desirable to record MCGs outside shielding. So far, device process techniques have been developed and consequently the individual HTSC device (e.g., Josephson junction, SQUID-ring, etc.) has been thought to be applicable in practical use. However, there still exists a difficulty with the HTSC SQUID design. It is difficult to prepare a HTSC differential pick-up coil with acceptable performance (due to immaturity of technologies for connection, film deposition on extremely large substrates, and so on). This has prevented practical SQUID fabrication. Therefore, a wide variety of first and second order gradiometers have been demonstrated, either deposited on a single chip or involving a flux transformer on a substrate inductively coupled to a SQUID on a separate chip [14,15,16,17]. SQUID operation has been studied in an unshielded environment [18,19,20,21].

In an alternative approach, electronically subtracting the outputs of two or more flux-locked SQUID magnetometers has formed gradiometers [20,22]. The signals of the reference magnetometers are added to the signal of the sensing magnetometer with coefficients adjusted such that their effective sensitive direction and sensitivity exactly compensates the sensing magnetometer. These coefficients are adjustable, so the difference in the measurement environments or sensitivity deviation from chip to chip can be compensated. In addition, various digital signal processing methods can be applied to determine the coefficients and to reduce the noise signal.

The completed new HTSC-SQUID magnetocardiograph worked well at 77 K as an "Open-SQUID" in ordinary circumstances (a magnetic shield is not required) and detected at least the magnetocardiogram signal.

3.1 Open-SQIUD Magnetocardiography Equipment

A discrete dc-SQUID was composed of an HTSC magnetometer and a flux transformer, which were encapsulated face to face in a nitrogen ambient sealed 37 mm diameter package. A magnetometer was made of a 200 nm thick $HoBa_2Cu_3O_{7-y}$ step-edge junction on a $SrTiO_3$ substrate. The hole size is 8 μm by 100 μm and the washer is 2 mm by 2 mm. The step edge junctions are located inside the washer and the step height is 200 nm. Figure 12a shows a schematic view of the SQUID pattern. The geometrically determined inductance of the SQUID is 30 pH. The flux transformer consisted of a single layer pickup coil (15 mm × 15 mm) and an input coil (2 mm × 1.5 mm), as shown in Fig. 12b. The thickness of $HoBa_2Cu_3O_{7-y}$ is 200 nm with a 10 nm CeO_2 buffer layer. The effective area of the SQUID with the flux transformer is about 0.2 mm^2.

Four SQUID elements were installed in a specially designed fixture as shown in Fig. 13. A top SQUID chip serves as a detector (for both the magnetocardiogram signal and magnetic environment noise signal) and the remaining three are an orthogonal reference triplet for noise canceling. The distance between the detector and reference, the so-called baseline, was 80 mm. This baseline distance was important to get high efficiency of the subtraction between the detector's signal and the reference signal(s). The environment magnetic field noise is almost homogeneous over the sensor head, while the magnetic field from the heart decreases rapidly with increase of distance inversely proportional to the cube of the distance. So increasing the baseline decreases the MCG signal component included in the reference signal. However, the difference of the magnetic environment noise fields between the detector and reference magnetometers becomes large because of the fluctu-

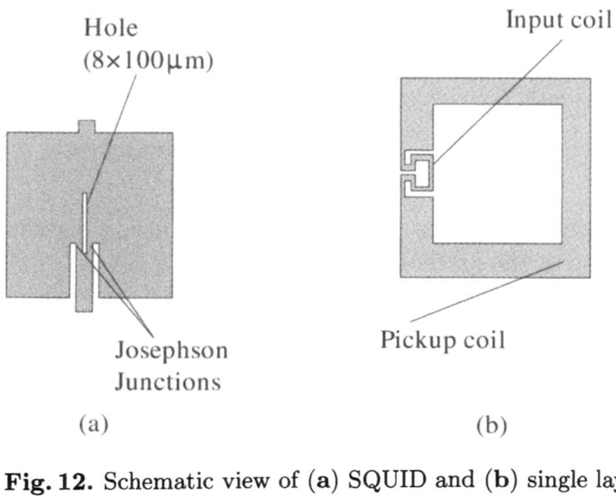

Fig. 12. Schematic view of (**a**) SQUID and (**b**) single layer flux transformer

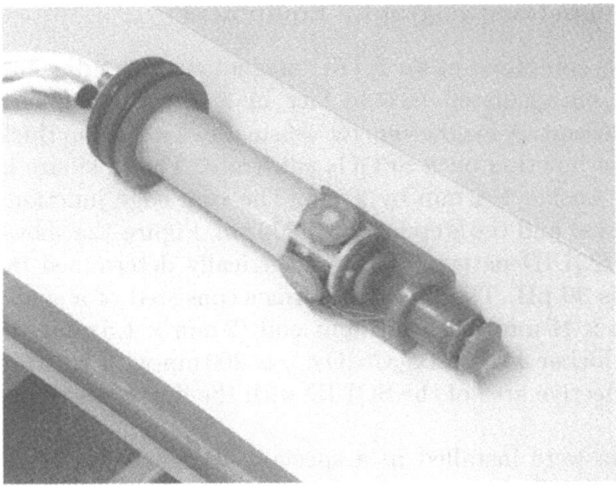

Fig. 13. Experimental set-up of four magnetometers in a specially designed suscepter

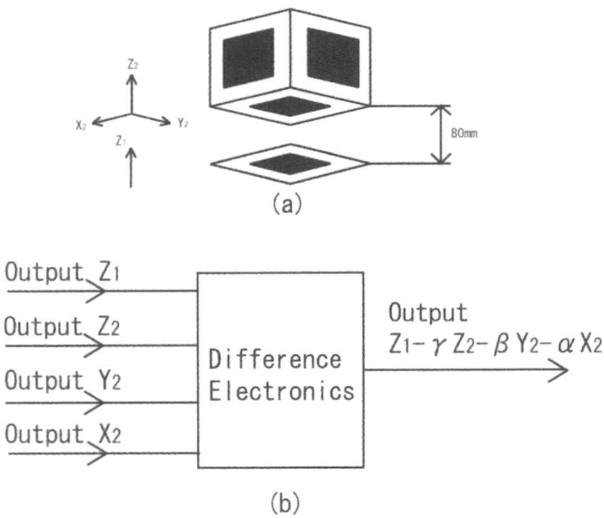

Fig. 14. Schematic of the four SQUID vector gradiometer system; (a) the gradiometers configuration and (b) the balance adjustment electronics

ation of the field. The baseline distance is determined by considering this situation. The configuration of the gradiometer is shown in Fig. 14.

The signals of the three reference magnetometers are added with coefficients adjusted such that their effective sensitive direction and sensitivity exactly compensate the sensing magnetometer. The output S of the gradiometer in a homogeneous field $\boldsymbol{B}$ can be expressed using their sensitivities η_i and their sensitive directions $\boldsymbol{x}_i$ with coefficients a_i,

$$S = \sum_{i=1}^{4} a_i \eta_i \boldsymbol{x}_i \cdot \boldsymbol{B}. \tag{1}$$

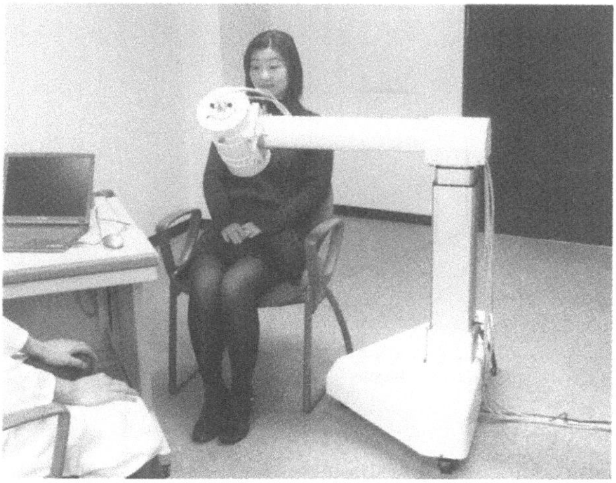

Fig. 15. Open-SQUID MCG detection system. A patient is sitting on a chair and need not lie down

Therefore if the coefficients a_2, a_3 and a_4 are adjusted to fulfill

$$a_2\eta_2\boldsymbol{x}_2 + a_3\eta_3\boldsymbol{x}_3 + a_4\eta_4\boldsymbol{x}_4 = -a_1\eta_1\boldsymbol{x}_1, \tag{2}$$

which corresponds to the representation of $-a_1\eta_1\boldsymbol{x}_1$ in the base $\boldsymbol{x}_2$, $\boldsymbol{x}_3$, $\boldsymbol{x}_4$, the balance should ideally approach infinity, meaning that an arbitrary homogeneous field $\boldsymbol{B}$ does not give a signal S. The base $\boldsymbol{x}_2$, $\boldsymbol{x}_3$, $\boldsymbol{x}_4$ formed by the reference magnetometers does not have to be orthogonal.

In a practical case the coefficients were adjusted manually to minimize the noise component of the output signal. The residual external noise components arising from any misalignment of each SQUID chip introduced unintentionally and from the special discrepancy between the top SQUID detector and the canceller were compensated in this manner. At least at present there still remains a problem of quality scatter from chip to chip. This scatter was also compensated in the same manner as stated above.

The completed SQUID holder (shown in Fig. 13) was set in a liquid nitrogen Dewar as shown in Fig. 15 where the dewar itself was 45 degree tilted for greater utility. The sizes of the dewar and total system including pedestal are ∅165 mm × 300 mm and 1000 mm × 800 mm × 1200 mm, respectively.

Figure 16 shows the noise spectra of the magnetometer and of the gradiometer in a shield-less environment. The outputs of the sensing and the reference magnetometers were filtered with a commercially available 60 Hz notch and 0.2–50 Hz band-pass filters. The gradiometer effectively suppressed the environmental noise. In this measurements, there was strong magnetic noise at 40 Hz of unknown origin. Though the adjustment was not so perfect because of the delicate operation this noise was suppressed about −27 dB. The rest of the residual noise was from the commercial electric lines; the noise

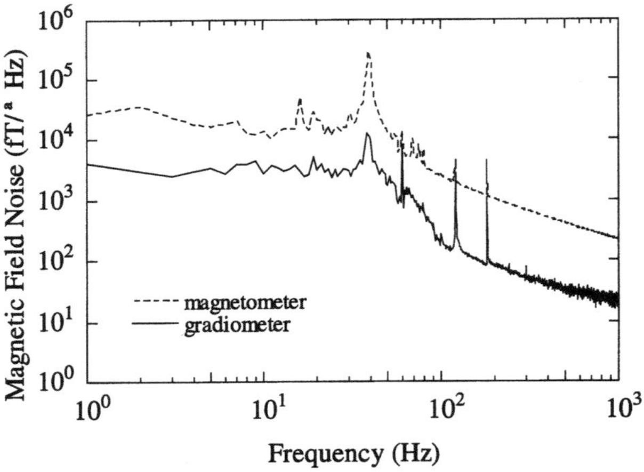

Fig. 16. Noise of the manually compensated gradiometer and of the magnetometer in an unshielded environment

spectral peaks were at 60 Hz and second and third harmonics of 120 and 180 Hz (Kansai area). Because the field produced by the commercial power lines is not homogeneous in the measurement room, the elimination ratio is somewhat smaller compared with other noise.

The raw MCG of a healthy person (a male 22 years old) taken by this Open-SQUID is shown in Fig. 17a. Because of the inhomogenity of the field from the electric power line, it could not be eliminated by first order gradiometer, even though the notch filter was used. We attempted to process the raw data by using the digital software filter. The results are shown in Fig. 17b and Fig. 17c, which are 60 Hz filtering and, 60, 120 and 180 Hz filtering traces, respectively. It is clear that the data were improved and by the software filtering.

The intrinsic noise of the individual SQUID chip (including the electronics noise) is as high as 0.5 $pT/Hz^{1/2}$. Moreover, the bubbling of liquid nitrogen may add to the noise level. These internal noises unfavorably limit the resolution at a relatively low level. However, the remaining external noise level in the Open-SQUID exceeded them. In other words, the existing internal noise is not an issue to be discussed more at least at present.

The MCG trace in Fig. 17c was smoothed by simple integration. A result of 20 s integration is given in Fig. 18, where one can see a good MCG waveform from a sound body.

3.2 Adaptive Noise Canceling Process

The adaptive filter (ADF) technique was introduced for adjustment of compensation parameters and afterwards to noise canceling [23]. The usual method of estimating a signal corrupted by additive noise is to pass the

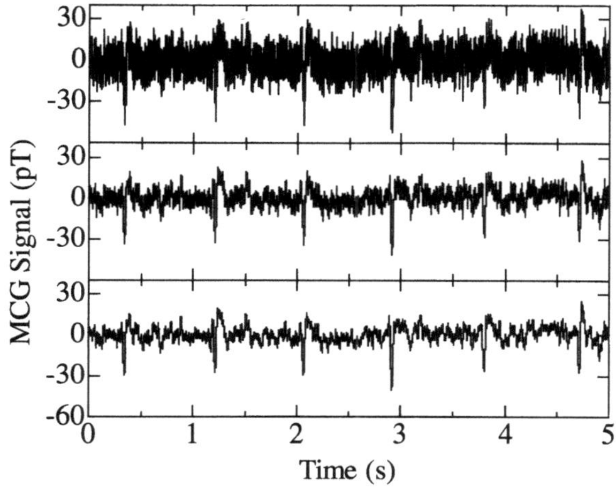

Fig. 17. MCG data taken by the completed Open-SQUID in the ordinary environment. (**a**) raw signal, (**b**) 60 Hz digital filtering and (**c**) 60, 120 and 180 Hz digital filtering

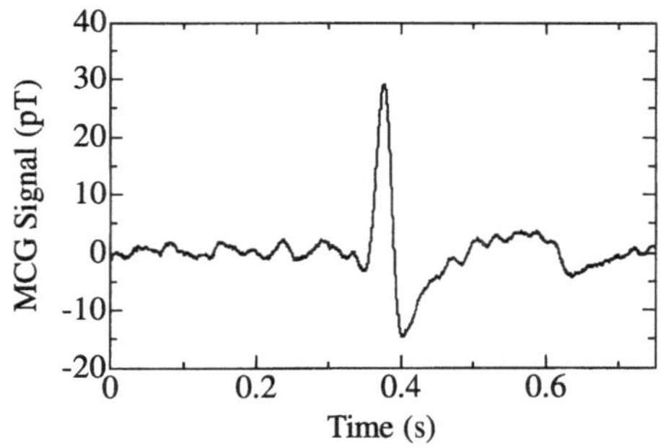

Fig. 18. Processed MCG waveform of the Open-SQUID. The data in Fig. 17c were simply integrated for 20 s

composite signal through a filter that tends to suppress the noise while leaving the signal relatively unchanged. Filters used for the foregoing purpose can be fixed or adaptive. The design of fixed filters must be based on prior knowledge of both the signal and the noise, but adaptive filters have the ability to adjust their own parameters automatically and their design requires little or no prior knowledge of signal or noise characteristics.

Noise canceling is a variation of optimal filtering that is highly advantageous in many applications. It uses an auxiliary or reference input derived from one or more sensors located at points in the noise field where the signal is weak or undetectable. This input is filtered and subtracted from a primary input containing both signal and noise. As a result, the primary noise is attenuated or eliminated by cancellation.

At first glance, subtracting the noise from a received signal would seem dangerous. If done improperly it could result in an increase in output noise power. If, however, an appropriate adaptive process controls filtering and

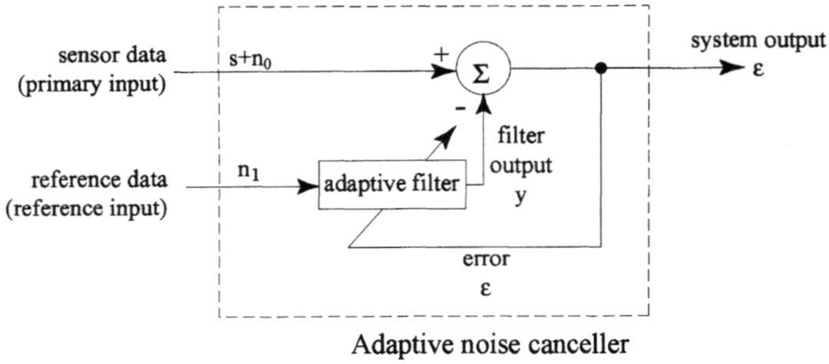

Fig. 19. Adaptive noise canceling concept

subtraction, noise reduction can in many cases be accomplished with little risk of distorting the signal or increasing the output noise level. In circumstances where adaptive noise canceling is applicable we can often achieve a degree of noise rejection that would be difficult or impossible to achieve by direct filtering.

3.2.1 The Concept of Adaptive Noise Canceling

The basic noise canceling situation is illustrated in Fig. 19. A signal is transmitted over a channel to a sensor (corresponding to Z_1 in Fig. 14) that receives the signal plus an uncorrelated noise, n_0. The combined signal and noise, $s + n_0$, form the "primary input" to the canceller. A second sensor (corresponding to Z_2, X_2 and Y_2 in Fig. 14) receives a noise n_1, which is uncorrelated with the signal but correlated in some unknown way with the noise n_0. This sensor provides the "reference input" to the canceller. The noise n_1 is filtered to produce an output y that is a close replica of n_0. This output is subtracted from the primary input $s + n_0$ to produce the system output, $s + n_0 - y$.

Let $\boldsymbol{X}_k = [x_k \, x_{k-1} \, \Lambda x_{k-L}]^T$ be a time series of reference data and $\boldsymbol{W}_k = [w_{0k} w_{1k} \Lambda w_{Lk}]^T$ be the impulse response of the adaptive filter, where L is the filter length. Then the k-th system output ϵ_k is calculated by subtracting the filter output $y = \boldsymbol{W}_k^T \boldsymbol{X}_k$ from the raw data d_k, and is expressed as

$$\epsilon_k = d_k - \boldsymbol{W}_k^T \boldsymbol{X}_k. \tag{3}$$

This ϵ_k also represents an error signal between raw data and filter output. In the system the reference data is processed by an adaptive filter that automatically adjusts its own impulse response through a least mean squares (LMS) algorithm that responds to an error signal dependent, among other things, on the filter's output. Thus with the appropriate algorithm the filter can work under a changing set of conditions and can readjust itself continuously and quickly to minimize the error signal,

$$\boldsymbol{W}_{k+1} = \boldsymbol{W}_k + 2\mu\epsilon_k \boldsymbol{X}_k, \tag{4}$$

where μ is the gain constant that regulates the speed and stability of adaptation. This canceling process can be calculated by a personal computer and the optimal adjustment parameters can be obtained without requiring manual operation.

3.2.2 MCG Measurement with Adaptive Noise Canceller

The MCG of a health person was measured with the Open-SQUID system including the adaptive noise canceller in a shield-less environment. The obtained MCG waveforms are shown in Fig. 20, where Fig. 20a was manually compensated and Fig. 20b was obtained with an adaptive noise canceller. It is clear that significant improvement of the signal-to-noise ratio of the MCG waveform was achieved.

The output Fourier spectra of the Z_1 magnetometer, electronics gradiometer and adaptive noise canceller are shown in Fig. 21. The noise level of the adaptive noise canceller is lower than those of other outputs in the low frequency region ($\sim$ 30 Hz). The high frequency noise level of the adaptive noise canceller is larger than that of the electronics gradiometer and is nearly equal to the input level because the adaptive noise canceller tends to adjust the coefficients to suppress signals in a wide power density region.

The data acquisition and the adaptive noise canceling have been done separately, namely in an off-line process, to get the data in Fig. 20. However, because of the small amount of calculation, the required time for the adaptive noise canceling process for two minutes' data is only a few seconds. Therefore, real-time processing will be possible if the signal processing algorithm is integrated in the data acquisition program of the measurement system.

3.3 Active Noise Control System for DC Fluctuations

Another problem of the MCG measurement in unshielded environments is a very slow but large fluctuations of the magnetic field. This fluctuation occasionally prevents the SQUID FLL circuit from locking and interrupts the MCG signal acquisition. Figure 22 shows an example of this DC fluctuation. Because the dynamic range of the FLL circuit corresponds to ± 170 nT, when the variation of the DC level exceeds the dynamic range no signal is obtained except for a DC value. The signal in Fig. 22 is obtained with no filtering process such as high-pass filter or notch filter. In the case of measurement with a high-pass filter, the obtained signal will vary near zero level, but if the magnetic field exceeds FLL dynamic range, the output signal will settle and shows exactly zero. In any case, the sequential measurement is discontinued.

To overcome this situation an active noise controller using Open-SQUID system is proposed. The system setup is depicted in Fig. 23. The FLL output of the reference SQUID Z_2 returns to both feedback coil 1 and 2 to counteract the magnetic field. Thus a magnetic negative feedback loop is constructed.

The output signal of the sensor SQUID Z_1 with working active noise control system is shown in Fig. 24. The DC fluctuation range is well-suppressed

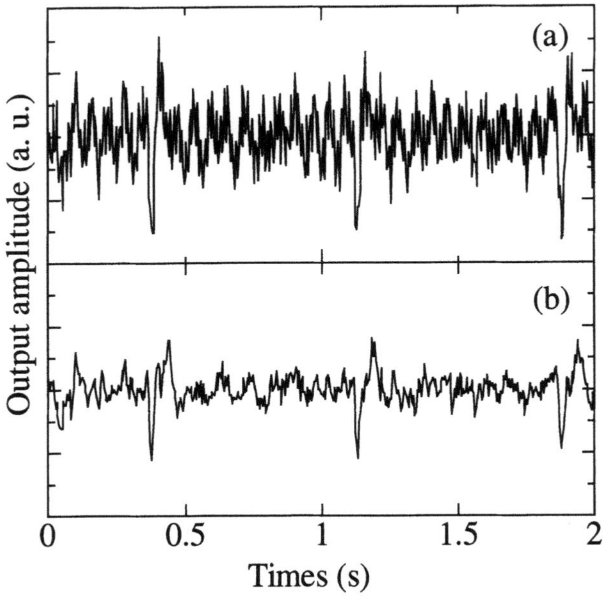

Fig. 20. The unaveraged signal sequences of the electronic gradiometer. (**a**) raw data and (**b**) processed data through an adaptive filter (noise canceling)

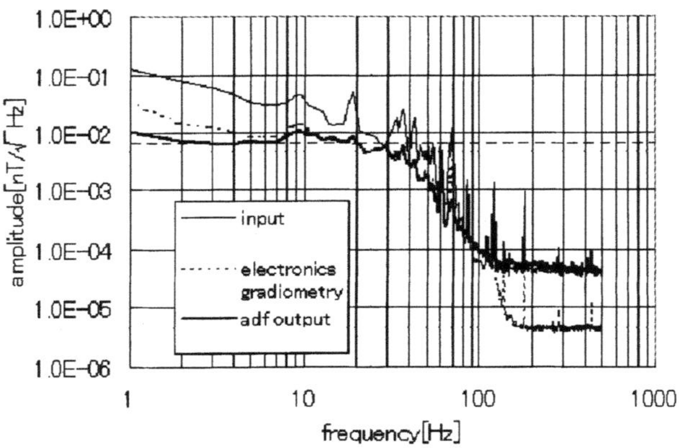

Fig. 21. Fourier spectra of magnetometer Z_1 (input), electronics gradiometer and adaptive noise canceling (adf output). The *thin dashed line* denotes a noise level of a 7 $pT/Hz^{1/2}$

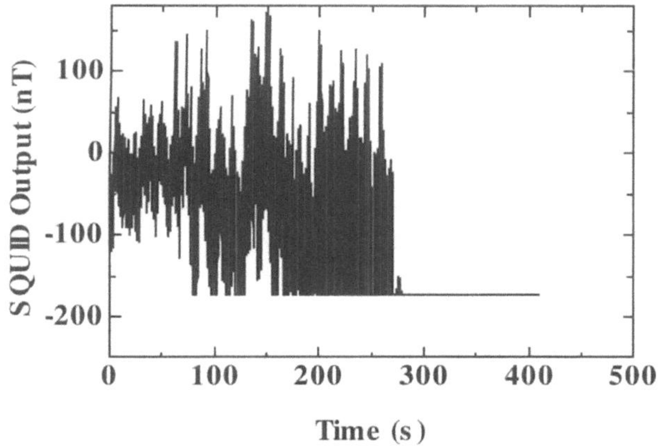

Fig. 22. The DC level fluctuation of a magnetic field from the environment

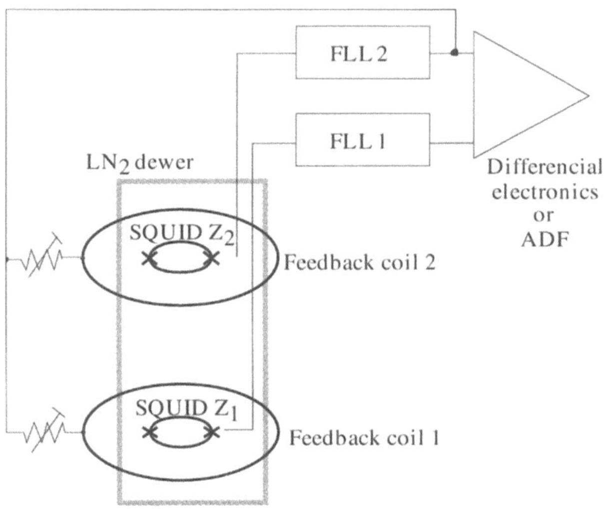

Fig. 23. Schematic of the active noise control attached to the Open-SQUID system. The final noise-canceled signal is produced by the differential electronics or the adaptive noise canceller

compare with a normal output shown in Fig. 22. Moreover, the magnetic noise from the electric power line at 60 Hz is reduced by this negative feedback circuit. This active noise control system provides a comfortable measurement environment during which long time date acquisition can be done without a interruption. It is worth noting that the effect of suppressing the magnetic noise from the power line is effective especially for the adaptive noise canceller. Because the adaptive noise canceling algorithm has the characteristic that it suppresses only the high power signal if it exists, and other low power

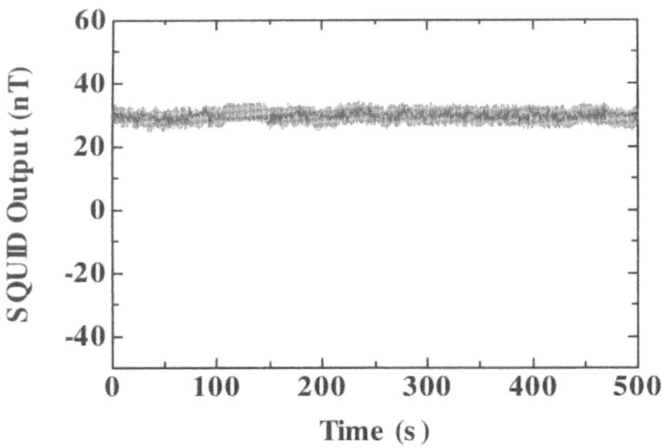

Fig. 24. SQUID output signal with active noise control system under shield-less conditions. No filter was used at measurement

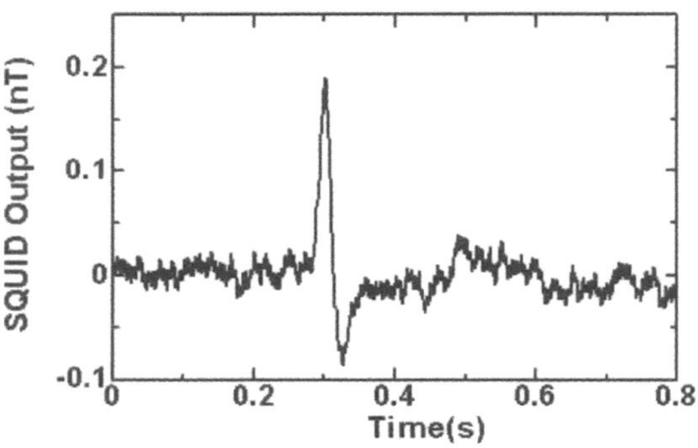

Fig. 25. MCG signal using the Open-SQUID with adaptive noise canceller and active noise control. The averaging time is 25 s

signals are not processed, reducing the high power noise from the power line by the active noise controller makes the adaptive noise canceller process the low power noise region.

Finally, Fig. 25 shows the MCG waveform obtained from the Open-SQUID magnetocardiography system with both the adaptive noise canceller and the active noise control circuit working under shield-less conditions. A clear MCG signal was obtained.

3.4 Summary

A shield-less HTSC SQUID magnetocardiography called the "Open-SQUID" system using four magnetometers has been proposed and developed. Its high flexibility of signal processing and of combination with each SQUID output signal processing offers various noise reduction techniques. With ordinary digital filtering and simple integration for only a 20 s interval a smooth MCG curve was obtained in an unshielded environment. The adaptive noise canceller made the adjustment of subtraction coefficients automatically. The active noise control system works as a magnetic negative feedback loop suppressing the DC fluctuation of the environmental magnetic field. As a result long term data acquisition was possible.

4 Biological Immunoassays

A highly sensitive SQUID magnetometer has been applied to biological immunoassays [24,25,26,27,28]. In the immunoassays, an antigen, such as pathogenic bacteria, cancer cell or environmental injuries, is detected with its antibody. In order to magnetically detect the binding reaction between antigen and its antibody, the antibody is labeled with a magnetic marker made of magnetic nanoparticles. The binding reaction can be detected by measuring the magnetic field from the magnetic marker. In this magnetic method, it is important to measure a very weak magnetic signal in order to detect a very small quantity of the antigen since the signal field is proportional to the quantity of antigen. A highly sensitive SQUID magnetometer is very useful for this purpose.

4.1 Measurement Principle

Until now two methods have been developed for SQUID immunoassays. One is to measure the Néel relaxation of the magnetic nanoparticles after a pulse field is applied [24,25,26]. The other is to measure the magnetic field produced by the magnetization of the magnetic nanoparticles when a static field is applied [27,28]. In this paper we present a system configuration for the latter method and show some experimental results of the biological immunoassays.

In Fig. 26a, the measurement principle of the binding reaction between an antigen and its antibody is schematically shown. Here the antibody is labeled with a magnetic marker. The magnetic marker that consists of γ-Fe_2O_3 nanoparticles is embedded in the core made of polymer and the antibody is attached around the surface of the core. The diameter of γ-Fe_2O_3 is usually reduced down to 10 nm in order not to have remanance, i.e., to be superparamagnetic. The diameter of the core is typically 50 nm (MACS; Miltenyi Biotec, Germany).

The sample to be measured is an assembly of antibodies labeled with the magnetic nanoparticles. The magnetic field B_{ex} is applied in parallel to the SQUID magnetometer in order to magnetize the nanoparticles, as shown in Fig. 26b. Then a signal field B_s is produced from the nanoparticles [29,30,31]. The vertical component of the signal field is collected with the pickup coil of the SQUID and is detected as a signal flux Φ_s. Since the signal flux Φ_s

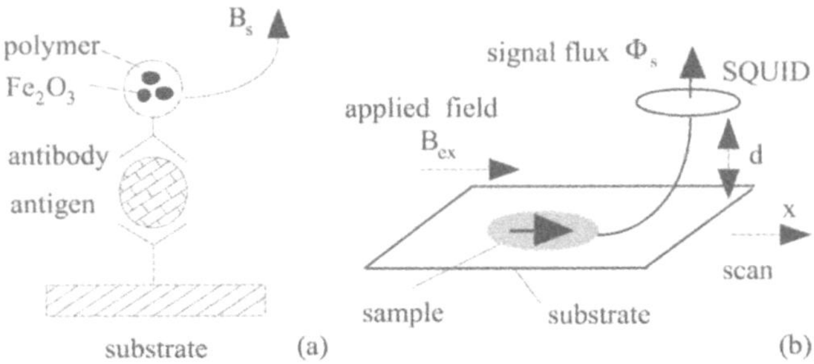

Fig. 26. Principle of immunoassay with the SQUID magnetometer. (**a**) Binding reaction between an antigen and its antibody. The antibody is labeled with the magnetic marker. (**b**) Schematic diagram of the measurement system

is proportional to the quantity of antigen, we can magnetically measure the binding reaction between an antigen and its antibody.

A theoretical expression for the magnetic flux Φ_s detected by the SQUID is given by

$$\Phi_s = K(d)\mu_0 mN \quad \text{with} \quad \mu_0 m \approx 3B_{ex}\frac{4\pi}{3}\left(\frac{a}{2}\right)^3 , \tag{5}$$

where μ_0 is the permeability of vacuum, N is the total number of the markers in the sample, a is the diameter of the marker and m is the magnetic moment of the single marker when the external field B_{ex} is applied. The moment m depends on both the applied field B_{ex} and the volume of the marker, but depends little on the relative permeability μ_s of Fe_2O_3 for the case $\mu_s \gg 1$.

As can be seen in (5), the signal flux increases in proportion to B_{ex} and a^3. Therefore, we can increase the signal flux Φ_s by increasing the applied field B_{ex}. A large marker size, a, is also preferable to increase Φ_s though the maximum size of a will be limited in the application. The value $K(d)$ in (5) represents the efficiency of collecting the magnetic flux produced from the nanoparticles. It depends on the distance d between the sample and the SQUID as well as the sizes of the sample and the pickup coil of the SQUID.

4.2 Measurement System

In Fig. 27a a measurement system is schematically shown [32]. The system consists of the SQUID magnetometer, field and compensation coils to apply the magnetic field, a Dewar to set the SQUID close to the room temperature sample, a motor-driven slider to move the sample and cylindrical magnetic shields with a shielding factor of 100. In the following we explain the main components of the system.

(a) Dewar. The signal flux Φ_s detected by the SQUID depends strongly on the distance d between the SQUID and the room temperature sample. In

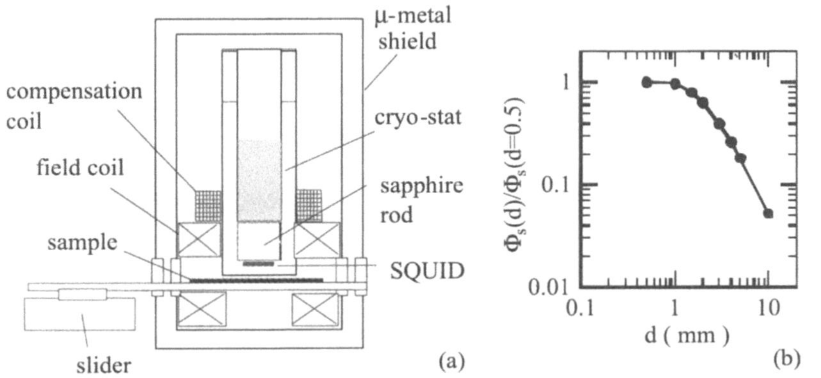

Fig. 27. (**a**) Schematic diagram of the high T_c SQUID system for immunoassays. (**b**) Relationship between the magnetic signal Φ_s and the sample to SQUID distance d. The *solid line* shows the calculated result of $K(d)$ in (5)

Fig. 27b calculated results on the relationship between Φ_s and d are shown. In the calculation we assume the case where the magnetic nanoparticles with $N = 10^8$ and $a = 50$ nm are uniformly placed in an area of 3 mm × 4 mm. The magnetic field from each nanoparticle with moment m is calculated and the magnetic flux Φ_s that interlinks the pickup coil of 3 mm × 6 mm is calculated. As shown in Fig. 27b, the signal flux Φ_s decreases with increasing distance d. However, the decrease is small when $d < 2$ mm. Therefore, we should choose $d < 2$ mm in the present case.

In order to realize the distance $d < 2$ mm we use a Dewar similar to that used in the SQUID microscope [33,34]. The SQUID is mounted in vacuum at the end of a sapphire rod that is thermally anchored to a can of liquid nitrogen. A thin sapphire window with a thickness of 0.3 mm separates the SQUID and the room temperature sample and maintains the vacuum. The distance between the SQUID and the sapphire window is adjusted to be less than 1 mm. As a result, the distance between the sample and the SQUID is typically 1.5 mm in the present experiment.

(b) SQUID Magnetometer. In Fig. 28a a direct-coupled SQUID magnetometer fabricated with 30 degree bicrystal junctions is schematically shown [35,36]. The outer size of the pickup coil is 6 mm × 3 mm. The pickup coil consists of 4 parallel loops. The line width of each loop is 160 μm, and a flux dam is inserted in each loop, i.e., the line width of the loop on top of the grain boundary of the substrate is reduced to 20 μm. The effective area of the magnetometer is $A_{\text{eff}} = 0.12$ mm^2 for a uniform magnetic field.

In Fig. 28b the flux noise spectra of the magnetometer are shown for three cases. The symbol (a) in the figure represents the flux noise when the SQUID is measured in liquid nitrogen inside three μ-metal and two cryo-parm shields. Therefore, the result denoted by (a) represents the intrinsic noise of

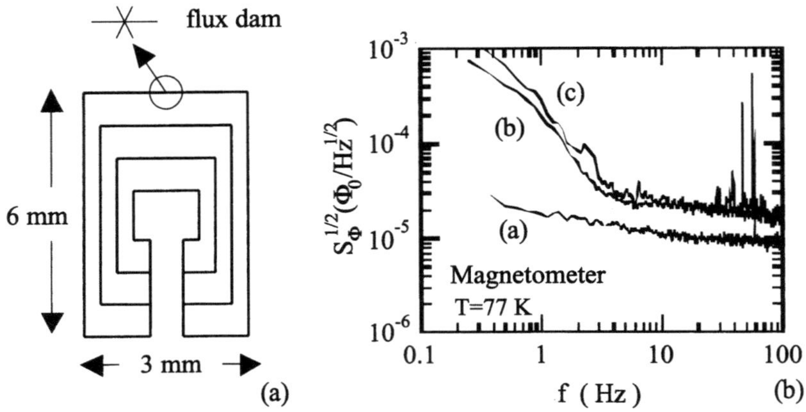

Fig. 28. (**a**) Schematic figure of the high T_c SQUID magnetometer. (**b**) Noise performance of the magnetometer system. Three results measured under different conditions are shown. The *symbols* (a), (b) and (c) in the figure denote the measurement conditions (see text)

the SQUID magnetometer. As shown, the flux noise is $S_\Phi^{1/2} = 10\mu\Phi_0/\mathrm{Hz}^{1/2}$ in the white noise region. The noise becomes white down to a frequency of $f = 5\,\mathrm{Hz}$. Below $f = 5$ Hz the noise slightly increases with decreasing a frequency and becomes $S_\Phi^{1/2} = 23\mu\Phi_0/\mathrm{Hz}^{1/2}$ at $f = 1\,\mathrm{Hz}$.

The result denoted by (b) represents the flux noise when the magnetometer is mounted on the sapphire rod of the Dewar. In this case the noise increases compared to case (a). Especially we observe a large increase of the noise in the low frequency region. For example, the flux noise of $S_\Phi^{1/2} = 200\mu\Phi_0/\mathrm{Hz}^{1/2}$ is measured at $f = 1$ Hz in case (b). This value is about 10 times larger than case (a). This flux noise corresponds to the field noise of $S_\mathrm{B}^{1/2} = S_\Phi^{1/2}/A_\mathrm{eff} = 3.5\ \mathrm{pT/Hz}^{1/2}$. This large low frequency noise will be due to the residual environmental noise: since the shielding factor in case (b) is estimated to be 100, residual environmental noise will remain.

The results denoted by (c) represent the flux noise when an external field of $B_\mathrm{ex} = 0.8$ mT is applied with the field and compensation coils and the motor-driven slider switched on, i.e., under practical measurement conditions. As shown, the noise spectra are almost the same for cases (b) and (c). This result means that the field coil and the slider do not increase the system noise. From the noise spectrum denoted (c) we can estimate the performance of the present system. Since the measurement bandwidth is $0.2\,\mathrm{Hz} < f < 5\,\mathrm{Hz}$, we can estimate the amplitude of the noise flux Φ_n by integrating the noise spectrum S_Φ in this region. From the result denoted (c), we estimated the minimum detectable amplitude of magnetic flux as $\Phi_\mathrm{n} = 0.6m\Phi_0$.

As shown in Fig. 28b, the system noise is worse than the intrinsic noise of the SQUID, especially in the low frequency region due to the residual en-

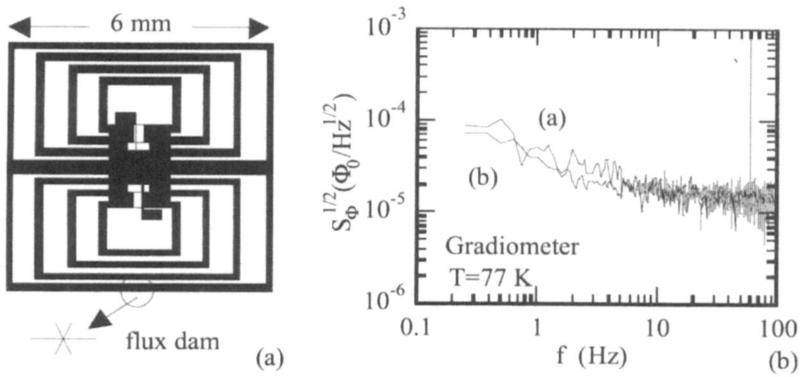

Fig. 29. (**a**) Schematic figure of the direct-coupled high T_c SQUID gradiometer. (**b**) Noise performance of the gradiometer system. The *symbols* (a) and (b) in the figure denote the measurement conditions (see text)

vironmental noise. In order to reduce the effect of the residual environmental noise, we replace the magnetometer by a single-layer gradiometer as shown in Fig. 29a. In Fig. 29b the flux noise spectrum of the system is shown when the gradiometer is mounted on the sapphire rod of the Dewar. The symbols (a) and (b) correspond to the cases of $B_{\mathrm{ex}} = 0$ and $B_{\mathrm{ex}} = 0.8$ mT, respectively. As shown, the low frequency noise is much improved in this case. The flux noise at $f = 1$ Hz is $S_\Phi^{1/2} = 40\mu\Phi_0/\mathrm{Hz}^{1/2}$. From the noise spectrum we can estimate the minimum detectable amplitude of the flux as 0.12 $m\Phi_0$ when the measurement bandwidth is $0.2\,\mathrm{Hz} < f < 5\,\mathrm{Hz}$. This value is 5 times smaller than that obtained in the case of the magnetometer. Therefore, we can improve the system noise by using the gradiometer in the present case.

(c) Field and Compensation Coils. In order to magnetize the magnetic nanoparticles, an external field B_{ex} is applied in parallel to the SQUID with the field coil. Since the signal field B_{s} produced from the nanoparticles is proportional to B_{ex}, it is desirable to apply a large field B_{ex}. If the field B_{ex} is applied perfectly in parallel to the SQUID, the field does not degrade the SQUID performance [29,30]. However, it must be noted that small vertical component of B_{ex} couple to the SQUID due to imperfect mechanical alignment. This vertical component of B_{ex} degrades the SQUID performance. For example, if the mechanical misalignment is 0.5% and an external field of $B_{\mathrm{ex}} = 1$ mT is applied, the vertical component of 5 μT couples to the SQUID magnetometer. In this case it can be calculated that a circulating current of $I_{\mathrm{cir}} = 5$ mA flows in the outermost loop of the pickup coil in the present magnetometer shown in Fig. 28a. This value is large enough to open the flux dam that is inserted in the pickup coil, and flux entry through the flux dam occurs. Since the flux dam needs a long time for completion of the flux entry, we have to wait a long time for the start of the measurement, e.g., 10 min [37].

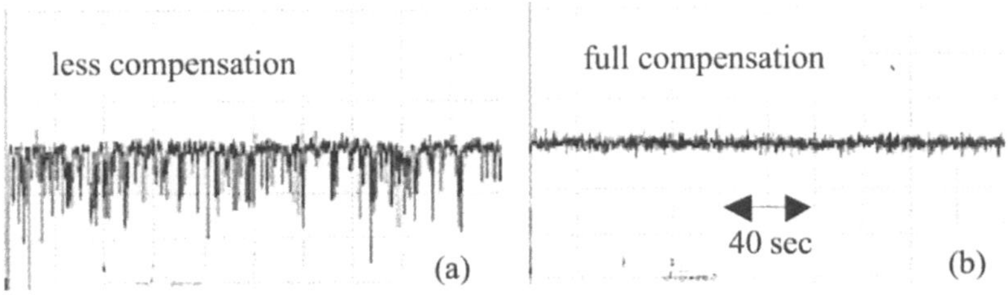

Fig. 30. Effect of the compensation coil. (**a**) Output of the FLL when the vertical component of B_{ex} is not sufficiently compensated. (**b**) Output of the FLL when a compensation coil is used to cancel the vertical component of B_{ex}. The output of the FLL is band passfiltered from $f = 0.2$ Hz to $f = 5$ Hz

In Fig. 30a the output of the FLL, which is measured before completion of the flux entry through the flux dam, is shown. Here the output of the FLL is band passfiltered from 0.2 Hz to 5 Hz. As shown, many voltage pulses corresponding to the flux entry are observed for long a time. During this time we cannot start the measurement.

Therefore it is important to compensate the vertical component in order to decrease the circulating current due to the vertical component. For this purpose, a compensation coil is added as shown in Fig. 27a. Using this compensation coil, the field that couples to the SQUID can be reduced down to 10 nT when the external field is $B_{ex} = 1$ mT. In Fig. 30b the output of the FLL is shown when the compensation is done. As shown, no voltage pulse is observed. This result can be compared with that shown in Fig. 30a when the compensation is not sufficient.

As the applied field B_{ex} becomes higher, the requirement on the compensation becomes more serious. In order to relax the requirement on the compensation, we also developed an electrical switch that can control the opening and the closing of the pickup coil. This switch can be used instead of the flux dam. The circulating current I_{cir} due to the vertical component of B_{ex} can be reduced quickly to zero by opening the pickup coil with the switch. As a result, we can avoid voltage pulses due to flux entry even when the compensation of the vertical component of B_{ex} is insufficient. Details of the switch are described in Sect. 6.2.

4.2.1 Experimental Results

(a) Measurement of Magnetically Labeled Antibody. Using the present system, we first measured the magnetic field from a solution of antibodies that are labeled with magnetic marker (MACS, Miltenyi Biotec, Germany), as shown in Fig. 26a. The marker consists of γ-Fe_2O_3 nanoparticles that are embedded in the core made of polymer, and the antibody is attached around the surface of the core. The diameter of the core is typically 50 nm. In Fig. 31a, a typical waveform measured with the present system is shown

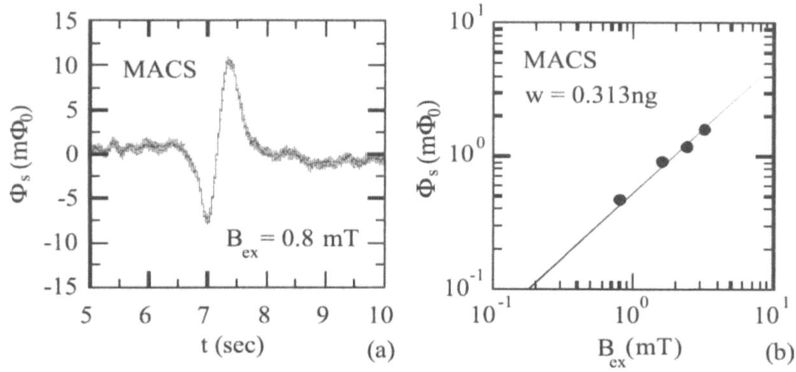

Fig. 31. (**a**) Measured waveform of the detected signal Φ_s when the sample is moved with a speed of 10 mm/s. (**b**) Relationship between the detected signal and the applied field B_{ex}

when an external field of $B_{ex} = 0.8$ mT is applied. Since the sample is moved in the x direction with a speed of 10 mm/sec as shown in Fig. 26b, we can obtain a large signal when the sample is passed beneath the magnetometer, as shown in Fig. 31a. In the measurement, the output voltage of the FLL circuit is band passfiltered from $f = 0.2$ Hz to $f = 5$ Hz in order to reduce the external noise. The peak value of the output signal in Fig. 31a gives the signal flux Φ_s from the nanoparticles.

In Fig. 31b the relationship between the detected signal Φ_s and the external field B_{ex} is shown. As expected, the signal flux Φ_s increases in proportion to the external field B_{ex}. Therefore it is effective to use a large external field B_{ex} in order to increase the signal flux Φ_s. At present, however, the system noise increases a little (about 1.5 times) when the field is increased up to $B_{ex} = 3.2$ mT. This increase in noise will be caused by the vertical component of the external field, and hence, will be solved by improving the compensation coil or by using the switch mentioned in Sect. 6.2.

In Fig. 32 the detected flux Φ_s is shown when the concentration of the solution is changed for different magnetic markers. Here the horizontal axis represents the total weight of the marker, i.e., the sum of the weights of the polymer and the magnetic nanoparticles. Circles show the result when the marker called MACS is used and an external field of $B_{ex} = 0.8$ mT is applied. As shown, good linearity is obtained between the magnetic signal and the weight of the markers in the solution. In this case the minimum detectable weight of the markers is 300 pg.

The open triangles in Fig. 32 show the results when the marker called Nanomag 250 (Micromod, Germany) is used. This marker is large compared to the MACS and its diameter is $a = 250$ nm. When an external field of $B_{ex} = 0.8$ mT is applied, the detected signal is much larger than the case of MACS. This difference will be caused by the different magnetic properties of

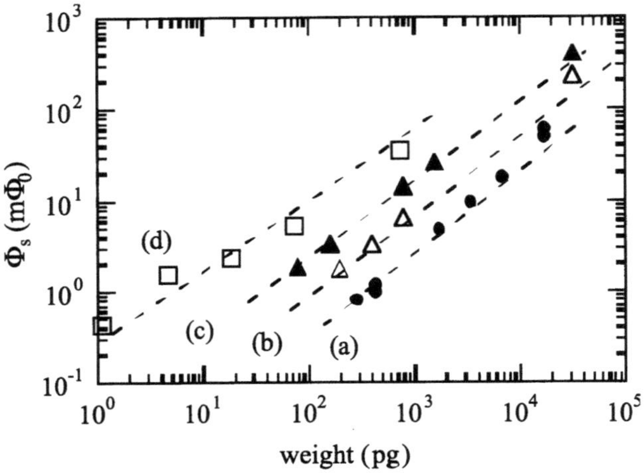

Fig. 32. Relationship between the detected signal Φ_s and the weight of the magnetic nanoparticles for different markers. The symbols show the markers: (**a**) MACS with $B_{ex} = 0.8\,mT$, (**b**) Nanomag 250 with $B_{ex} = 0.8\,mT$, (**c**) Nanomag with B_r and (**d**) Toda 25 with B_r (see text)

the nanoparticles used for the markers. Therefore, there remains much room for optimization of the magnetic nanoparticles for the present purpose.

The marker called Nanomag 250 can keep a remnant magnetic moment when a field of 0.1 T is applied, unlike usual markers. Therefore, we use the remnant field B_r to detect the magnetic marker. In Fig. 32 closed triangles show the magnetic signal from the remnant moment when the weight of the markers is changed. As shown in Fig. 32, the magnetic signal becomes much larger when we use the remnance. Although we cannot treat less than 70 pg of marker at present due to the difficulty of sample preparation, it will be possible to detect much smaller amounts of the marker.

It is known that the remnance disappears exponentially due to thermal noise when the diameter of the nanoparticles becomes less than 10 nm. If the diameter is larger than 10 nm, on the other hand, remnance begins to appear. In order to check this size effect, we measured the remnant magnetic moment of Fe_3O_4 nanoparticles with $a = 10\,nm$ and 25 nm (TODA, Japan). An external field of 0.1 T is applied with a permanent magnet and the remnant magnetic moment is measured. As expected, we observed a little remnance for the case of $a = 10\,nm$, while a large remnant magnetic moment of 9.2 emu/g is obtained for the case of $a = 25\,nm$.

In Fig. 32 rectangles show the magnetic signal Φ_s from the remnant moment when the weight of the Fe_3O_4 nanoparticles with $a = 25\,nm$ is changed. As shown, we obtain a large magnetic signal and can detect nanoparticles as small as 1 pg. Therefore, it is strongly desired to develop marker that is optimum for the present purpose.

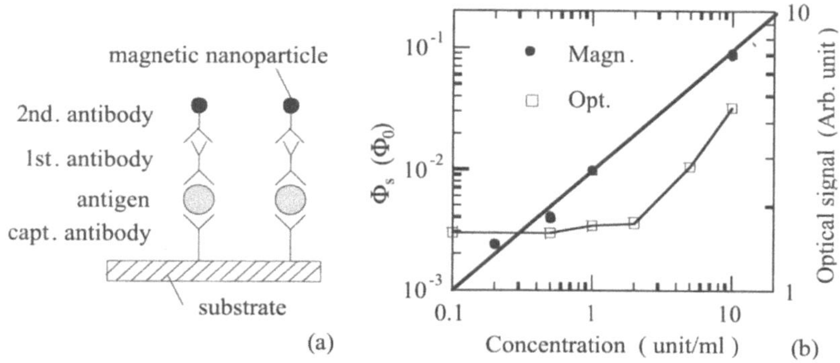

Fig. 33. (**a**) Schematic diagram of the immunoassay using the magnetic marker. (**b**) Relationship between the detected signal Φ_s and the concentration of antigen. The result is compared with that of the conventional optical method

(b) Measurement of Antigen-Antibody Binding Reaction. Using the present system, we conducted an experiment to detect biological antigen-antibody reactions. The sample used in the experiment is schematically shown in Fig. 33a, which was prepared using the following standard procedure. First, a glass slide (Nalge Nunc Int., USA) was coated with capturing antibody (mouse anti-human interferon β monoclonal antibody: Yamasa, Japan) in an area of diameter 8 mm. Next, serially diluted antigen (human interferon b: Pestka Biomed, USA) was added and incubated at 37°C for 3 h. The amount of solution including the antigen is 200 μl. Then, primary detecting antibody (rabbit anti-human interferon β polyclonal antibody: PAESEL, Germany) was added and incubated at 37°C for 1 h. Finally, secondary detecting antibody labeled with magnetic nanoparticles (MACS: goat anti-rabbit IgG) was added and incubated at 37°C for 1 h. The magnetic field from the magnetic nanoparticles was measured using the present system. The scanning speed of the slider is increased to 20 mm/s in the present experiment since the size of the sample is large, i.e., 8 mm in diameter.

In Fig. 33b closed circles show the relationship between the magnetic signal Φ_s and the quantity of the bound antigen. The vertical axis on the left shows the magnetic flux detected by the SQUID magnetometer. The horizontal axis shows the concentration of the bound antigen (interferon β) in units/ml. Here, one unit/ml of interferon means the quantity necessary to produce a cytopathic effect of 50%. As shown, reasonable linearity is obtained between the magnetic signal and the concentration. At present we can detect antigen-antibody reaction at concentrations as low as 0.2 unit/ml.

For comparison, antigen-antibody reaction was also measured using the conventional optical method, i.e., with an analyzer called ELISA. In this case, peroxidase labeled goat anti-rabbit IgG (Bio-Rad, USA) was used as the secondary detecting antibody for optical detection. After the antigen-antibody reaction was terminated, tetramethylbenzidine (TMB) was added in order to produce fluorescence from the peroxidase attached to the secondary detecting antibody. The degree of reaction was measured by detecting the amount of fluorescence.

In Fig. 33b open rectangles show the results, where the vertical axis on the right represents the strength of the optical signal. As shown, a linear response of the optical detector is obtained above concentrations of 2 unit/ml. Below 2 unit/ml concentration, the optical response becomes almost independent of the concentration due to the noise of the optical system. Therefore, the sensitivity of the optical method is limited to about a 2 unit/ml concentration.

As shown in Fig. 33b, the sensitivity of the present system is about 10 times better than that of the optical method. Since the sensitivity of the present system is not limited by the noise of the SQUID, we can expect that the sensitivity of the present system can be improved by more than one order of magnitude when the system is sophisticated, as discussed before. If the sensitivity is improved, we can expect to measure antigen-antibody reaction at concentrations as low as 0.01 unit/ml.

4.2.2 Conclusion

A high T_c SQUID system is developed for application to biological immunoassay. In this application, Fe_2O_3 nanoparticles are used as the magnetic marker in order to label the antibody, and the binding reaction between an antigen and its antibody is detected by measuring the magnetic field from the marker. The design and setup of the system are described. The minimum detectable amplitude of the magnetic fluxes are $0.6\,\mathrm{m}\Phi_0$ and $0.12\,\mathrm{m}\Phi_0$ for the measurement bandwidth from 0.2 Hz to 5 Hz when we use the magnetometer and the gradiometer, respectively. The system noise does not increase when a magnetic field of 1 mT is applied in parallel to the SQUID. An experiment to measure the antigen-antibody reaction shows that the sensitivity of the present system is 10 times better than that of the conventional method using an optical marker. Since the system performance strongly depends on the magnetic properties of the nanoparticle, it is important to develop a magnetic marker suitable for the present system, e.g., nanoparticles having remnance. It will be easy to develop a system that is 100 times more sensitive than the conventional optical method.

5 Monitoring Environmental Magnetic Field Related to Earthquakes

Recently SQUIDs are being applied to geophysical measurements such as transient electromagnetic geophysical exploration [38,39], archeometry [40], investigation of shallow subsoil [41], detection of seismic shaking [42] and environmental magnetic field measurement [43]. It is expected that the extraordinary high sensitivity of SQUIDs bring forth a breakthrough in these applications. However, the measurement of magnetic field from the point of view of geophysical application needs special considerations about the SQUID system. This is one of the barriers for use of SQUIDs in field experimentation. Efforts to overcome these difficulties have been started. In this section, measurement of the environmental magnetic field related to earthquakes is described.

5.1 ULF Variation as a Precursory Phenomenon of an Earthquake

Seismic activities and the volcanic activities are assumed to accompany an abrupt change of stress distribution in the crust of the earth. It has been reported that various changes of the gravity, earth magnetic field, level and chemical composition of underground water, etc. are induced by changes in the crust [44]. Variations of the underground electromagnetic field and ground resistivity are a possible field of application of magnetic field measurement using SQUIDs for the purpose of investigation of crustal change.

Prediction of the earthquake and volcanic eruption is required from the viewpoint of mitigation of disasters. Many precursory phenomena of earthquake and volcanic eruption have been observed and electromagnetic phenomena are expected to be a candidate for the predictor [45,46,47]. In 1982, anomalous electromagnetic field changes in the low frequency (LF) and very low frequency (VLF) bands were reported as precursory phenomena of an earthquake [48]. It seems very probable that precursory anomalous changes also appear in the ultra low frequency (ULF) and the extra low frequency (ELF) bands. Higher frequency electromagnetic radiation generated deep in the earth's crust should be damped due to skin effects. Thus, monitoring of ULF variations is considered to be more effective for the investigation of earthquake occurrence because we can expect to detect anomalous variations at greater distances from the hypocenter.

There is a report by *Fraser-Smith* et al. about the obvious precursor ULF magnetic field anomaly measured prior to the occurrence of a large earthquake (the magnitude 7.1 Loma Prieta earthquake of 17 October 1989) [49]. The measurement system with an induction coil was set about 7 km from the epicenter. A substantial increase in the noise background started about 10 days before the earthquake and an exceptionally high level of activity, about $60\,\mathrm{nT}/\,\mathrm{Hz}^{1/2}$ at most, in the range 0.01–0.5 Hz started about three hours before the earthquake. Precursor phenomena in the ULF band were also observed related to the Spitak earthquake ($M_\mathrm{s} = 6.9$) on 7 December 1988 [50] and the Guam earthquake ($M_\mathrm{s} = 7.1$) on 8 August 1993 [51]. Associated with volcanic eruption, *Fujinawa* et al. observed anomalous underground electric field variations of the ULF band a few days prior to, and a month after, a minor volcanic eruption of Mt. Mihara on Izu-Oshima island in Japan [52]. Recently, electric field and magnetic field variations were observed before of the eruptions on 18 and 26 August 2000 at Miyake Island in Japan [53,54].

An anomalous electromagnetic radiation preceding the Ito seismic swarm of 1989 was observed by underground electric field measurement [55]. It is remarkable that the anomalous radiation was observed associated with a rather small magnitude earthquake compared with Loma Prieta earthquake. The anomalous electric field variations were also observed before the swarm starting at Hodaka on 7 August 1998 [56]. An example of the anomalous variations is shown in Fig. 34a. The figure shows that the anomalies appeared in the frequency range of DC (0–0.7 Hz), ULF (0.07–0.7 Hz) and VLF (1–9 kHz). Similar variations observed before and after the earthquake of magnitude 4.1 on 12 August 1998 are also shown in Fig. 34b.

Reports have been accumulated to provide evidence showing that particular anomalous ULF variations correlate with earthquakes and volcanic eruption occurrence. However, we need to elucidate the generation mechanism of the ULF variations in order to use them with confidence for the prediction of earthquakes or eruptions. There are numerous hypotheses on the generation mechanism [57,58,59]. *Fujinawa* et al. presented a hypothesis taking into account recent electromagnetic field observations, laboratory experiments on micro-cracks in rock [60] and theoretical analysis [61]. They stated that underground water is concerned in generation through electrokinetic interaction between fluid motion and electric current [62]. An electromagnetic field can be generated by ion transport in fluids. The coupling between fluids and electric current is described by (6).

$$\begin{bmatrix} I_{\mathrm{e}} \\ J_{\mathrm{f}} \end{bmatrix} = \begin{bmatrix} L_{11} & L_{12} \\ L_{21} & L_{22} \end{bmatrix} \begin{bmatrix} -\nabla \Phi \\ -\nabla P \end{bmatrix} , \tag{6}$$

where I_{e} is the electric current density, J_{f} the fluid volume velocity, Φ the electric potential, P the pore pressure and L_{ij} the generalized coefficients of conductivity [61]. They developed equations for the transient electromagnetic field by introducing Maxwell's equations into (6). It was shown by means of the formation that the typical pulse-like signals in the ULF bands associated with the seismic swam can be explained by confined water pressure change assuming a simple conductivity structure model.

A verification of the hypothesis is to fined the generator of the anomalous ULF radiation by direct inversion from measured electromagnetic signal data. Magnetic field measurement at multiple points may provide useful data for generator estimation based on the analogy of biomagnetic source localization [63].

5.2 Requiremente on the Measurement System of ULF Radiation

A more sensitive magnetic sensor may be helpful to further study of the generator of ULF activity. A sensitivity better than 10 pT is desired [54]. A SQUID has the merits of higher sensitivity, small size and wide frequency range operation compared with a fluxgate magnetometer and induction coil

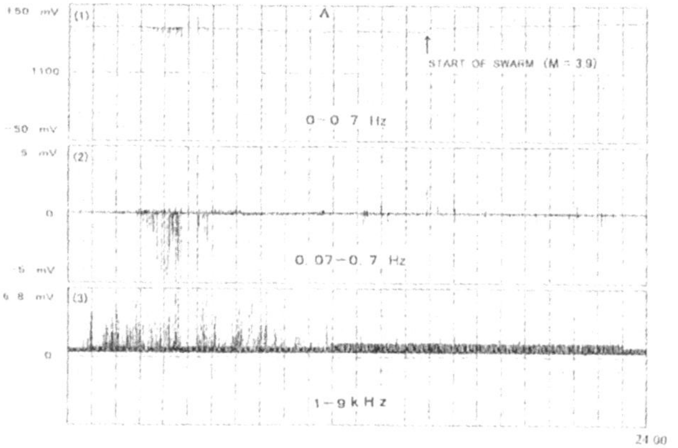

(a)

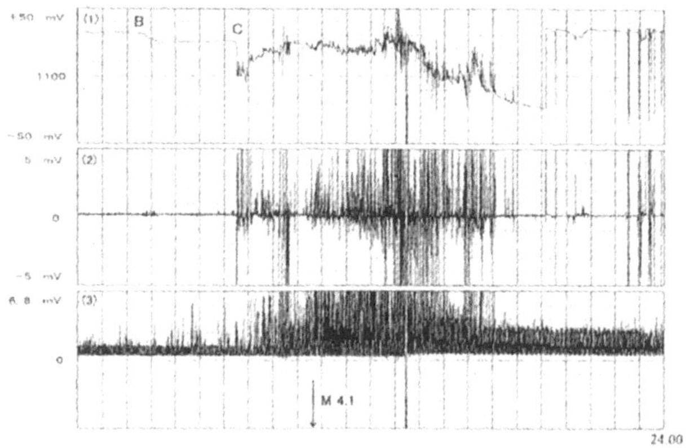

(b)

Fig. 34. Anomalous electric field variation in the ULF band related to the swarm in Hodaka. (**a**) The anomalous electric field variations were observed about 10 h before the swarm starting on 7 August 1998. (**b**) The anomalous electric field variation before and after the earthquake on 12 August 1998

sensor. The conversion factor from magnetic flux density to output voltage of SQUIDs is independent of the environmental temperature because SQUIDs are operated in liquid nitrogen. The feature is useful for field measurement. A study was started to monitor the ULF band magnetic field using HTS-SQUID by *Kasai* et al. in 2001 [43].

Several points should be taken account for the measurement system for ULF band magnetic field. *Mortyanov* et al. predicted a relationship between detectable ULF activity, magnitude of earthquake and distance from epicenter by a model computation [64]. *Nagao* et al. arranged data on major earthquakes up to date to make clear the relationship. The results agreed well with the predicted one. The results show that observation of the ULF activity associated with the earthquake is easier as the measurement point is closer to the epicenter and the earthquake is larger. For measurement near a crustal activity area the system is desired to be portable since crustal changes and volcanic eruptions generally occur far from town. Operation by battery and remote control of operation and acquisition of data is also desired.

Noise exclusion is very important for analysis on the electromagnetic data [65]. For instance, it is reported that the ratio of vertical and horizontal components (Z and H components) Z/H is useful in discrimination of the radiations presumably of seismic origin from space plasma waves [51]. From the results of a forward analysis of a model using an electric dipole in space, the atmosphere and underground, it was shown that the ratio was small when the generator was in space and was close to or exceeded 1 when the generator was underground. The ULF magnetic field measured by the fluxgate magnetometers at the Guam earthquake was analyzed using the ratio and confirmed the usefulness of the ratio [54,66]. For measurement of Z and H components, the system is desired to have three sensors set orthogonally. A vector measurement system is preferred for estimation of generator location.

Long term monitoring of the environmental magnetic field in the field is necessary to obtain useful data because crustal or volcanic activity occurs intermittantly. The system has to work stably over the long term.

The items for the system are simply arranged as follows.

(1) high magnetic field sensitivity in the ULF band
(2) vector measurement
(3) portable size
(4) stable operation over several weeks
(5) remote control

5.3 SQUID System for ULF Magnetic Field Measurement

As the first stage system for observation of the ULF magnetic field, a portable system was constructed to serve the present objective that SQUIDs could be operated in the field over the long term [43]. The system was composed of an HTS-SQUID magnetometer, electronics for SQUID operation, a personal computer for operation control, a digital audio tape (DAT) recorder for acquisition of data, a Dewar and an electromagnetic shield case made of brass as shown in Fig. 35.

The HTS-SQUID magnetometer was used as the sensor and set to measure the vertical component of the magnetic field. It is well known that since SQUID gradiometers are effective in reducing ambient magnetic field noise, they are used for onsite non-destructive testing of aircraft, bridges, etc. However, the generator of the ULF magnetic field is supposed to be under ground

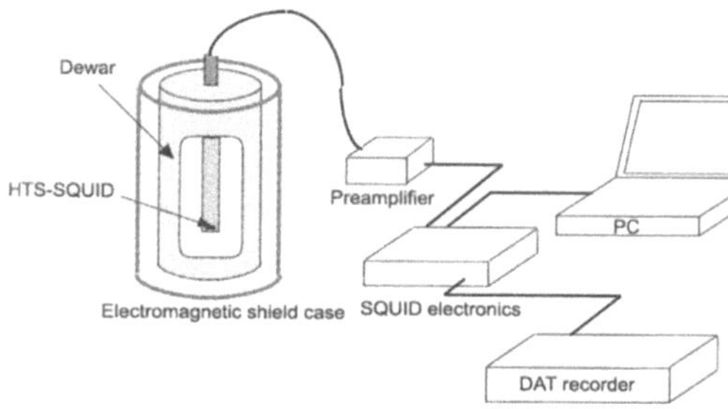

Fig. 35. Schematic view of the portable SQUID system for ULF magnetic field monitoring

and far from the measuring site. A very long baseline of gradiometer is necessary to obtain the ULF signal induced by the generator. An electric gradiometer constructed by using two or more sets of the portable system enables this.

The used HTS-SQUID magnetometer was the SEIQUID II (Sumitomo Electric Hightechs Co.). The SQUID was operated by the direct offset integrated technique (DOIT) to obtain a high slew rate that will be necessary for the field operation of the SQUID. A high-pass filter of 0.2 Hz, a low-pass filter of 100 Hz and a notch filter of 50 Hz (power line frequency) were used in the SQUID system. The sensitivity of the SQUID at 77 K measured in a μ-metal shield case was $2\,\mathrm{pT/Hz}^{1/2} (f > 5\,\mathrm{Hz})$. The coefficient of output voltage of SQUID electronics to magnetic flux density, α, was estimated by adding an AC magnetic field with a field coil. The output voltage increased linearly with added magnetic field up to 5 nT and the α is estimated to be 10 nT/V in this region. The frequency response of the SQUID output was measured in the frequency range below 50 Hz. The lower cut off frequencies of the response is 0.6 Hz and the response is constant up to 40 Hz.

A portable Dewar that keeps liquid nitrogen over long term was constructed. The shape and size of the Dewar is schematically shown in Fig. 36a. The keeping ability of the liquid nitrogen of the Dewar was checked experimentally. After filling the Dewar with liquid nitrogen, the total weight of the Dewar itself and the remained liquid nitrogen was measured by a weighting machine. The residual liquid nitrogen in the Dewar was calculated from the weight. The reduction rate of liquid nitrogen is shown in Fig. 36b. It suggests that the measurement using a SQUID should be possible for two weeks running at least. The total weight is about 15 kg when it is filled. The size and the weight are acceptable for carrying it by only one person. The Dewar was covered by an electromagnetic shield case to protect the SQUID from rf-magnetic field noise during the measurement.

The capability of the system was tested in the facilities of the National Research Institute for Earth Science and Disaster Prevention in Tsukuba. The system worked well without unlocking of SQUID operation at a point

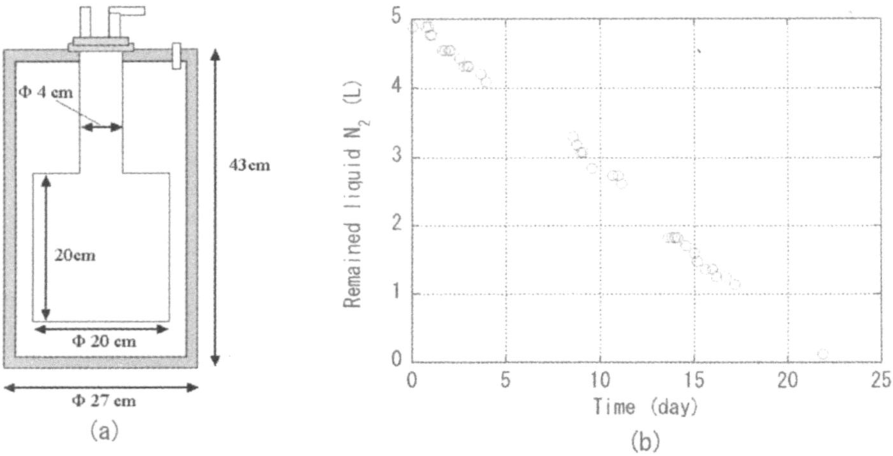

Fig. 36. High performance Dewar. (**a**) Shape and size of the Dewar. The total weight is about 15 kg when it is filled. (**b**) Capability for keeping liquid nitrogen

only 100 m distant from high voltage (about 6×10^4 V) power cable. The measured power line noise via a notch filter of 50 Hz was over 30 nT at point about 300 m distant from the high voltage power cable. The magnetic field change of the power line frequency measured without the 50 Hz notch filter shows that the slew rate of this system is higher than 30 mT/s.

5.4 Field Measurement

Field monitoring with the portable system was performed at the foot of Mt. Bandai (Fig. 37) from the end of October to the beginning of November 2000. Mt. Bandai is an active volcano and volcanic earthquakes often occurred in 2000. The system was set in a wooden cabin in the field (Fig. 38). The Dewar was filled with liquid nitrogen on site and the system was operated using a lamp wire. The system worked well for about 36 h from 7:30 p.m. on the 31st of October to 8:00 a.m. on the 2nd of November. Data on the magnetic field B_z below 100 Hz was recorded on 19 digital audio tapes with a sampling frequency of 200 Hz.

The environmental magnetic field level was constant within 3 nT, mainly power line noise (50 Hz), during calm periods. Some large magnetic field variations B_{av} were recorded and sometimes it exceeded the dynamic range of the SQUID system (15 nT). The evolution of the anomalous magnetic field variations was intermittent and sometimes continued 20 min or more. They resemble those of the anomalous electromagnetic field changes in the ELF/VLF band related to seismic swarms [51]. The magnetic field variation in the 36 h is shown in Fig. 39. The expression of the variation was prepared as follows. The obtained magnetic field data were filtered using a digital low-pass

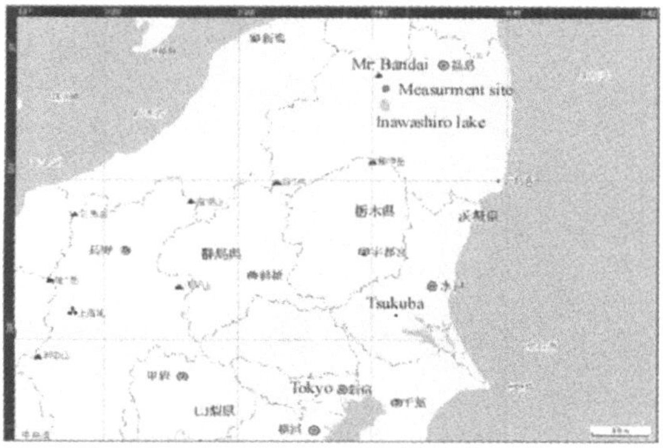

Fig. 37. Map of Mt. Bandai. Mt. Bandai is about 200 km from Tokyo

Fig. 38. Photograph of the wooden cabin where the SQUID system was set

filter of 45 Hz to eliminate the power line noise. The amplitude of the filtered data was averaged over one minute. Anomalous magnetic field variations are clearly seen in several periods in Fig. 39. An example of the waveform of the anomalous ULF magnetic field, pulse-like waveforms, is shown together with the spectrum in Fig. 40. The main frequency components of the anomalous magnetic field in every period with anomalous variations were between 5 Hz and 20 Hz.

As mentioned above, there are many generators of ULF magnetic field. The possibility of being a source of the anomalous magnetic field was investigated for the following generators. There was no abnormal power transmission (Tohoku Electric Power Corporation) there during the whole measurement period and there were no thunderbolts (Flankrin Japan Corporation) within 8 km around Mt. Bandai. There seemed to be no correlation between a train passing through the station of electric railway, about 3 km distant from the measurement site, and the anomalous magnetic field changes.

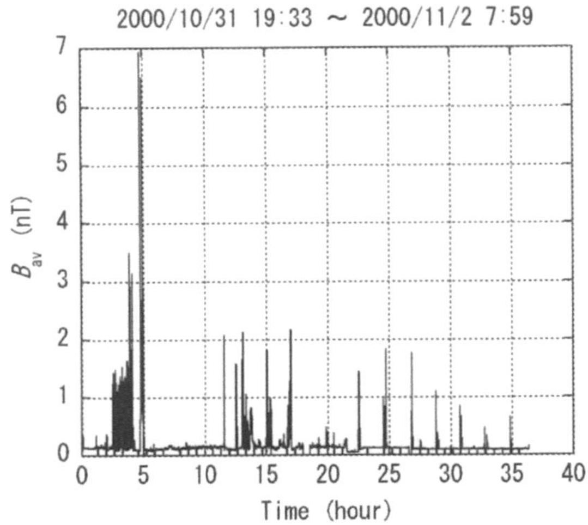

Fig. 39. The results of the long term magnetic field measurement. B_{av} is the avarage of the amplitude of the magnetic field B_z over one minute

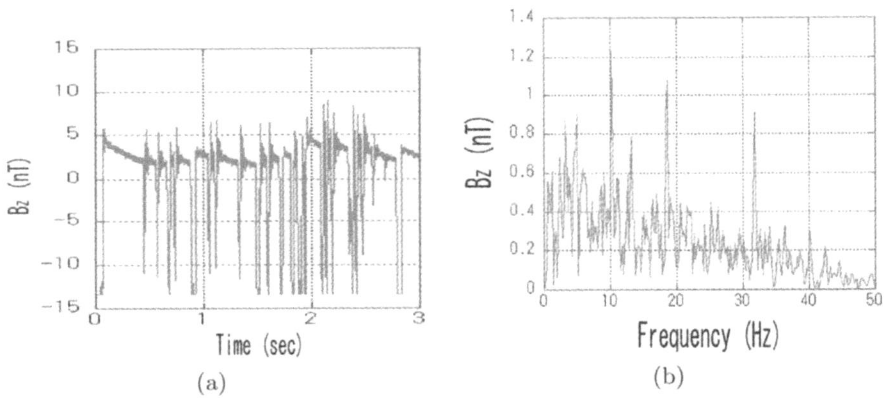

Fig. 40. Example of the anomalous ULF magnetic field. (**a**) Time domain waveform. (**b**) Spectrum of the waveform

An earthquake occurred related to volcanic activity at 3:14 a.m. on 1 November at Bandai ($M = 1.7$, Japan Meteorological Agency). The epicenter was about 3 km from the magnetic field measurement site. The seismic waveforms of the three directions (North-South, East-West and Up-Down) measured at the Wakamatsu meteorological observatory in the Mt. Bandai are shown together with the spectrum in Fig. 41 [62]. No anomalous magnetic field variation appeared at the time. For comparison, the spectrum of the magnetic field is shown in Fig. 42. It was supposed that the ground swing was not enough to induce significant SQUID output change by comparison of spectra.

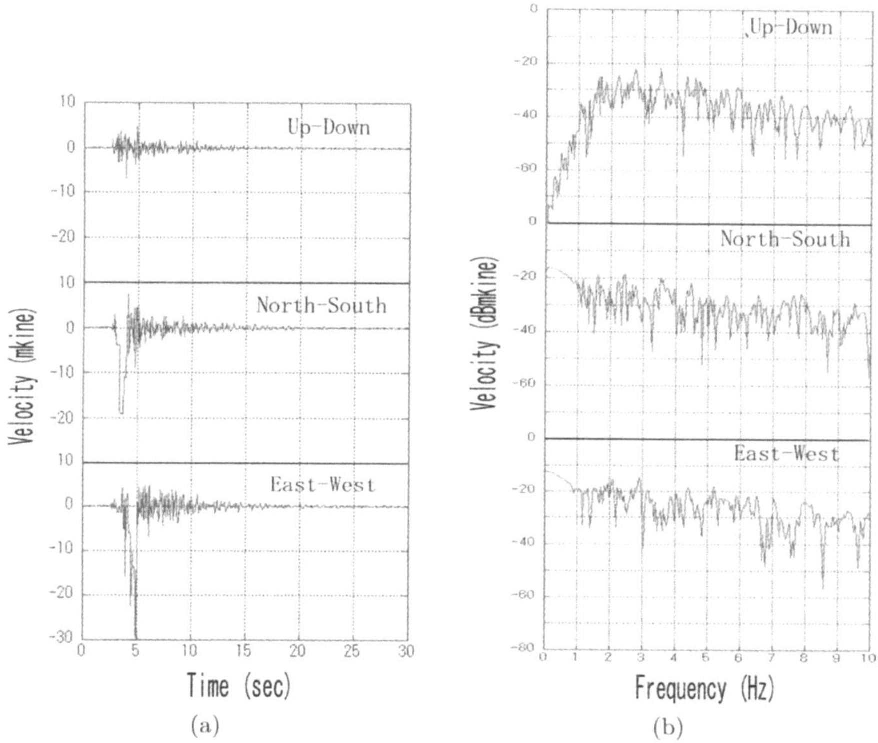

Fig. 41. Waveform of the earthquake. (**a**) Time domain waveform. (**b**) Spectrum of the waveform

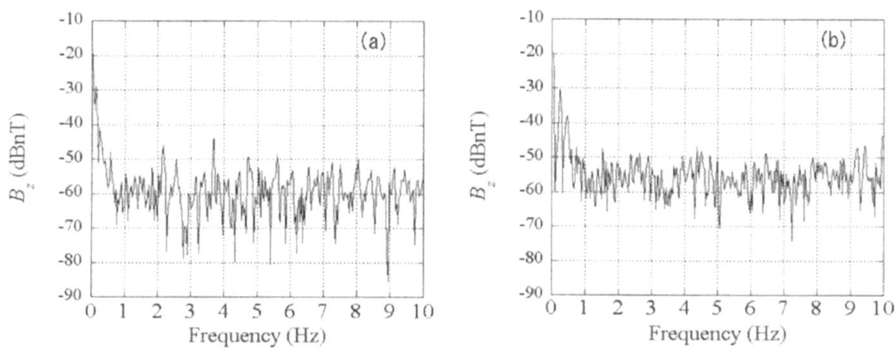

Fig. 42. Spectrum of the magnetic field (**a**) when the earthquake occured and (**b**) in a calm period

5.5 Future Problem

Anomalous ULF magnetic field variation were observed before and after a weak earthquake. However, the relation between the anomalous field variation and the earthquake is not sure. Further studies about the generator are necessary. Simultaneous measurement of ULF magnetic field, seismograph, electric potential and other measurements will be done.

The improvement of the system for environmental magnetic field measurement is progressing. A vector HTS-SQUID system was made and its utility investigated at Mt. Bandai. The whole system, electronics for SQUID operation, computers and data logger, were provided electricity by car batteries. A data logger was developed for long term data acquisition. A preliminary study for remote sensing using a pocket telephone was carried out.

6 Laser-SQUID Microscope for LSI Chip Defect Analysis

We use emission microscopy, IR-OBIRCH (infrared optical beam induced resistance change) and electron beam testing for nondestructively localizing electrical defects in the failure analysis of LSIs. However, a weak point of these methods is the electrical contact to the outside of the LSI chips or the boards that the LSI chips are mounted on. Pattern defect inspection methods do not require electrical contact, but the defects found are not always causes of electrical failures.

We propose a new technique that can detect electrical defects nondestructively without electrical contact to the outside [67,68]. The position of this new method, scanning laser-SQUID microscopy ("laser-SQUID" for short) with respect to nondestructive methods is shown in Fig. 43.

The technique detects magnetic fields induced by OBIC (optical beam induced current). DC-SQUIDs (superconducting quantum interference devices) are used to detect magnetic fields because they are the most sensitive detectors of magnetic flux currently available (Fig. 44). We chose HTS (high-T_c superconducting) DC-SQUID because of its relative ease of operation. The intensity of magnetic flux detected by the SQUID magnetometer is imaged as black and white corresponding to the $+z$ and $-z$ components of the magnetic flux while scanning a laser beam across the sample (Fig. 45). The spatial resolution of the laser-SQUID is much better than conventional SQUID microscopy because it is limited by the laser beam diameter and not the size of the SQUID detector and the distance between the detector and a sample. A demonstration of pn junction imaging using a prototype laser-SQUID showed that the spatial resolution is about 1.3 μm. This is more than 15 times better than the resolution obtained using a similar type of microscopy in the past [69,70]. Furthermore, this is 25 to 40 times better than conventional

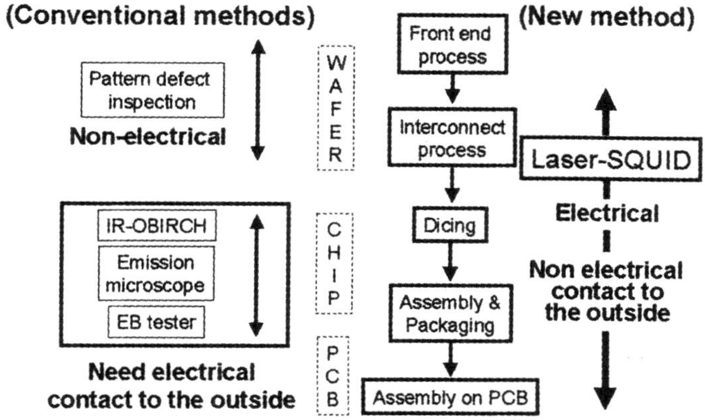

Fig. 43. Position of the Laser-SQUID in nondestructive methods

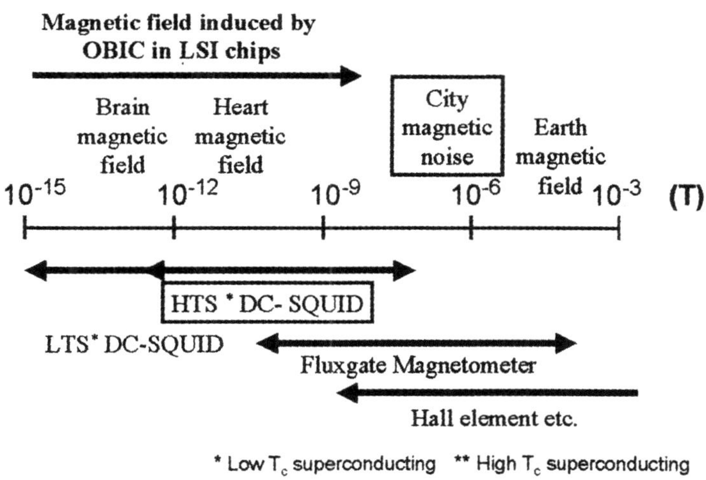

Fig. 44. Magnetic field intensity and detector sensitivity

SQUID microscopy [71,72] and also better than SPM (scanning probe microscope) based SQUID microscopy [73] or micro DC-SQUID microscopy [74] (Fig. 46).

6.1 The Laser-SQUID System

The essence of the laser-SQUID is shown in Fig. 47. As a laser beam irradiates a chip, the beam induces a current that produces a magnetic field detectable by a SQUID magnetometer. In many cases a transient current decays much faster than the DC-SQUID time response of > 1 μs. Only a slow-decay current can be detected by the DC-SQUID magnetometer. When at least one closed

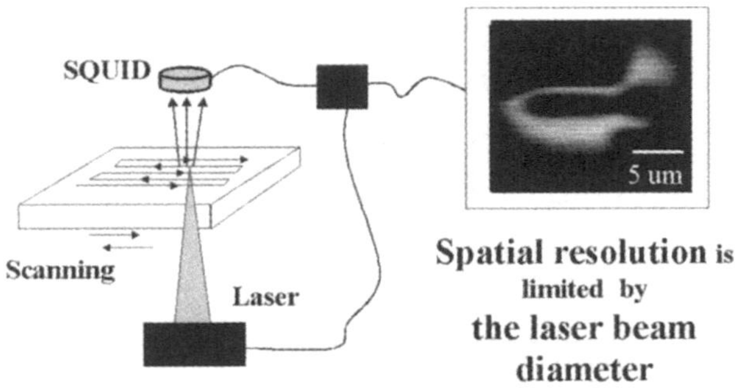

Fig. 45. Spatial resolution of laser-SQUID

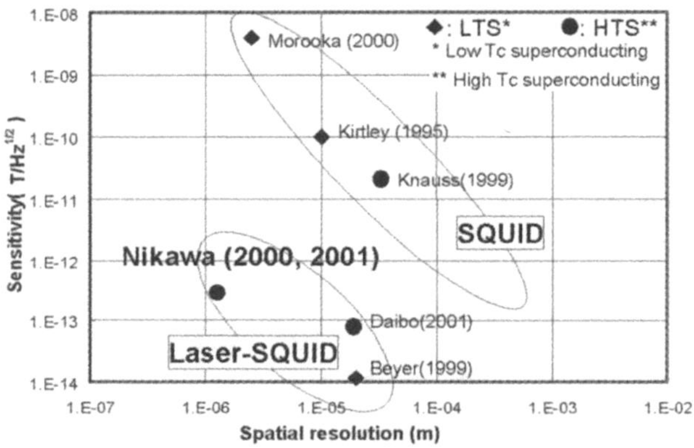

Fig. 46. Spatial resolution vs. sensitivity

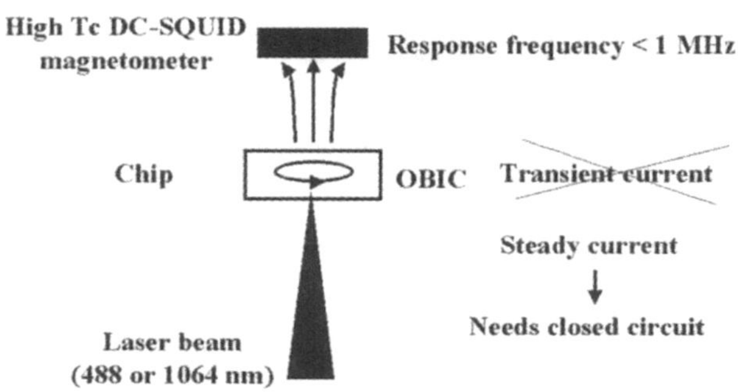

Fig. 47. Essence of the laser-SQUID

circuit exists, then steady-state current flow and DC-SQUID can detect the magnetic flux.

The OBIC and Seebeck effects generate laser beam induced current. Since the current is usually much stronger in the OBIC than in the Seebeck effect, we used the OBIC effect to demonstrate the laser-SQUID. The laser-SQUID

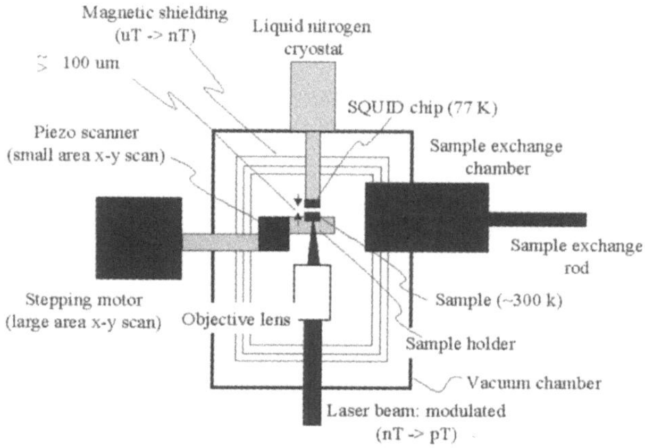

Fig. 48. Main part of the prototype

can detect short, open, and high resistance defects that are connected to pn junctions by comparing the magnetic flux difference between a good chip and a chip under test. Defect localization is facilitated by the CAD (computer aided design) layout data.

6.2 Prototype System Setup

A prototype system setup is shown in Fig. 48. The magnetically shielded SQUID detector, sample holder and objective lens are placed in a vacuum chamber with the sample changed through a vacuum exchange port. The sample is scanned using a piezo scanner (maximum $100 \times 100\,\mathrm{mm}^2$) or stepping motor (maximum $7 \times 7\,\mathrm{mm}^2$). The SQUID detector is cooled to about 77 K using a liquid nitrogen cryostat. The distance between the sample and the SQUID chip can be set to about 100 μm by a z-stage and the laser beam is introduced from the bottom side. We can use a 488 and a 1064 nm wavelength laser. The intensity of the laser beam is modulated with an acousto-optic modulator or a mechanical chopper and the resulting modulated SQUID output signal is captured by a lock-in amplifier. The output of the lock-in amplifier is used for imaging. The combination of magnetic shielding and "laser modulation/lock-in" approach make pT (pico Tesla) order magnetic signals detectable with μT order magnetic noise.

6.3 Basic Demonstration Using 488 nm Laser

We used a Ti-Si/poly-Si line test structure for demonstration. Gate oxide shorts caused by ESD (electrical static discharge) produced an OBIC signal because the shorts made current paths from a Ti-Si/poly-Si line to the Si substrate through pn junctions. The closed circuit required for steady-state current generation was constructed by connecting several probe pads

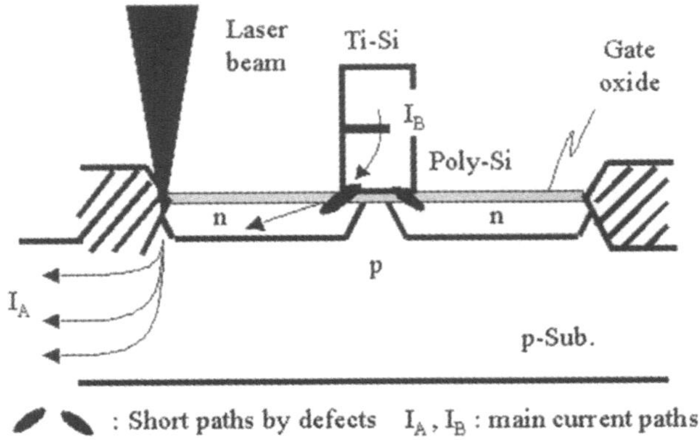

Fig. 49. Current paths in the cross section at the dot line in Fig. 50b

to a bonding pad using a FIB (focused ion beam). The resulting current paths in the chip are shown in Fig. 49 as I_{A} and I_{B}. The rest of the closed circuit was constructed by externally shorting the chip. This simulated a gate-short site localization on a chip that is mounted on a board. An Ar laser (488 nm) irradiated the chip from the front side.

A laser-SQUID image is shown in Fig. 50a and a laser scanning microscope (LSM) image of the same area is shown in Fig. 50b. Figure 49 shows the cross section along the dotted line in Fig. 50b where the probing pad was omitted.

The image in Fig. 50a shows that OBIC signals do not come from the bottom of pn junctions. This is because the DC-SQUID magnetometer we used detects only the z-component of magnetic flux. The long minority carrier diffusion length (Fig. 51) generated the broad bright contrast in the image in Fig. 50a. The narrow bright contrast in the image occurs where the pn junction was covered with a probing metal pad. Here only the light scattered at the edge of the probe pad generated the OBIC signal. This allowed us to measure the spatial resolution of the laser-SQUID image. The intensity profile across the bright narrow line in Fig. 50a is shown in Fig. 50c. The observed spatial resolution, defined as FWHM (full width at half maximum), was about 1.3 μm.

6.4 Backside Failure Identification before Bondpad Patterning

We demonstrated defective chip identification before bondpad patterning by examining many chips from several wafers that were sampled from a wafer fabrication process after second level metal deposition (before patterning). Figure 52a shows many features with black and white contrast in the laser-SQUID image. For this image we used a 1064 nm laser from the backside of the chip (Fig. 53). The distance between the chip surface and the SQUID was 0.5 mm. After imaging Fig. 52a, two first level metal lines were cut by FIB

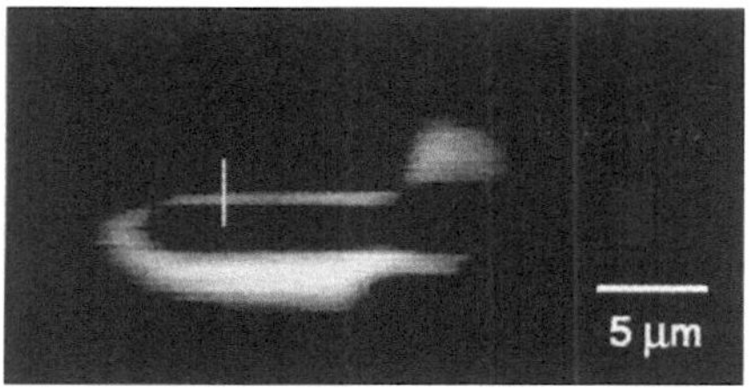

(a) Laser-SQID image

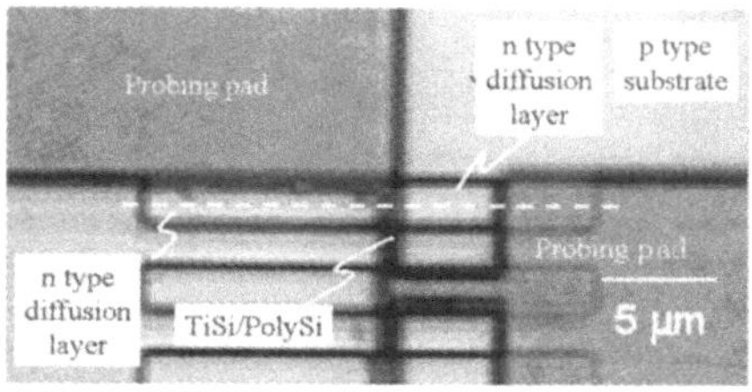

(b) Laser scanning micorscope image

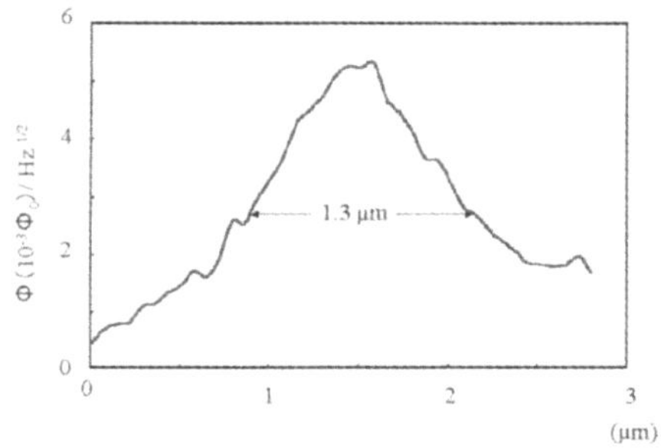

(c) Profile showing the spatial resolution

Fig. 50. Basic demonstration result

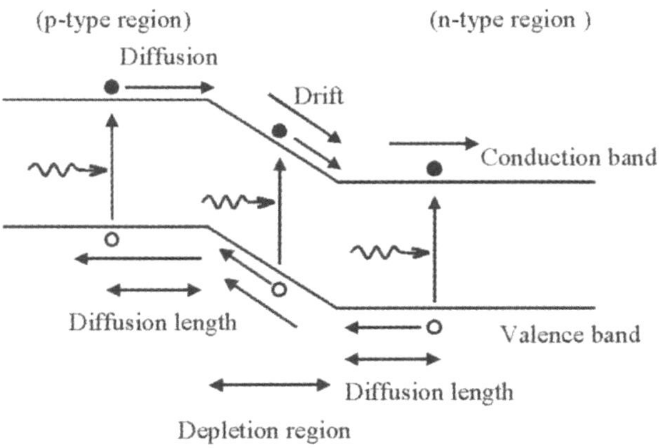

Fig. 51. OBIC generation at a pn junction

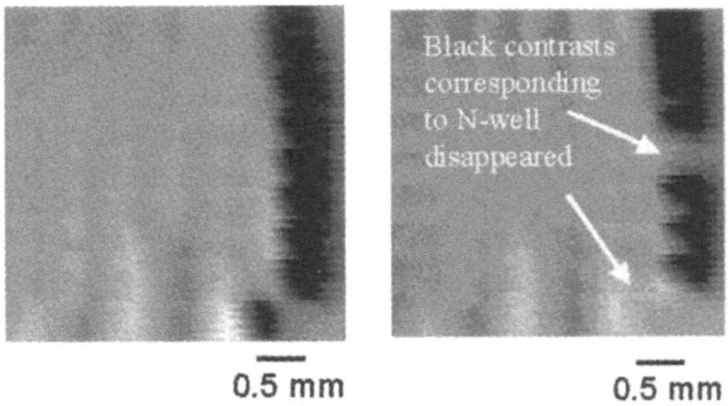

(a) Befor FIB cut (b) After FIB cut

Fig. 52. Laser-SQUID images before and after FIB cut

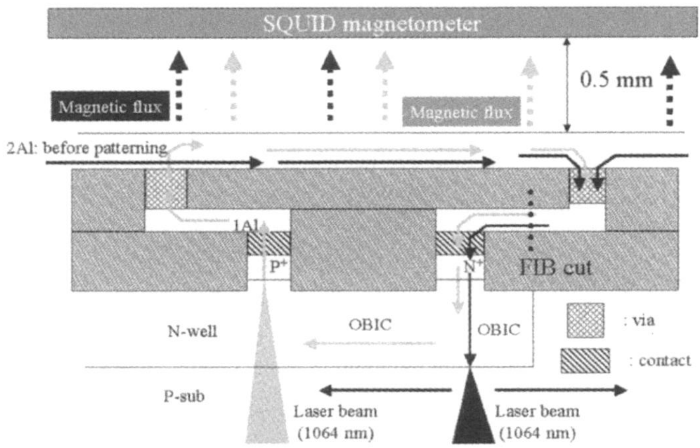

Fig. 53. Laser beam, pn junction, current, magnetic flux: two cases

as shown in Fig. 53. This "defective" chip image is shown in Fig. 52b. When compared with Fig. 52a, we see that two black-contrast areas corresponding to n-wells disappeared.

6.5 Defective Chip Identification after Bonding and Packaging

To demonstrate defective chip identification after bonding, all the pins of packaged LSI chips were shorted and then observed by the laser-SQUID. The ICs were packaged in a 100-pin QFP (quad flat package). Plastic materials over the chip surfaces were removed before laser-SQUID observation.

Figure 54 shows the laser-SQUID images of defective and good chips. The defective chip image in Fig. 54a and the good chip image in Fig. 54b are clearly different. This distinct difference between a defective chip image and a good chip image was not seen with other electrical connections such as

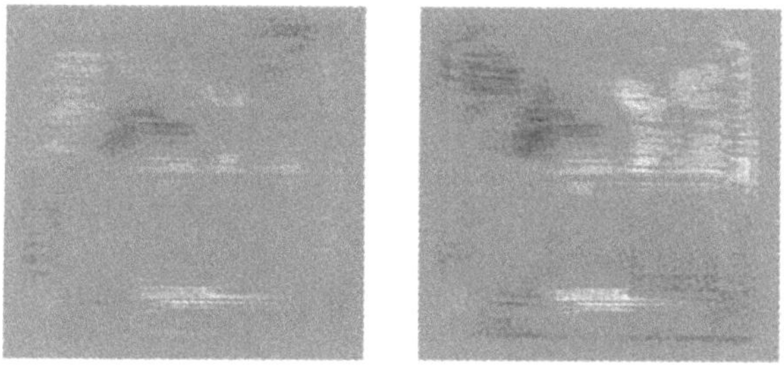

(a) defective chip (b) good chip

Fig. 54. Laser-SQUID images: all pins shorted (100 pin QFP, 1064 nm laser, 3.5 mm sq. area scan)

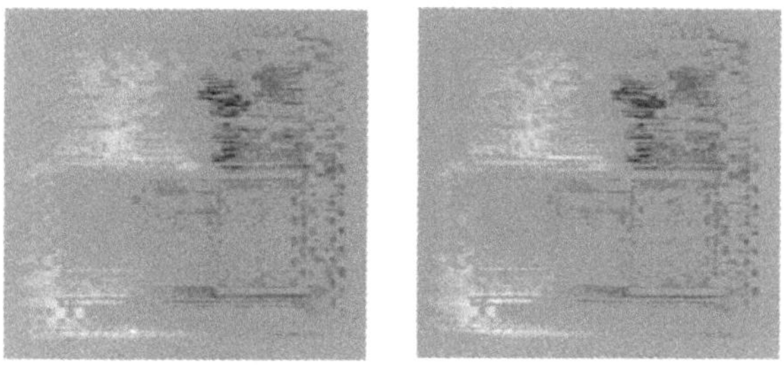

(a) defective chip (b) good chip

Fig. 55. Laser-SQUID images: V_{DD}-GND short (100 pin QFP, 1064 nm laser, 3.5 mm sq. area scan)

a V_{DD}-GND short (Fig. 55) or an all-pin open condition (Fig. 56). We also observed these two chips with conventional OBIC between V_{DD} and GND and found no difference between the two OBIC images (Fig. 57). The laser-SQUID contrast in the all-pin-short condition suggests that the difference between defective and good chips might be due to an ESD failure. Defective parts in an ESD failure are typically near input and output pads. Therefore, the current paths could change in the all-pin-short condition, but they could not change in a V_{DD}-GND-short condition or an all-pin-open condition.

6.6 Localization from Whole Chip Area to Micrometer Area

One of the most important functions in a defective site localizaion tool is to localize from the whole chip area down to a micrometer area. We demonstrated this function by using the same chip that demonstrated the spatial resolution in the previous section. Figure 58 shows the localization results for the defective gate short with decreasing fields of view. The view decreased from the whole chip down to a $50 \times 50\,\mu\mathrm{m}^2$ area. A $25\,\mathrm{mm}^2$ area was first

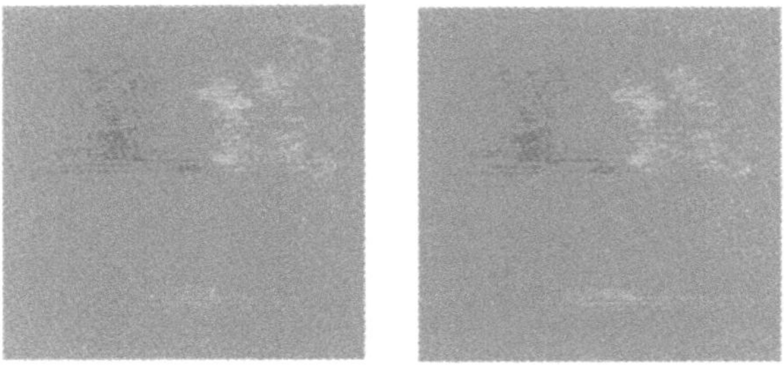

(a) defective chip (b) good chip

Fig. 56. Laser-SQUID images: all pins open (100 pin QFP, 1064 nm laser, 3.5 mm sq. area scan)

(a) set up (b) defective chip (c) good chip

Fig. 57. Conventional OBIC images: V_{DD}-GND (100 pin QFP, 1064 nm laser, 3.5 mm sq. area scan)

scanned using $x-y$ stepping motors as seen on the left side of Fig. 58. Two bright spots can be seen in the image. A $500 \times 500\,\mu m^2$ area that included these two bright spots was scanned next, also using the $x-y$ stepping motors (middle of Fig. 58). Some detail in these two bright spots can be seen in the image. A $50 \times 50\,\mu m^2$ area was then scanned around the brighter of the two sites (right side of Fig. 58). A similar shape and contrast is seen in Fig. 50a. In this particular case defect site localization was easy because the defect was near the pn junction. In many cases, however, you may need the help of CAD layout data.

6.7 Conclusions

We developed the laser-SQUID for LSI chip inspection, monitoring and analysis. We demonstrated a prototype laser-SQUID and showed

- Spatial resolution of about 1.3 μm
- Ability to detect a gate-shorted chip mounted on a board
- Ability to detect a defective site from the whole chip area down to a μm^2 area

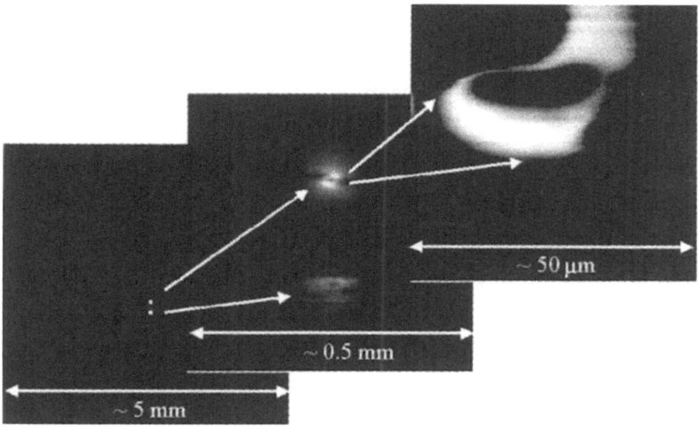

Fig. 58. Localization on a chip on a board

- Identification of a defective chip before bonding-pad patterning
- Identification of a defective chip also after bonding pad patterning

These results suggest that the laser-SQUID has many applications for in-line monitoring, in-line inspection and failure analysis.

7 Small-Scale HTSC Digital Applications

High-T_c superconductor (HTSC) Josephson junctions are naturally overdamped, which means that their I–V curves do not show hysteresis, and the junctions in single flux quantum (SFQ) circuits must be ovedamped junctions [75]. Tunnel-type low-T_c superconductor (LTSC) Josephson junctions, on the other hand, are underdamped and require some shunt resistance between the two electrodes of each junction for erasing the hysteresis in their I–V curves [76]. This lowers the characteristic voltage ($I_c R_n$ product), which results in lower operating speeds and complicates the layout and the fabrication process. The $I_c R_n$ product of HTSC junctions can also be expected to be larger than that of LTSC junctions because it intrinsically depends on the gap voltage of the superconductor. Moreover, devices made from HTSC materials have lower cooling costs as well as higher operating frequencies.

However, the higher operating temperature results in more thermal noise. The energy barrier between two flux states in an SFQ gate is very low. A rough estimation [77] shows that for typical critical currents of the order of 10^{-4} A, this energy barrier is of the order of 10^{-19} J. Thus fluctuations may increase the spontaneous switching between flux states. The probability of SFQ gate errors caused by thermal noise has been investigated theoretically and experimentally [78,79].

Parasitic inductance is unavoidable in practical Josephson junction layouts and contacts of SFQ digital circuits. The line sheet inductance of HTSC circuits is twice that of LTSC circuits. Moreover, smaller inductance elements are used in HTSC circuits to ensure that the LI_c product of each SFQ loop remains constant against larger I_c. The parasitic inductance is therefore a more serious problem for HTSC SFQ circuits than for LTS circuits.

A number of tests of SFQ circuits using HTSC Josephson junctions have been reported [80,81,82,83,84], but these circuits were small scale because the fabrication technology for HTSC junctions is still in a primitive stage. The stronger thermal noise and the parasitic inductance have also hindered researchers from increasing the scale of HTSC digital circuits. Samplers and analog-to-digital converters (ADC) have been developed for HTSC digital applications. A sampler is a measurement system for observing repeated electrical signal waveforms with high frequency. ADC is one of the signal processing circuits. HTSC ADCs also have high resolution as well as a wide bandwidth and may find application in wireless communications and in radar systems.

7.1 Sampler

Since a Josephson junction has a well defined current threshold for switching from the superconducting to the resistive state and can generate an ultra short current pulse, it can be used as a sampler switching element. Moreover, the high speed and sensitivity of Josephson junctions make them attractive for use in a sampler.

Much research in the 1980s was devoted to developing Josephson samplers using LTSC latching circuits [85,86]. For example, in 1987, Hypres Inc., developed a Josephson sampler system that consisted of a sampler chip, a liquid-helium cooling system and a semiconductor peripheral system [87]. Unfortunately, these LTSC samplers have seen only limited commercial application because the cost of operating the liquid helium cooling system is very high. An HTSC sampler, on the other hand, would be more suited to commercial application because it could be cooled with a single-stage cryocooler, which is compact and easy to operate.

7.1.1 Sampler Circuit

Figure 59 shows the principle used to measure a signal current waveform with the Josephson sampler circuit [85]. At a given moment a pulse current I_p and a feedback current I_f are added to a signal current $I_s(t)$. When the sum of the three currents exceeds the threshold current of the Josephson gate, the gate turns on and an output voltage appears. The $I_{fmin}(I_s)$ value, which is the minimum I_f to turn on the gate, can be determined by repeating the above operation with various I_f values. Comparing $I_{fmin}(I_s)$ with $I_{fmin}(0)$, which is the I_{fmin} for $I_s = 0$, we can obtain the I_s value at the moment determined from I_p. The whole I_s waveform is measured by detecting I_s at various I_p supply moments.

Figure 60 shows the circuit design of the original HTSC sampler circuit [88]. The circuit consists of five over damped junctions and inductance

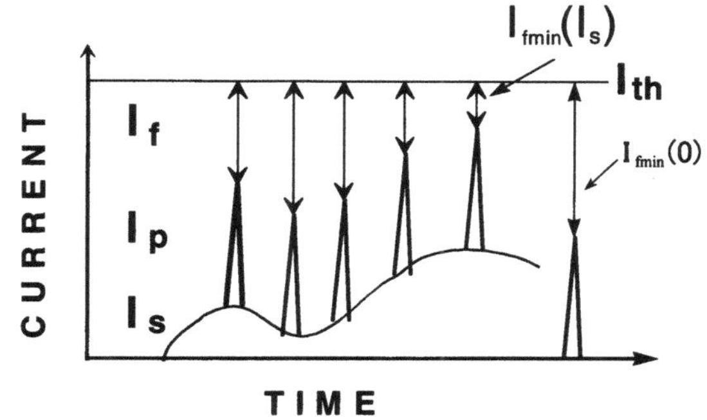

Fig. 59. Measurement principal of the Josephson sampler

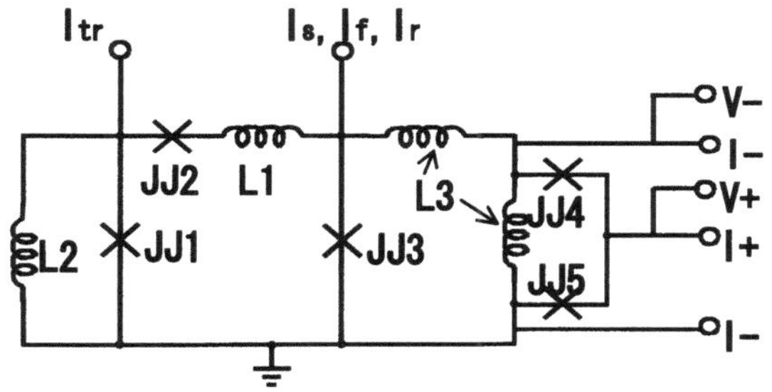

Fig. 60. Circuit diagram of the original HTSC sampler. The design parameters are $I_c(\mathrm{JJ1}) = I_c(\mathrm{JJ3}) = 0.5\,\mathrm{mA}$, $I_c(\mathrm{JJ2}) = I_c(\mathrm{JJ4}) = I_c(\mathrm{JJ5}) = 0.25\,\mathrm{mA}$, L1 = 3.7 pH and L2 = L3 = 5.0 pH. $V+$, $V-$ and $I+$, $I-$ are, respectively, the voltage and current terminals of the readout SQUID

elements. A current pulse I_p due to SFQ is generated by JJ1 and JJ2 at the moment the trigger current I_{tr} rises, and this pulse propagates to a comparator junction (JJ3) where it combines with a signal current I_s and a feedback current I_f. When the sum of the three currents exceeds the threshold value, JJ3 turns on and an SFQ is stored in the superconducting loop that contains JJ3 and L3. Thus, the momentary turning-on of JJ3 is converted into a circulating current corresponding to the SFQ in the loop. The stored SFQ induces an output voltage at the readout SQUID which consists of JJ4 and JJ5. The stored SFQ in the loop is reset using an opposite direction reset current I_r at the end of each sampling cycle.

7.1.2 Fabrication Process

The superconductor that has been most widely used in HTSC digital circuits is $YBa_2Cu_3O_x$ (YBCO) because of its reproducibility of T_c, robustness to water and large critical current density. There are a number of suitable substrates on which YBCO films can be grown: $SrTiO_3$ (STO), MgO, $LaAlO_3$, $NdGaO_3$, YSZ (yttria-stabilized zirconia), Sr_2AlTaO_6 (SAT),

Sr_2AlNbO_6(SAN), and $(La_{0.3}Sr_{0.7})(Al_{0.65}Ta_{0.35})O_x$ (LSAT). Among them STO has been the most popular substrate material for SFQ digital circuits because its lattice constant and thermal expansion coefficient are close to those of YBCO. The original sampler circuit was fabricated on the STO substrate. The main drawback with STO is its large dielectric constant, approximately 10^3 in films at 77 K, which makes it unsuitable for high-frequency applications. The substrate was replaced by MgO, whose dielectric constant is 10. Although STO thin film was used as an insulation layer between YBCO electrodes at first, it was also replaced by CeO_2 thin film which has a lower dielectric constant of 16.

Of the various types of Josephson junctions developed so far, ramp-edge type junctions [89], shown schematically in Fig. 61, seem to be the most promising for digital circuit applications because of their small dimensions, the potential controllability of the junction critical current and junction resistance values and ease of wiring. Interface-engineered ramp-edge junctions (IEJ) developed by *Moeckly* et al. [90] attracted much attention because their reproducible fabrication process is quite suitable for digital circuit applications. This process has no barrier deposition step: the barrier is formed simply by structural modification using ion bombardment and vacuum annealing. *Satoh* et al. modified the process in which the edge of the base YBCO film is formed by normal ion milling and the film is then heated to deposition temperature for the counter electrode in O_2 without being exposed to air. Their modified interface junctions (MIJ) also showed reproducible I_c with a 1σ spread in I_c of 8% for 100 junctions [91]. The MIJ was used as the Josephson junction in the sampler circuit.

Development of a circuit process integrating reproducible Josephson junctions into epitaxial multilayers is required before high-performance HTSC SFQ circuits can be implemented. In particular, a superconducting ground plane is required to keep circuit inductance low enough that the SFQ pulse can generate enough current in the load inductor. Ramp-edge junctions with a 450 nm thick stacked ground plane were first made by *Miura* et al. [92] and several HTSC Josephson junctions with stacked ground planes have been developed [93,94,95]. However, these ground planes are buried under the junctions and the thickness of the buried ground planes had to be kept below 200 nm because thicker ground planes would have resulted in more surface roughness, which reduces junction quality. To overcome this drawback, the sampler circuit was given a stacked structure, called a HUG (HTSC circuit with an upper-layer ground plane) structure, where the YBCO ground plane is the top layer shown in Fig. 62 [96]. The advantage of this structure is that better quality junctions can be made directly on the smooth substrate compared with methods where the junction is over a thick ground plane. Moreover, the upper ground plane can be made as thick as desired. Each YBCO layer of the HUG structure was verified to have a current density close to that of an as-grown YBCO film. The measured resistance of the

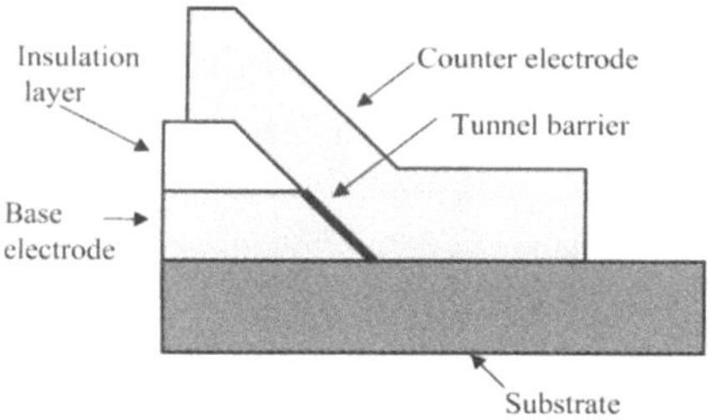

Fig. 61. Schematic cross section of a ramp-edge Josephson junction

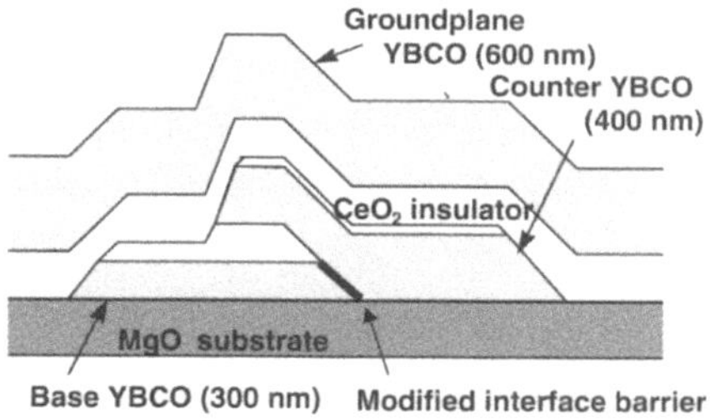

Fig. 62. Schematic cross section of the HUG structure

trilayer across the 400 nm thick STO film was over 1 MΩ in a range from 4.2 K to 300 K over an area of 100 μm × 100 μm which is sufficient for circuit operation. The high temperature process used in forming the ground plane does not affect the junction quality, such as the $I_c R_n$ product and excess current. At 30 K, experimentally estimated L values were around 1 pH with a 600 nm thick ground plane, while L without a ground plane was 2.8 pH.

7.1.3 Current Measurement System

The HTSC sampler is able to measure current waveforms directly with picosecond and microampere resolutions [97]. Semiconductor samplers and electro-optic (E-O) samplers are well known for characterizing the temporal shape of high speed electrical signals. However, semiconductor samplers measure voltage and E-O samplers measure electrical field. To measure current using these samplers, the electrical impedance of the measured part has to be known. The demand for a current measurement capability to evaluate circuit designs and electromagnetic compatibility (EMC) will increase as the operation frequency of semiconductor large-scale integrated circuits (LSIs) enters the gigahertz frequency regime. However, since the impedance of the wiring in an LSI under test is generally unknown because of the LSI's complex layered structure and via holes, current flowing through the wiring cannot be measured using semiconductor or E-O samplers.

Figure 63 shows the measurement principle of this system [98]. I_s flowing through the device under test (DUT) generates a magnetic field. The magnetic field induces a shielding current I_{cir} in the superconducting loop, which includes the comparator junction of the HTSC sampler. Since I_{cir} is proportional to I_s, I_s can be measured by the HTSC sampler without contact. A schematic cross section of the current probe is shown in Fig. 64. The sampler chip is mounted on the bottom of a chip holder head. The sampler chip is cooled to its operating temperature by a single stage GM-pulse tube cryocooler. Because the pulse tube cooler has no moving parts in its cold section, the electromagnetic noise level and vibration level in the cold section are low enough for current probing. The chip holder head is housed in a vac-

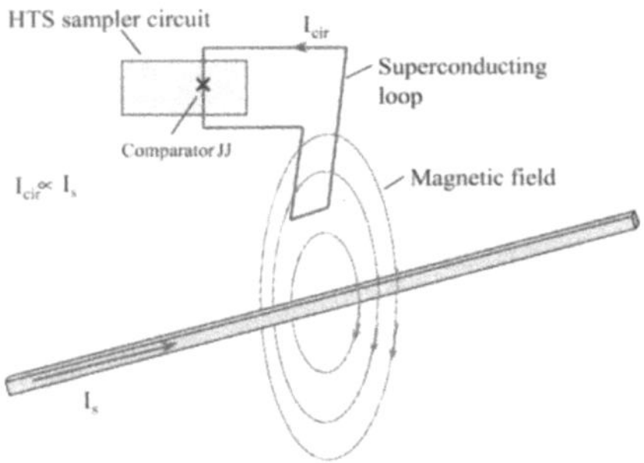

Fig. 63. Measurement principle using the HTS sampler and a superconducting loop

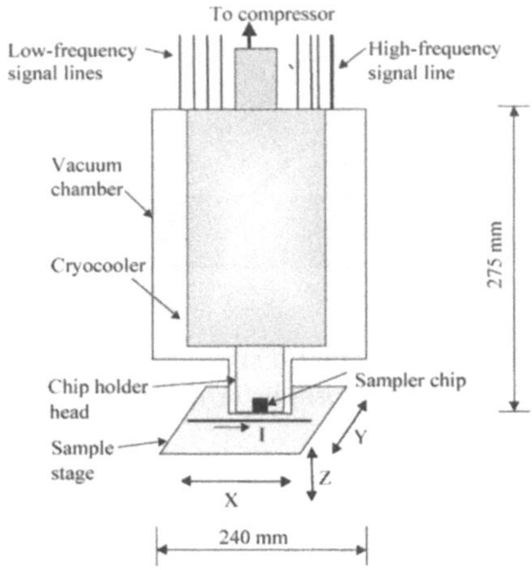

Fig. 64. Schematic cross section of the current probe for the sampler current waveform measurement system [98]

uum chamber and the distance between the sampler chip and DUT at room temperature can be reduced to 2 mm. A high speed line for the trigger current and seven low speed lines are wired from the room temperature control unit to the chip holder head. The heat transfer through the signal lines was estimated to be 0.5 W at 40 K.

Figure 65 shows a photograph of the HTSC sampler current measurement system. One can measure a current waveform by setting the DUT under the current probe and outside the vacuum chamber. During the measurements the current probe is covered with a magnetic shield box made of μ-metal. Figure 66 shows the current measurements produced by the system. When the input frequency was 7 GHz the input sinusoidal waveforms were reconstituted with the correct period.

7.1.4 Digital Signal Waveform Measurement System

Next-generation ultra high speed optical transmission technologies for terabit/s networks based on 40 Gbps optical communication technologies are being developed. In these networks, 40 Gbps optical digital signals are converted to electrical signals at both ends of an optical fiber. The waveforms of the 40 Gbps electrical digital signals need to be measured in order to develop and maintain the related components. The samplers need to have a bandwidth at least three times wider than the 40 Gbps digital waveform (120 GHz bandwidth). The widest bandwidth of commercially available semiconductor samples is, however, only 65 GHz. The bandwidth of the HTSC sampler is expected to be more than 120 GHz below 40 K, according to an estimation

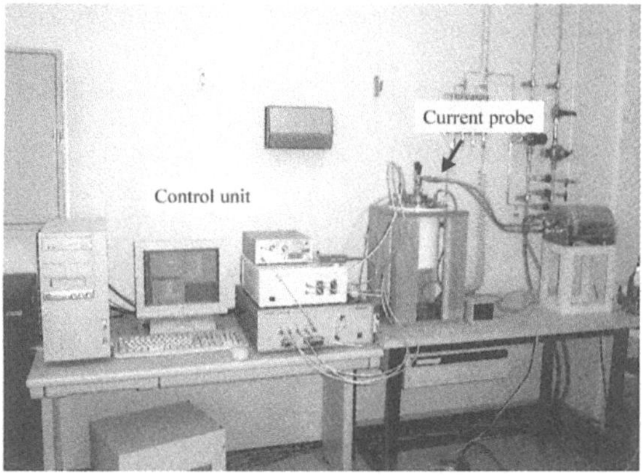

Fig. 65. Photograph of the HTS sampler current measurement system [98]

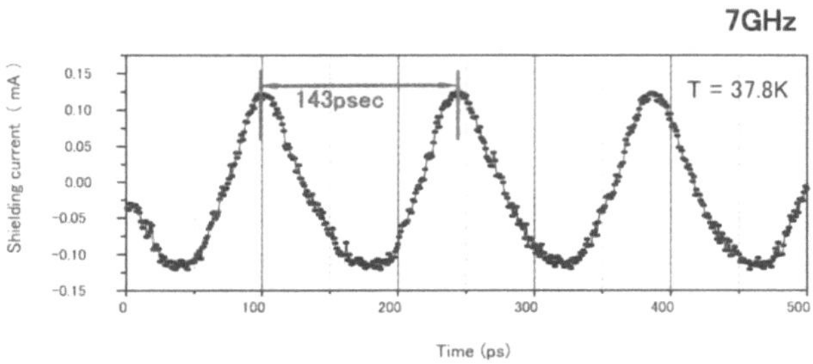

Fig. 66. Measured sinusoidal current waveform produced by the prototype system with input frequency of 7 GHz

using the $I_c R_n$ products of the MIJ [91]. A digital waveform measurement system for observing the 40 Gbps digital signals was attempted. As a first step, a prototype system whose target bandwidth was 40 GHz was developed [99].

In the prototype system, a sampler chip is cooled down to its operating temperature by a single stage Stirling cryocooler and connected to a 40 GHz bandwidth assembly line. Figure 67 shows the equivalent circuit of the sampler chip, which consists of a sampler circuit and two impedance transformation circuits for a signal and a trigger input line. The dimensions of the cryocooler unit are 28 cm × 35 cm × 50 cm. The lowest temperature achieved by the system was 35 K, in spite of the two high frequency lines and the seven low frequency lines being connected to the room temperature control unit. The heat transfer was estimated to be 0.8 W at 40 K. When the cold head temperature was 40 K, the input power was 80 W.

Figure 68 shows the prototype of the HTSC sampler digital signal waveform measurement system and Fig. 69 shows a reconstituted digital waveform of 2.95 GHz = 5.9 Gbps input. The solid line in Fig. 69 represents the waveform measured by the HTSC sampler system and the dotted line is that

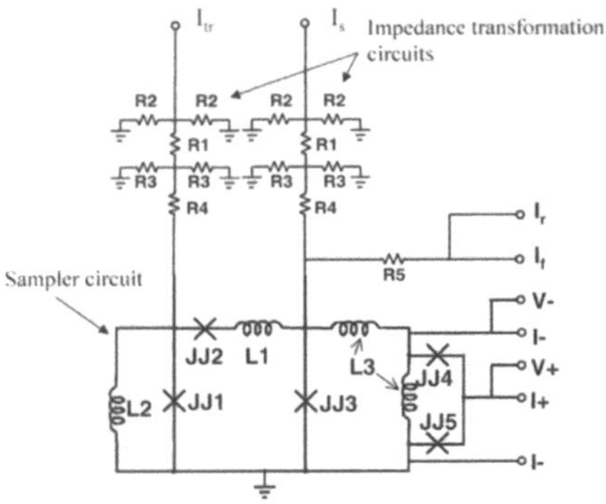

Fig. 67. Equivalent circuit of the sampler chip for the digital waveform measurement system [99]

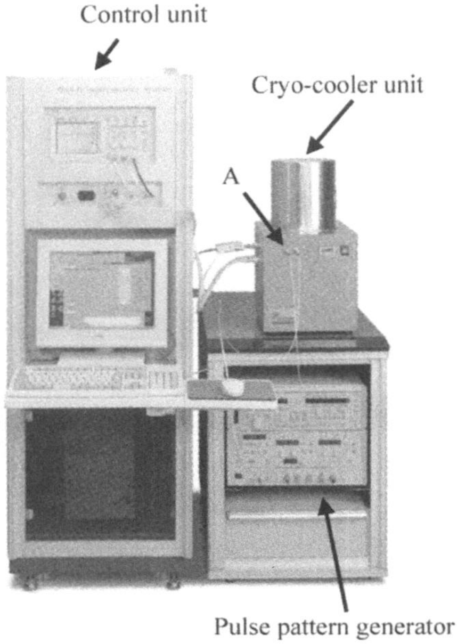

Fig. 68. Prototype of an HTS sampler digital signal waveform measurement system. Digital signals to be measured are generated by a pulse pattern generator and incerted into the sampler from the port A [99]

measured by a semiconductor sampling oscilloscope (Tektronix 11801C with an SD-32 50-GHz bandwidth sampling head). The reconstituted waveform produced by the HTSC sampler system has the correct period of 339 ps and its shape is roughly coincident with the reconstituted one produced by the semiconductor sampler. These results show that the HTSC sampler system can measure the 5.9 Gbps digital waveform coming from outside the system.

7.2 Analog to Digital Converter

7.2.1 Superconducting Sigma-Delta Modulator

Figure 70 shows the block diagram of a first order Sigma-Delta (Σ-Δ) modulator [100]. An input signal is integrated and the result is sampled by a clocked comparator. The digitized signal is subtracted from the input signal in a feedback loop. This type of feedback causes the averaged output of the comparator to be exactly the input signal. The deviation of the quantized output of the comparator from the analog input signal can be described as noise and is called quantization noise. The quantization noise is shifted to higher frequencies. It is obvious that the signal-to-noise ratio can be improved tremendously

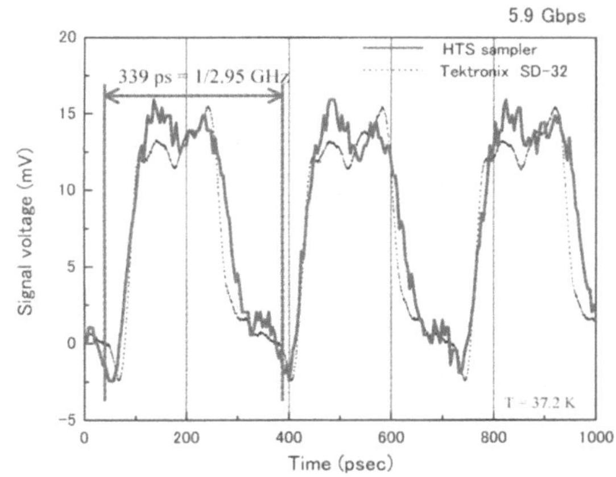

Fig. 69. Digital waveforms reconstituted by the HTS sampler system (*solid line*) and a semiconductor sampling oscilloscope (*dotted line*). In the HTS sampler measurement, a 1 ps delay between every sampling point and a 1 μA step for increasing it were used [99]

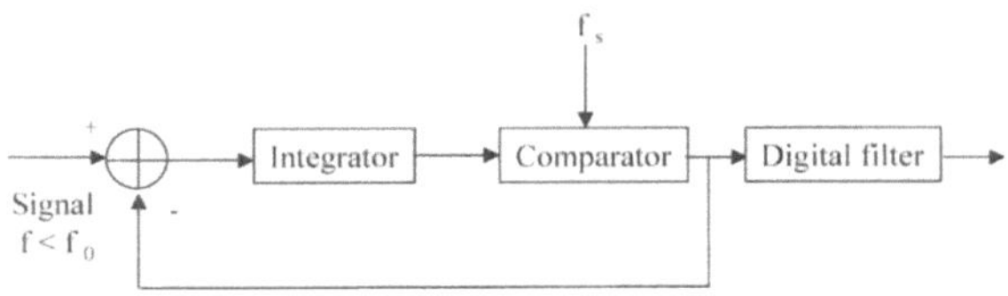

Fig. 70. Block diagram of a first order Σ-Δ modulator with a digital filter [100]

by filtering the frequencies above the signal bandwidth. The Σ-Δ architecture is the preferred architecture for ADCs with large dynamic range [101]. This oversampling approach implemented by semiconductor devices is used in audio applications where signals at kHz frequencies are sampled at MHz by a Σ-Δ modulator and the resulting bit stream digitally filtered to provide a resolution of 18–20 bits. The semiconductor Σ-Δ AD converter is limited to megahertz sampling and digital filtering. A superconductor Σ-Δ AD converter, however, can perform gigahertz sampling and apply the advantages of digital filtering to megahertz bandwidth signals. Moreover, flux quantization in a superconducting loop provides a quantum mechanically precise feedback mechanism unavailable with other technologies. Precision feedback is essential to the performance of Σ-Δ ADCs [102].

7.2.2 Implemented HTSC Σ-Δ Modulators

The integrator, comparator and feedback loop of the first order Σ-Δ modulator are easily implemented by using Josephson junctions as shown in Fig. 71. *Forrester* et al. fabricated a simple HTSC Σ-Δ modulator with 15% Co-doped YBCO barrier ramp-edge junctions in an epitaxial multilayer process

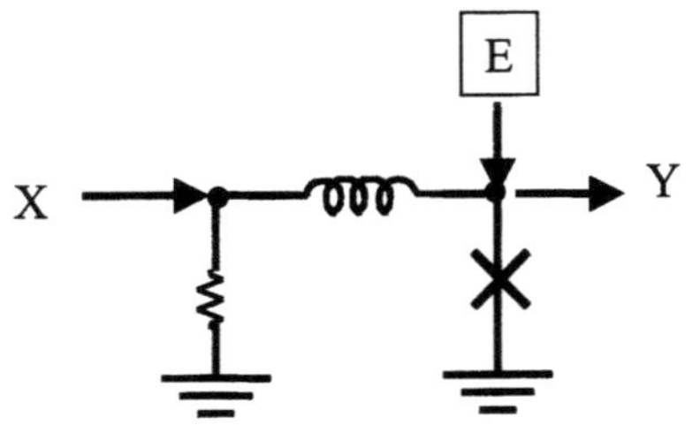

Fig. 71. A superconducting Σ-Δ modulator to convert analog current source X to digital output Y [102]

utilizing three YBCO layers, two epitaxial insulators and an integrated Au resistor [103]. They measured its performance at 35 K by inputting a 5.01 MHz signal and sending the output bit stream to a spectrum analyzer that measured the relative amplitude of the unwanted harmonics which determine the spur-free dynamic range (SFDR). With a 27 GHz sampling rate a SFDP of greater than 75 dB was measured. This value is comparable to that of an LTS modulator [104].

Ruck et al. also reported an HTSC Σ-Δ modulator [105]. The circuit was fabricated on a STO bicrystal substrate. The YBCO/STO/YBCO trilayer was fabricated by laser deposition. The bottom layer served as a superconducting ground plane and the Josephson junctions were formed in the upper layer. Pd/Au thin film was used for the integrator resistance. The circuit consisted of a dc/SFQ converter, a JTL, a comparator, a L/R integrator and an output stage with ten junctions. Correct operation of the modulator was confirmed through dc measurements at 34 K. The linearity of the modulator was studied by measuring the harmonic distortions of a 19.5 kHz sine wave input signal and a minimum resolution of 5 bits could be estimated from the recorded spectrum. This accuracy was limited by the noise of the preamplifier. After the feedback inductance was cut the current feedback loop operated correctly.

A Σ-Δ modulator with 13 IEJs was fabricated and its dc characteristics were tested at 20 K by *Saito* et al. [106]. Figure 72 shows an optical micrograph and schematic diagram of their Σ-Δ modulator. The dc-characteristics (Fig. 73) show that VJ1 is independent of the input current and the sum of VJ3 and VJ4 is constant in the full range of input signal. The former result indicates that there was no interference between the sampling and input signals. The latter one suggests that the number of SFQ pulses is conserved and the sampling frequency reaches 100 GHz.

7.2.3 An Application of the HTSC ADC

Mobile wireless communication network traffic is rapidly increasing and new radio standards and advanced services are being developed to augment or supplant conventional services. This rapid change shortens the useful lifetime of handsets and base stations. Replacement of old base stations with new ones

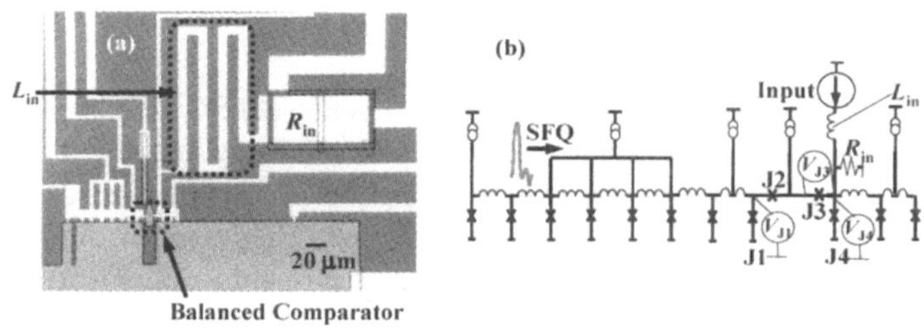

Fig. 72. (**a**) Optical micrograph of Σ-Δ modulator and (**b**) schematic equivalent circuit [106]

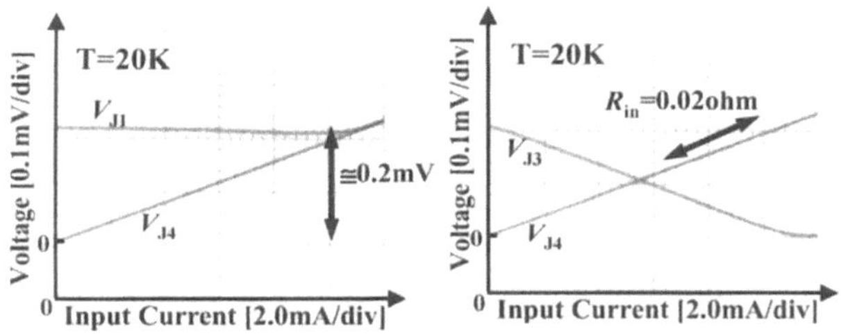

Fig. 73. Measured dc-characteristics of the Σ-Δ modulator [106]

wastes much time and costs money, so future base stations will need to be easily upgradable. A software defined radio (SDR) communication system [107] has software defined functions, e.g., functions the can be altered by software download or exchange identity cards. High speed ADCs with large dynamic ranges are needed because SDR would ultimately require a direct ADC for GHz frequency carriers with a signal bandwidth above 60 MHz and an accuracy of 14 to 16 effective bits [108]. This capability is far beyond the limits of current semiconductor ADC technology. Only the superconducting ADC, which combines Σ-Δ modulator and a digital filter, is capable of satisfying such a stringent requirement. A decimation filter is a kind of digital filter that receives the output data stream from a Σ-Δ modulator and removes the high frequency noise from the output of the modulator. The type of decimation filter is defined by its transformation format. In the sinc filter [109], the stages are connected in series and only the first stage is required to operate at the modulator's sampling clock frequency, which reduces the size of the high speed operation circuit. In the case of a 100 MHz bandwidth and a 14 bit Σ-Δ modulator, the number of Josephson junctions comprising the first and second stages is roughly one thousand, and lower stages can be constituted by semiconductor circuitry. One thousand junctions is a good target for the next generation of HTSC circuits.

7.3 Summary

Samplers and ADCs have been developed for small scale HTSC digital applications. For practical implementation, the sampler's bandwidth must be enlarged and the Σ-Δ modulator's frequency characteristics and accuracy must be enhanced. ADCs using the Σ-Δ modulator will also need a digital filter. The author believes that these devices will see practical application in the near future because of their low cooling costs.

Acknowledgements

The authors would like to thank Shoji Inoue (TDI Co. Ltd.), Hideo Itozaki (Sumitomo Electric), Tatusoki Nagaishi (Sumitomo Electric Hightecs), Hirotoshi Terada (Hamamatsu Photonics), Toshihiko Nagamura, Shigenobu Yamanaka and Yoshio Kamitsuji (Unisoku), Harumi Hifumi (NEC Yamaguchi), Kazuzo Fujito (NEC Kyushu) Shuichi Tahara, Mutsuo Hidaka, Wataru Hattori, Hiroshi Tsuda, Yumi Kakuhara, Hiromi Fukui, Masami Yanagisawa and Tohru Tsujide (NEC) for their contributions to this study. The sampler work is supported by the New Energy and Industrial Technology Development Organization (NEDO) through ISTEC as Collaborative Research and Development of Fundamental Technologies for Superconductivity Applications.

References

1. J. Clarke: Nature **372**, 501 (1994)
2. J. P. Wikswo, Jr.: IEEE Trans. Appl. Supercond. **5**, 74 (1995)
3. A. I. Bragainski, H. J. Krause: Physica C **335**, 179 (2000)
4. R. Kötitz, H. Matz, L. Trahms, H. Koch, W. Weitschies, T. Rheinlander, W. Semmler, T. Bunte: IEEE Trans. Appl. Supercond. **7**, 3678 (1997)
5. K. Enpuku, T. Minotani, T. Gima, Y. Kuroki, Y. Itoh, M. Yamashita, Y. Katakura, S. Kuhara: Jpn. J. Appl. Phys. **38**, L1102 (1999).
6. Y. R. Chemla, H. L. Grossman, Y. Poon, R. McDermott, R. Stevens, M. D. Alper, J. Clarke: Proc. Nat. Acad. Sci. (USA) **97**, 14268 (2000)
7. C. P. Folley, K. E. Leslie: Exploration Geophys. **29**, 30 (1998)
8. U. Kalberkamp, U. Matzander, K. D. Husemann, G. Panaitow, E. Zimmermann, Y. Zhang: Appl. Supercond. **5**, 205 (1998)
9. K. Nikawa: IEICE Trans. Electron. **E85-C**, 746 (2002)
10. D. Drung: Supercond. Sci. Technol. **4**, 377 (1991)
11. D. Drung, E. Dantsker, F. Ludwig, H. Koch: Appl. Phys. Lett. **68**, 1856 (1996)
12. T. Hirano, T. Nagaishi, H. Itozaki: Supercond. Sci. Technol. **12**, 759 (1999)
13. H.-P. Muller, P. Godde, K. Czerski, M. Oeff, R. Agrawal, P. Endt, W. Kruse, U. Steinhoff, L. Trahms: Phys. Med. Biol. **44**, 105 (1999)
14. Y. J. Tian, S. Linzen, F. Schmidl, L. Dorrer, R. Weidl, P. Seidel: Appl. Phys. Lett. **74**, 1302 (1999)
15. E. Dantsker, O. M. Froehlich, S. Tanaka, K. Kouznetsov, J. Clarke, Z. Lu, V. Matijasevic, K. Char: Appl. Phys. Lett. **71**, 1712 (1997)

16. A. Kittel, K. A. Kouznetsov, R. McDermott, B. Oh, J. Clarke: Appl. Phys. Lett. **73**, 2197 (1998)
17. Y. Zhang, G.. Panaitov, S. G. Wang, N. Wolters, R. Otto, J. Schubert, W. Zander, H. -J. Krause, H. Soltner, H. Bousack, A. I. Braginski: Appl. Phys. Lett. **76**, 906 (2000)
18. Y. Tavrin, Y. Zhang, M. Muck: A. I. Braginki, and C. Heiden: Appl. Phys. Lett. **62**, 265 (1993)
19. J. Borgmann, G. Ockenfuss, R. Otto, J. Scjibert, W. Zander, A. I. Braginski, P. David: Rev. Sci. Instrum. **68**, 2730 (1997)
20. D. F. He, H.-J. Krause, Y. Zhang, M. Bick, H. Soltner, N. Wolters, W. Wolf, H. Bousack: IEEE Trans. Appl. Supercond. **9**, 3684 (1999)
21. F. Ludwig, A. B. M. Jansman, D. Drung, M. O. Lindstrom, S. Bechstein, J. Beyer, J. Flokstra, T. Schurig: IEEE Trans. Appl. Supercond. **11**, 1315 (2001)
22. K. Sakuta, T. Fujii, K. Ogawa, H. Tamai, T. Kobayashi, M. Tonouchi, H. Itozaki, T. Nagaishi, F. Kamitani: Physica C **378–381**, 1391 (2002)
23. B. Widrow, S. D. Stearns: *Adaptive Signal Processing*, (Prentice-Hall, New Jersey 1985), pp. 302
24. R. Kötitz, H. Matz, L. Trahms, H. Koch, W. Weitschies, T. Rheinlander, W. Semmler, T. Bunte: IEEE Trans. Appl. Supercond. **7**, 3678 (1997)
25. A. Haller, H. Matz, S. Hartwig, T. Kerberger, H. Atzpadin, L. Trahms: IEEE Trans. Appl. Supercond. **11**, 1371 (2001)
26. Y. R. Chemla, H. L. Grossman, Y. Poon, R. McDermott, R. Stevens, M. D. Alper, J. Clarke: Proc. Nat. Acad. Sci. (USA) **97**, 14268 (2000)
27. K. Enpuku, T. Minotani, T. Gima, Y. Kuroki, Y. Itoh, M. Yamashita, Y. Katakura, S. Kuhara: Jpn. J. Appl. Phys. **38**, L1102 (1999)
28. K. Enpuku, T. Minotani, M. Hotta, A. Nakahodo: IEEE Trans. Appl. Supercond. **11**, 661 (2001)
29. S. Kumar, R. Matthews, S. G. Haupt, D. K. Lathrop, M. Takigawa, J. R. Rozen, S. L. Brown, R. H. Koch: Appl. Phys. Lett. **70**, 1037 (1997)
30. K. Schlenga, R. McDermott, J. Clarke, R. E. de Souza, A. Wong-Foy, A. Pines: Appl. Phys. Lett. **75**, 3696 (1999)
31. S. Tanaka, A. Hirata, Y. Saito, T. Mizaguchi, Y. Tamaki, I. Sakita, M. Monden: IEEE Trans. Appl. Supercond. **11**, 665 (2001)
32. K. Enpuku, T. Minotani: IEICE Trans. Electron. **E84-C**, 43 (2001)
33. E. F. Fleet, S. Chatraphorn, F. C. Wellstood, L. A. Knauss: IEEE Trans. Appl. Supercond. **9**, 4103 (1999)
34. T. S. Lee, E. Dantsker, J. Clarke: Rev. Sci. Instrum. **67**, 4208 (1996)
35. J. Beyer, D. Drung, F. Ludwig, T. Minotani, K. Enpuku: Appl. Phys. Lett. **72**, 203 (1998)
36. F. Ludwig, D. Drung: Appl. Phys. Lett. **75**, 2821 (1999)
37. K. Enpuku, D. Tokimizu, D. Kuroda, S. Hijiya: IEICE Trans. Electron **E85-C**, 681 (2002)
38. T. Nagaishi, H. Itozaki, H. Katayama, H. Noguchi: Extended Abstracts 8th Int. Supercond. Electronics Conf. (2001) p. 471
39. V. Zakosarenko, A. Chwala, J. Ramos, R. Stolz, V. Schultze, H. Luetjen, J. Blume, T. Schueler, H-G. Meyer: IEEE Trans. Appl. Supercond. **11**, 896 (2001)
40. A. Chwala, R. Ijsselsteijn, R. Stolz, V. Schultze, N. Ukhansky, H-G. Meyer: Extended Abstracts 8th Int. Supercond. Electronics Conf. (2001) p. 467

41. D. Drung, T. Radic, H. Matz, H. Koch: IEEE Trans. Appl. Supercond. **7**, 3283 (1997)
42. R. Stoltz, V. M. Zakosarenko, L. Fritzsch, N. Oukhanski, A. Chwarla, H.-G. Meyer: Extended Abstracts 8th Int. Supercond. Electronics Conf. (2001) p. 473
43. N. Kasai, Y. Fujinawa, H. Iitaka, K. Nomura, Y. Hatsukade, S. Sato, H. Nakano, T. Doi, T. Nemoto, A. Ishiyama: Supercond. Sci. Technol. **14**, 1135 (2001)
44. T. Rikitake: *Earthquake Prediction* (Elsevier, New York 1976) p. 357
45. S. K. Park, M. J. S. Johnstone, T. R. Madden, F. D. Morgan, F. Morrison: Rev. Geophys. **31**, 117 (1993)
46. M. Parrot, J. Achache, J. J. Berthelier, E. Blanc, A. Deschamps, F. Lefeuvre, M. Menvielle, J. L. Plantet, P. Tarits, J. P. Villain: Phys. Earth Planet. Inter. **77**, 65 (1993)
47. M. Hayakawa, Y. Fujinawa: *Electromagnetic Phenomena Related to Earthquake Prediction* (Terra Sci., Tokio 1994)
48. M. Van Exter, Ch. Fattinger, D. Grischkowsky: Appl. Phys. Lett. **55**, 337 (1989)
49. A. C. Fraser-Smith, A. Bernardi, P. R. McGill, M. E. Ladd, R. A. Helliwell, O. G. Villard, Jr: Geophys. Res. Lett. **17**, 1465 (1990)
50. O. A. Molchanov, Y. A. Kopytenko, P. M. Voronov, E. A. Kopytenko, T. G. Matiashvili, A. C. Frather-Smith, A. Bernardi: Geophys. Res. Lett. **19**, 1495 (1992)
51. M. Hayakawa, R. Kawate, O. A. Molchanov, K. Yumoto: Geophys. Res. Lett. **23**, 241 (1996)
52. Y. Fujinawa, T. Kumagai, K. Takahashi: Geophys. Res. Lett. **19**, 9 (1992)
53. Y. Fujinawa, T. Matsumoto, M. Ukawa, H. Iitaka, K.Takahashi, H. Nakano, T. Doi. T. Saito, N. Kasai, S. Sato: J. Appl. Geophys. (in press)
54. T. Nagao, K. Hattori, K. Sayanagi: J. Jpn. Soc. Appl. Electromag. Mech. **9**, 435 (2001) (in Japanese)
55. Y. Fujinawa, K. Takahashi: Nature **347**, 376 (1990)
56. K. Takahashi, Y. Fujinawa, T. Matsumoto, T. Nakayama, T. Sawada, H. Sakai, H. Iitaka: Tech. Note Nat. Res. Instit. Earth Science Disaster Prevention **204**, 1 (2000)
57. Y. Fujinawa, K. Takahashi, T. Matsumoto, H. Iitaka, S. Yamane, T. Nakayama, T. Sawada, H. Sakai: J. Appl. Geophys. **25**, 247 (2000)
58. S. K. Park, M. J. S. Johnston, T. R. Madden, F. D. Morgan, F. Morrison: Rev. Geophys. **31**, 117 (1993)
59. O. A. Molchanov, M. Hayalkawa: Geophys. Res. Lett. **22**, 3091 (1995)
60. M. A. Fenoglio, M. J. S. Johnston, J. D. Byerlee: J. Geophys. Res. **100**, 12951 (1995)
61. T. Ishido, H. Mizutani: J. Geophys. Res. **86**, 1763 (1981)
62. Y. Fujinawa, T. Matsumoto, K. Takahashi: J. Appl. Geophys. **49**, 101 (2002)
63. T. Yoshimoto, M. Kotani, S. Kuriki, H. Karibe, N. Nakasato: *Recent Advances in Biomagnetism* (Tohoku Univ. Press, Sendai 1999)
64. O. A. Molchanov, M. Hayakawa, V. A. Rafalsky: J. Geophys. Res. **100**, 1691 (1995)
65. M. Wyss: *Evaluation of Proposed Earthquake Precursors* 94 (AGU, Washington, DC 1991)

66. Y. A. Kopytenko, T. G. Matiashvili, P. M. Voronov, E. A. Kopytenko: *Electromagnetic Phenomena Related to Earthquake prediction* (Terra Sci., Tokyo 1994) p. 175
67. K. Nikawa, S. Inoue: Proc. LSI Testing Symposium 203 (2000) (in Japanese)
68. K. Nikawa, S. Inoue: Proc. Int. Reliability Phys. Symp. (Orlando, Florida, 30 April–3 May 2001) p. 289 (2001)
69. J. Beyer, H. Matz, D. Drung, Th. Schurig: Appl. Phys. Lett. **74**, 2863 (1999)
70. M. Daibo, T. Kotaka, A. Shikoda: Physica C **357–360**, 1483 (2001)
71. L. A. Knauss, B. M. Frazier, H. M. Christen, S. D. Silliman, K. S. Harshavardhan, E. F. Fleet, F. C. Wellstood, M. Mahanpour, A. Ghaemmaghami: Proc. Int. Symp. for Testing and Failure Analysis (ISTFA '99), 1999, p. 11
72. S. Chatraphorn, E. F. Fleet, F. C. Wellstood, L. A. Knauss, T. M. Eiles: Appl. Phys. Lett. **76**, 2304 (2000)
73. J. R. Kirtley, M. B. Ketchen, K. G. Stawiasz, J. Z. Sun, W. J. Gallagher, S. H. Blanton, S. J. Wind: Appl. Phys. Lett. **66**, 1138 (1995)
74. T. Morooka, K. Tanaka, A. Odawara, S. Nakayama, A. Nagata, M. Ikeda, K. Chinone: Physica C **335**, 157 (2000)
75. K. K. Likharev, V. K. Semenov: IEEE Trans. Appl. Supercond. **1**, 3–28 (1991)
76. H. Numata, M. Tanaka, Y. Kitagawa, S. Tahara: Extended Abstracts 7th Int. Supercond. Electronics Conf., Berkeley USA (1999) pp. 272–274
77. T. V. Filippov, Yu A. Polyakov, V. K. Semenov, K. K. Likharev: IEEE Trans. Appl. Supercond. **5**, 2240–2243 (1995)
78. V. K. Semenov, T. V. Filippov, Yu. A. Polyakov, K. K. Likharev: **7**, 3617–3621 (1997)
79. O. B. Oelze, B. Ruck, E. Sodtke, T. Filippov, A. Kidiyarova-Shevchenko, M. Kupriyanov, W. Prusseit: IEEE Trans. Appl. Supercond., **7**, 3450–3453 (1997)
80. Z. G. Ivanov, V. K. Kaplunenko, E. A. Stepantsov, E. Wikborg, T. Claeson: Supercond. Sci. Technol. **7**, 239–241 (1994)
81. S. Shokhor, B. Nadgomy, M. Gurvitch, V. Semenov, Yu. Polyakov, K. Likharev: Appl. Phys. Lett., **67**, 2869–2871 (1995)
82. G. J. Gerritsma, M. A. J. Verhoven, R. J. Wiegerink, H. Rogalla: IEEE Trans. Appl. Supercond. **7**, 2987–2992 (1997)
83. J. D. McCambridge, M. G. Forrester, D. L. Miller, B. D. Hunt, J. Pryzbysz, J. Talvacchio, R. M. Young: IEEE Trans. Appl. Supercond. **7**, 3622–3625 (1997)
84. T. Hashimoto, S. Inoue, T. Nagano, J. Yoshida: IEEE Trans. Appl. Supercond. **9**, 3821–3824 (1999)
85. S. A. Faris: Appl. Phys. Lett. **36**, 1005–1007 (1980)
86. P. Wolf, B. J. Van Zeghbroeck, U. Deutsch: IEEE Trans. Magn. **MAG-21**, 226–229 (1985)
87. S. R. Whitely, G. K. G. Hohenwarter, S. A. Faris: IEEE Trans. Magn. **MAG-23**, 899–902 (1987)
88. M. Hidaka, S. Tsai: IEEE Trans. on Appl. Supercond. **5**, 3353–3356 (1995)
89. J. Gao, Yu. M. Boguslavskij, B. B. G. Klopman, D. Terpstra, R. Wijbrans, G. J. Gerritsma, H. Rogalla: J. Appl. Phys. **72**, 575–583 (1992)
90. B. H. Moeckly, K. Char: Appl. Phys. Lett. **71**, 2526–2528 (1997)
91. T. Satoh, M. Hidaka, S. Tahara: IEEE Trans. Appl. Supercond. **9**, 3141–3144 (1999)
92. S. Miura, W. Hattori, T. Satoh, M. Hidaka, S. Tahara, J. S. Tsai: Supercond. Sci. Technol. **9**, A59–61 (1996)

93. W. H. Mallison, S. J. Berkowitz, A. S. Hirahara, M. J. Neal, K. Char: Appl. Phys. Lett. **68**, 3808–3810 (1996)
94. B. D. Hunt, M. G. Forrester, J. Talvacchio, J. D. McCambridge, R. M. Young: Appl. Phys. Lett. **68**, 3805–3807 (1996)
95. M. Hidaka, J. S. Tsai: IEEE Trans. Appl. Supercond. **5**, 3353–3356 (1995)
96. H. Terai, M. Hidaka, T. Satoh, S. Tahara: Appl. Phys. Lett. **70**, 2690–2692 (1997)
97. M. Hidaka, T. Satoh, M. Koike, S. Tahara: IEEE Trans. Appl. Supercond. **9**, 4081–4086 (1999)
98. M. Hidaka, N. Ando, T. Satoh, S. Tahara: IEEE Trans. Appl. Supercond. **11**, 986–989 (2001)
99. M. Hidaka, T. Satoh, M. Kimishima, M. Takayama, S. Tahara: IEEE Trans. Appl. Supercond. **11**, 267–270 (2001)
100. B. Ruck, Y. Chong, R. Dittmann, A. Engelhardt, E. Sodtke, M. Siegel: Supercond. Sci. Technol. **12**, 701–703 (1999)
101. J. C. Candy, G. C. Temes: *Oversampling A/D converters* (IEEE Press, Washington 1992)
102. J. X. Przybysz, D. L. Miller, E. H. Naviasky, J. H. Kang: IEEE Trans. Appl. Supercond. **3**, 2732–2735 (1993)
103. M. G. Forrester, B. D. Hunt, D. L. Miller, J. Talvacchio, R. M. Young: Extended Abstracts 7th Int. Supercond. Electronics Conf., Berkeley USA (1999) pp. 29–31
104. D. L. Miller, J. X. Przybysz, D. L. Meier, J. H. Kang, A. Hodge: IEEE Trans. Appl. Supercond. **5**, 2453–2456 (1995)
105. B. Ruck, Y. Chong, R. Dittmann, M. Siegel: Physica C **326-327**, 170–176 (1999)
106. K. Saitoh, Y. Soutome, T. Fukazawa, K. Takagi: Supercond. Sci. Technol. **15**, 280–284 (2002)
107. J. Mitola: IEEE Commun. Mag. **33**, 26–38 (1995)
108. E. B. Wikborg, V. K. Semenov, K. K. Likharev: IEEE Trans. Appl. Supercond. **9**, 3615–3618 (1999)
109. H. Hasegawa, T. Hashimoto, S. Nagasawa, H. Suzuki, K. Miyahara, Y. Enomoto: IEEE Trans. Appl. Supercond. **11**, 517–520 (2001)

Material Technology for Vortex Electronics

T. Kobayashi[1], S. Oda[2], O. Michikami[3], and T. Terashima[4]

[1] Graduate School of Engineering Science, Osaka University
1-3 Machikaneyama, Toyonaka, Osaka 560-8531, Japan
kobayashi@sup.ee.es.osaka-u.ac.jp

[2] Research Center for Quantum Effect Electronics, Tokyo Institute of Technology
2-12-1 O-okayama, Meguro-ku, Tokyo 152-8552, Japan
soda@pe.titech.ac.jp

[3] Faculty of Engineering, Iwate University
4-3-5 Ueda, Morioka, Iwate 020-8551, Japan
michikam@msv.cc.iwate-u.ac.jp

[4] Institute for Chemical Research, Kyoto University
Gokasho, Uji, Kyoto 611-0011, Japan
terashim@scl.kyoto-u.ac.jp

Abstract. high-T_c superconductor (HTSC) thin films are typically grown by mean of pulsed laser deposition (PLD), metalorganic chemical vapor deposition (MOCVD), sputtering or molecular beam epitaxy (MBE). This chapter reviews recent progress in the thin film growth technologies of HTSCs.

1 Introduction

In Sect. 2 sophisticated but very practical pulsed laser deposition (PLD) methods named "eclipse PLD" and "aurora PLD" have been developed for HTSC and related thin film formation. A droplet-free thin film can be obtained by employing the eclipse PLD method where a small shadow mask placed between the target and substrate facilitates capturing the droplets in the laser plasma plume almost completely. The aurora PLD, where a diverging magnetic field is applied from the substrate to the target, is characterized by an activation and/or enhanced ionization of the ablated particles during transport. It results in room temperature epitaxy of (100)NiO on (001)MgO. Using the water-cooled magnetic platelet located between the target and substrate, "eclipse-aurora PLD" worked well even the heated substrate. Much better quality in $SrTiO_3$ film was obtained by this improved PLD. Many basic findings suggested that the main mechanism of the aurora PLD is the electron recoil arising from the magnetic mirror effect which, in turn, gives rise to interaction of energetic electrons and atoms (or ions).

MOCVD is a suitable technology for large scale production of high quality vortex thin films. Important issues for commercial vortex application are the reproducibility of thin film growth and the reliability of vortex oxide films, which are subjects in Sect. 3. In order to improve the reproducibility of MOCVD we have investigated real-time monitoring and controlling

T. Kobayashi et al. (Eds.): Vortex Electronics and SQUIDs,
Topics Appl. Phys. **91**, 249–291 (2003)

of growth processes, which includes an ultrasonic sensor for precursor supply rate monitoring and in-situ ellipsometry of thin-film surfaces during deposition. Spectroscopic ellipsometry brings useful information on the initial growth processes of oxide thin films. Ga addition is an effective approach for solving the problem of degradation of the vortex interface caused by oxygen deficiency, because oxygen diffusion is suppressed by Ga substitution for Cu in $YBa_2Cu_3O_7$(YBCO).

In sputter deposition of HTSC materials with weak crystal lattices containing oxygen atoms, bombardment of a substrate surface with high energy particles from the target must be avoided. Representative sputtering methods in which the energetic particles are controlled are described in Sect. 4. The features of these methods are described and the possibility of large sized film deposition is discussed. It is shown that facing targets magnetron sputtering is a promising method for obtaining uniformity and enlargement of a thin film. The results of deposition of epitaxial $EuBa_2Cu_3O_7$ (EBCO) thin films on R-plane sapphire substrates using single target planar magnetron sputtering are shown. By optimizing the sputtering conditions of a CeO_2 buffer layer, a PBCO template layer and an EBCO thin film, high-quality *a*- and *c*-oriented EBCO thin films were obtained. Furthermore, it was shown that the activated oxygen plasma can be used for the recovery process of a degraded HTSC thin film.

In Sect. 5 ultrathin YBCO films and YBCO/$PrBa_2Cu_3O_7$(PrBCO) superlattices were epitaxially grown by MBE with monitoring of reflection high energy electron diffraction (RHEED) intensity oscillations. The interlayer coupling between the YBCO layers has been found to disappear when they are separated by a 2 unit cells thick PrBCO layer.

2 Pulsed Laser Deposition Method for HTSC and Related Oxide Film Formation

In recent superconducting electronics, oxide electronics, nitride electronics and so on, preparation of very high quality thin films has been requested. So far, several preliminary experiments on functional oxide devices have been reported intimating opening a new method. However, presently the oxide semiconductors still face the problem of lattice imperfections (trap problem) which makes us feel there still is a long way to go [1,2]. In addition, there is a demand for low temperature epitaxy. For instance, the antiferromagnetic NiO/$YBa_2Cu_3O_7$(YBCO) layered structure junction is a promising candidate for low noise SQUID fabrication. To establish this junction, the NiO cap layer should be grown at lower temperatures (as low as room temperature).

Since the discovery of high temperature superconducting oxides, the pulsed laser deposition (PLD) method has become a familiar technology for thin film preparation. The PLD method is very advantageous over the conventional thin film technologies in view of the high deposition rate, composition transfer from the target even for multielement compounds, etc. However, even the PLD was not sufficient to overcome the problems stated above. Besides, the PLD itself had a problem different from the above: the droplet problem. Via microexplosion at the molten target surface a number of small droplets are transferred together with the growth species and deposit on the film surface. In consequence, fatal damage is brought about to the film by coexistence of the droplets when one uses the film for an electronic application. To fulfill these requirements, new PLD techniques named the "eclipse method", the "aurora method" and their combination the "eclipse-aurora method" have been proposed and developed [3,4,5,6]. The eclipse PLD, where only a tiny shadow mask is placed between the target and the substrate, is used obtain droplet-free thin films. The shadow mask serves as a filter to eliminate the flying droplets in the laser plasma plume. On the other hand, the aurora PLD, wherein a diverging magnetic field is applied from the substrate side toward the target, is an effective method to provide high crystal quality thin film and low temperature film growth. The aurora method is characterized by enhanced ionization of the ablated particles during transport from the target to the substrate through interaction between the particles and the applied magnetic field. Owing to the aurora PLD ability, room temperature epitaxy of (100)NiO on (001)YBCO became feasible even when the eclipse shadow mask was used.

One can find, in the text, a set of very strange plume images, drastic enhancement of the plume emission in the aurora PLD (see Figs. 4–7). The enhanced activation of growth species during the transfer is thought to be responsible for the low temperature epitaxy. What is the main mechanism of the aurora effect? This question should be solved for further improvement of the PLD technique as well as for progress in laser ablation science.

So far, aurora PLD has faced the problem of thermal degradation of the permanent magnet due to radiant heat from the heated substrate holder when we grow films at elevated temperatures (more general case). Actually, the permanent magnet is indeed less resistive to the heat. An electromagnet can also provide a magnetic field in place of a permanent magnet. However, its drawback is a large geometrical volume of the magnet and less ability to provide large field divergence (related to the aurora mechanism).

Very recently, the mechanism of the enhanced plume emission in the aurora PLD method was closely investigated and, on the basis of new findings, an advanced aurora PLD technique (named eclipse-aurora PLD) was developed for practical use.

2.1 Eclipse PLD

In most cases of PLD experiments and applications one faces the problem of droplets which disperse on the entire film surface. The droplets are thought to originate from microexplosions of the molten target surface. *Ishibashi* et al. explored the laser ablation mechanism of the oxide target and they concluded

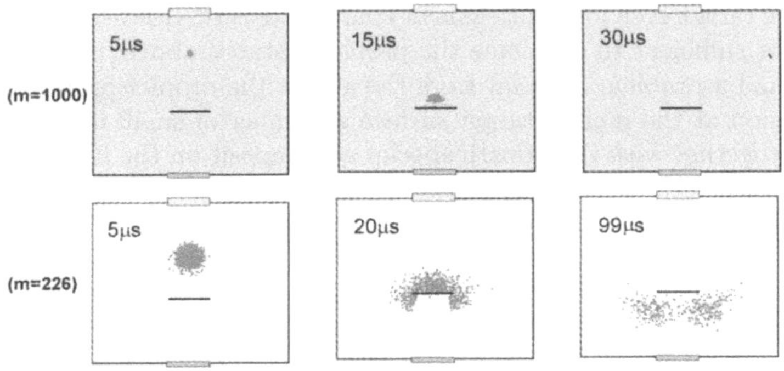

Fig. 1. Monte Carlo simulation of the particle transfer in the chamber. The spacing between the target and substrate is 5 cm. Oxygen gas of 0.2 Torr is fed inside the chamber. Scattering is due to collision between the particles and oxygen molecules

that thermal ablation dominates on the oxide targets instead of photochemical ablation [7]. In the thermal ablation process, the growth species departs from the target via boiling of the molten target surface and therefore, as is well known, microexplosion of the boiling melt takes place.

To solve the droplet problem we should first look at the dynamics of the flying particles (light particles: growth species, and heavy particles: droplets) obtained by numerical calculation (Monte Carlo simulation) [8]. In Fig. 1 the results for YBCO deposition are shown where the mass numbers of droplets and growth species are, respectively, 1000 and 226 (corresponding to Y_2O_3), both initial drifting energies are 100 eV and the background gas is oxygen at 0.2 Torr. As clearly seen in this figure, the droplets move almost straightforward because of less scattering due to their heavy mass. On the other hand, the growth species scatter frequently and strongly diffuse on the transfer path. Accordingly, small platelet (shadow mask) placed between the target and substrate can serve as a mass filter for eliminating the droplets. Eclipse PLD is based on the concept of the mass filter by the shadow mask.

By simply placing a tiny shadow mask (1 cm × 1 cm) between the target and the substrate one can set up the eclipse PLD apparatus. Droplet-free YBCO and related films are now routinely obtained by utilizing the eclipse PLD method as shown in Fig. 2.

2.2 Aurora PLD

The aurora PLD equipment is basically conventional except for introduction of the permanent magnet on the axis and at (and near) the substrate holder. The configuration of the components is schematically drawn in Fig. 3 where the eclipse shadow mask is also used [5]. The light source is an ArF excimer laser (wavelength 193 nm) whose pulse energy and repetition rate were, respectively, 4 J/cm^2 and 10 Hz. The laser beam irradiates the target in vacuum or oxygen ambient at pressures lower than 0.1 Torr (mostly 0.01 Torr).

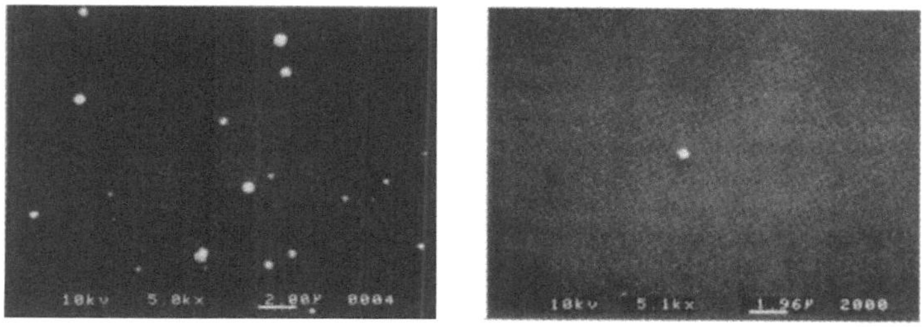

(a) (b)

Fig. 2. Scanning electron microscope images of YBCO films. (**a**) Conventional PLD, and (**b**) eclipse PLD. Oxygen pressure was 0.2 Torr

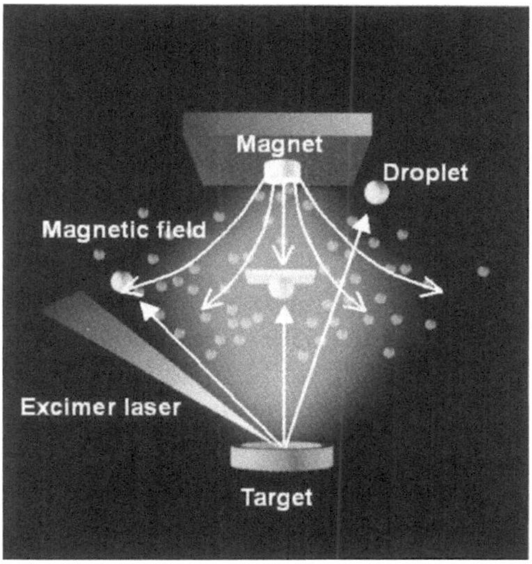

Fig. 3. Schematic drawing of the aurora PLD method. In this figure the eclipse shadow mask is used

The plume dynamics is measured by an ultrahigh speed framing streak camera (Hamamatsu Photonics, C4187 with a minimum window time as short as 50 ns). With the aid of an optical interference filter, measurements are done for the individual species (separately for atoms and ions). As to the plume spectral measurement, a rapid multi-channel analyzer (Hamamatsu Photonics, PMA-11) is used. Local spectra are also obtained using this apparatus. Film crystallinity, morphology and so on were characterized for several aurora conditions. Throughout this work, commercially available permanent magnets (Nd and SmCo system, 1.5 cm diameter) are used. The surface magnetic flux density is about 0.45 T. As we will discuss later, smaller size magnet platelet is much better for the aurora PLD.

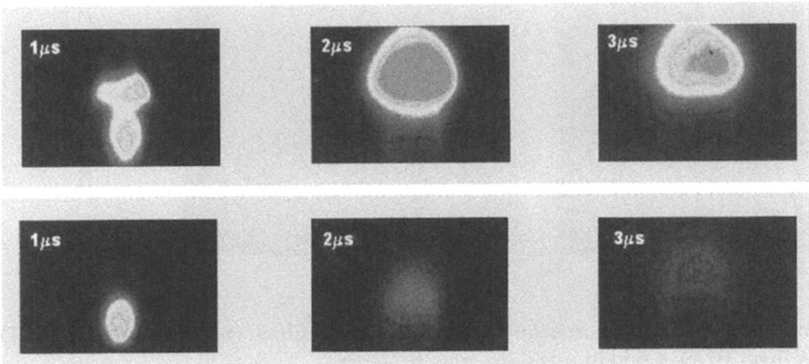

Fig. 4. Time resolved plume observation. *Top* images are taken from the aurora PLD and *bottom* images from conventional PLD

2.2.1 Fundamentals of Aurora PLD

A time sequence of the plume image change is given in Fig. 4, where an $SrTiO_3$ target is ablated in 0.01 Torr O_2 ambient and a permanent magnet is placed just at the position of the substrate holder. In this figure the temporal changes of the plume image are also shown as a reference. A marked difference with the aurora PLD is growth of the shining emission of the plume as the plume gets closer to the magnet. By inserting optical interference filters [center wavelengths corresponding to emissions from Sr^+ (422 nm) and Sr (461 nm)] at the camera head, wavelength-resolved plume images were taken and compared with results in Fig. 4, indicating that the shining emission of the aurora PLD comes from Sr^+ species.

Plume emission spectral measurements were done near the target and substrate sides for both the aurora and conventional PLDs using $SrTiO_3$ targets. Although the results are not given here, it was known that spectral peaks of Sr^+ near the substrate side increased to more than ten times higher than the initial level in the aurora PLD. Moreover, a peak from Ti^+ was also seen but showed no special change on application of the magnetic field. In other words, there must be a preferential activation for species in the aurora PLD.

The observed marked increase in Sr^+ emission intensity implies the possibilities that Sr atoms in the plume are intensively ionized and activated and/or existing Sr^+ ions are effectively activated due to the presence of the magnetic field. At the moment we cannot say which case is more appropriate from present data. Whichever the real case might be, in the following, we shall explore, the role of the applied magnetic field in aurora PLD.

To learn the aurora PLD mechanism (enhanced emission), we have checked the plume dynamics for various magnet configurations. The results are given in Fig. 5 where the permanent magnet location is indicated by the inset. Omitting description of the details of data analysis, we summarize the

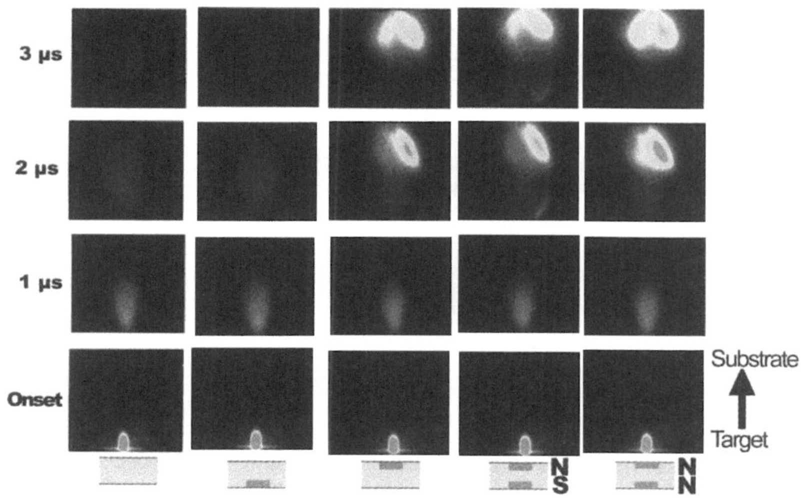

Fig. 5. Plume changes under various conditions of magnetic field application in aurora PLDs

main results on the plume change for variation of magnetic field application as follows:

(1) Magnetic field from the target side: No Change
(2) Magnetic field from the substrate side: Enhanced
(3) Magnetic field from both the substrate and target sides (N-S): Enhanced but rather weak
(4) Magnetic field from both the substrate and target sides (N-N): More enhanced.

As a general rule, species such as neutral atoms and ions are too heavy to directly feel the effect of the magnetic field. Furthermore, the used magnetic flux density was at most 0.45 T. Therefore, it is hard to explain the above mentioned observations (1)–(4) by a model of the direct interaction of particles (atoms and ions) and magnetic field. There must coexist energized electrons together with the atomic species in the plume. Electrons can respond to the applied magnetic field sufficiently because of their light mass. To what extent does the magnetic field bring about modulation of the electron dynamics? We performed a computer simulation of the electron transport in the case where the magnetic field is supplied from a magnetic dipole for simplicity and O_2 at 0.01 Torr is fed.

In Fig. 6 some of calculated results are shown and one can compare them with each other. In the figure, the target and substrate are located at the bottom and top, respectively, and the magnet location is indicated by a thin rectangular symbol. In any case, electrons depart from the target surface with an initial energy of 80 eV. (This energy value was used tentatively.) In

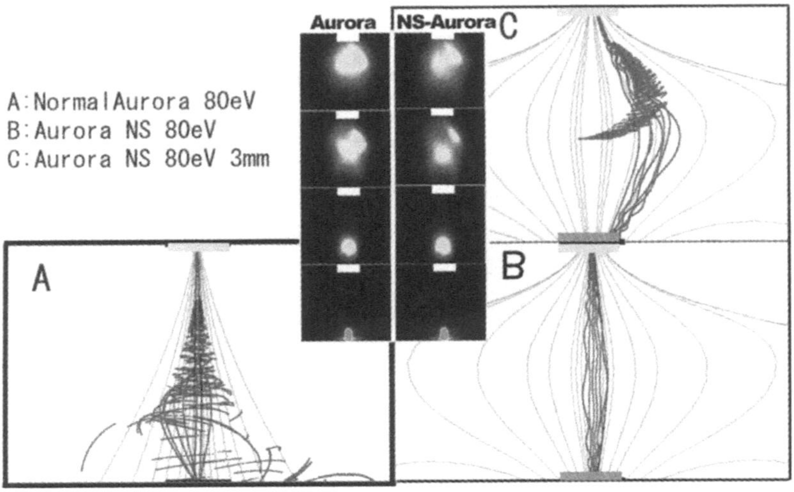

Fig. 6. Calculated electron motion in the magnetic field

case A (standard aurora PLD), electrons move toward the substrate but most of them begin to recoil (shown by the complicated trajectory) at a position some distance from the magnet surface. Here it should be noted that electrons initially moving on the central axis and near the axis zone seldom recoil. This kind of electron recoil behavior is known as a magnetic mirror effect, very familiar in the field of the plasma confinement, and it is related to the amount of div $\boldsymbol{B}$ (divergence of the flux density). According to this view, we arrive at the expectation that the magnetic force lines in cases B and C are rearranged to be more and more straight as compared to case A, and therefore recoil behavior can no longer take place as long as we use the initial conditions for electrons in the calculation. In the simulation of B, we cannot see any sign of a recoil trajectory. If we give a slight movement (3 mm) of the electron departure position from the target (case C), we have recoil behavior as shown in the figure. Although we have not yet calculated the case of the configuration of magnets N-N (or S-S), one can imagine electron recoil occurring more and more frequently based on the magnetic mirror effect model (and from a simple analogy to Fig. 6).

At this stage, we can suppose that electrons initially distribute in the leading edge of the plume, move together with the atoms and ions toward the substrate (magnet), interact with the magnetic field, and then some of them return by the mirror effect. When electrons are returning they can collide with neutral atoms and ionize them and/or activate the existing ions. In consequence, a combination of this activation model with the magnetic mirror effect reasonably explains the results given in Fig. 5.

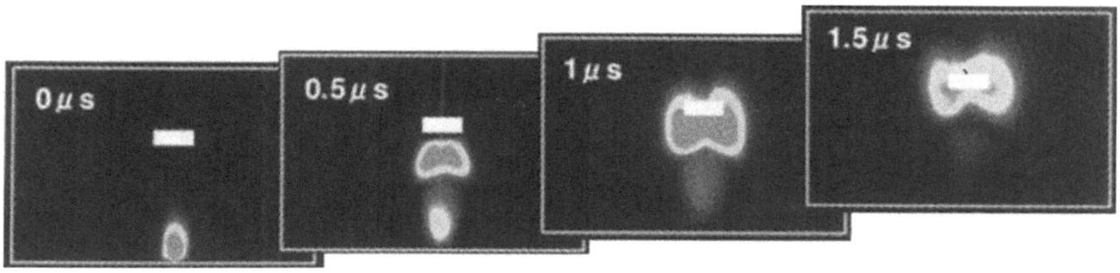

Fig. 7. Aurora PLD with the magnet at the shadow mask position

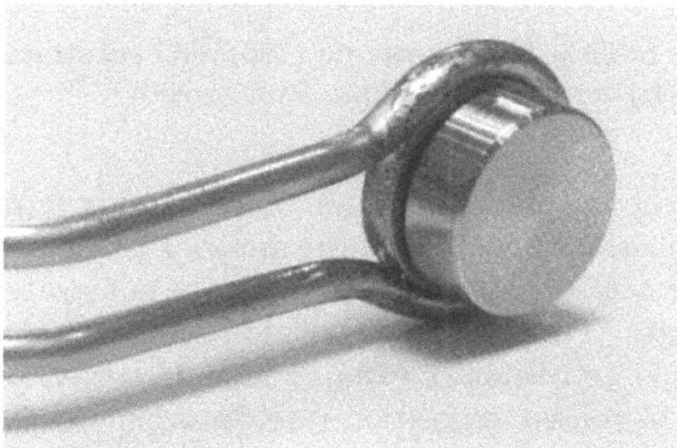

Fig. 8. Photograph of a water-cooled magnet for the eclipse-aurora PLD

2.2.2 Eclipse-Aurora PLD (Advanced Technology)

First, we placed the permanent magnet in contact with the substrate holder, which, in turn, easily damaged the magnet when the holder was heated up. As seen in Fig. 7, it was found that remote (but near the substrate side) placement of the magnet can also enhance the activation of particles.

In this case the magnet serves simultaneously as the shadow mask of the eclipse PLD, suggesting the feasibility of the eclipse-aurora PLD. Unfortunately, however, even the remote magnet still suffered from the damage due to heat radiation from the substrate holder when it was heated up to 600°C. Our final version in the present preliminary experiment was the eclipse-aurora PLD equipped with a water-cooled shadow magnet-mask between the target and substrate. Though data are not shown here, the cooled magnet-mask worked well at the elevated substrate temperatures (even at 600°C). The water-cooled magnet-mask is shown in Fig. 8.

2.2.3 On the Deposition Rate of Aurora PLD

As to the deposition rate of the eclipse-aurora PLD, from the experimental data not shown here, it can be definitely said, that the rate increases to about twice as high as that of conventional eclipse PLD (depending on the configuration of the magnet-mask, its diameter and ambient pressure). An increase in the rate can be understood if one takes the enhanced ionization model into account for the aurora PLD. Namely, more ionized species are

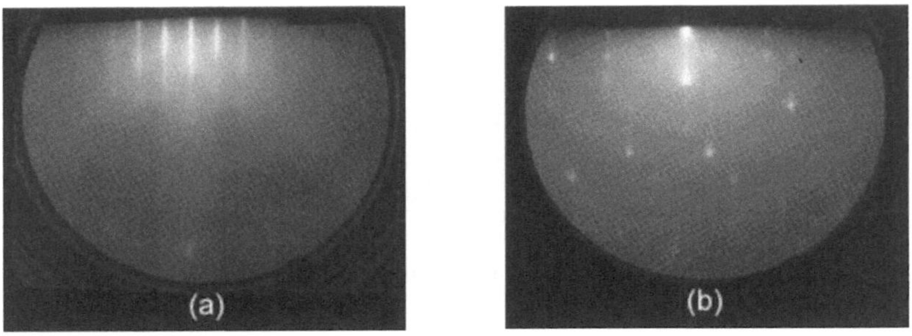

Fig. 9. RHEED patterns taken from $SrTiO_3$ films grown on (100)MgO substrate. (**a**) Eclipse-aurora PLD growth and (**b**) conventional eclipse PLD growth

likely to move forward along with the magnetic flux lines. Since the flux lines converge toward the magnet (or substrate position) in the aurora PLD, more species can be confined around the central axis and deposit on the substrate, resulting in a higher deposition rate. The enhanced emission of the plume could be explained also by enhanced activation of existing ions. However, in this case it is hard to explain the increased deposition rate. In other words, the increased growth rate supports the aurora effect by which the neutral ablation atoms are efficiently ionized via collision with the recoiled electrons.

2.3 Application of the Eclipse-Aurora PLD Method to Electronics Devices

Figure 9 is the first demonstration of RHEED (reflection high electron energy diffraction) patterns taken from $SrTiO_3$ films grown at 500°C in 0.01 Torr O_2 pressure on (100)MgO substrate with (eclipse-aurora PLD) and without (eclipse PLD) magnetic field applications. As clearly seen in this figure, well-ordered $SrTiO_3$ film grew under the eclipse-aurora condition (left image), which can be compared with conventional eclipse PLD growth (right image). The AFM observation revealed a very smooth surface of the eclipse-aurora PLD grown $SrTiO_3$ films shown in Fig. 10. A very rough film surface resulted unless the aurora effect was involved.

Antiferromagnetic NiO film is known to crystallize at room temperature when using several deposition methods, and epitaxially grow on lattice-matched substrates. However, as shown in Fig. 11a, polycrystalline NiO resulted on (100)MgO substrate (due to a large lattice mismatch) when the eclipse PLD was used at room temperature. A remarkable improvement of the room temperature NiO epitaxy was obtained by the eclipse-aurora PLD method. As seen in Fig. 11b, streak lines are observed in the RHEED pattern. This improvement is indeed a result of enhanced activation of growth species by the aurora effect. However, it should be noted that, at least at present, the epitaxial growth of (100)NiO is not obtained on the entire area of (100)MgO substrate. The same is true for the NiO/(001)YBCO layered structure. There

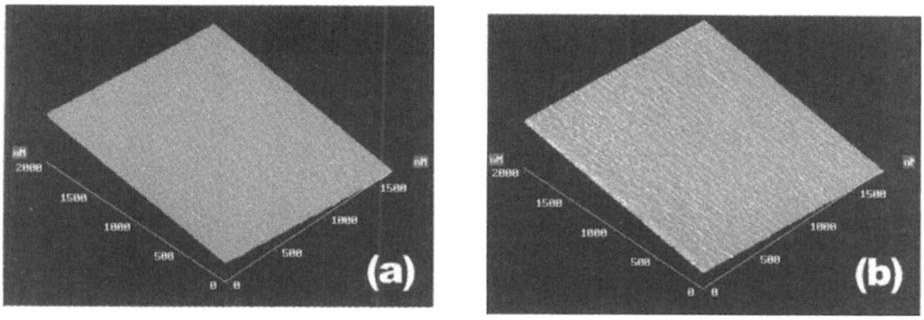

Fig. 10. AFM images taken from $SrTiO_3$ film surfaces grown on (100)MgO substrate. (**a**) Eclipse-aurora PLD growth and (**b**) conventional eclipse PLD growth

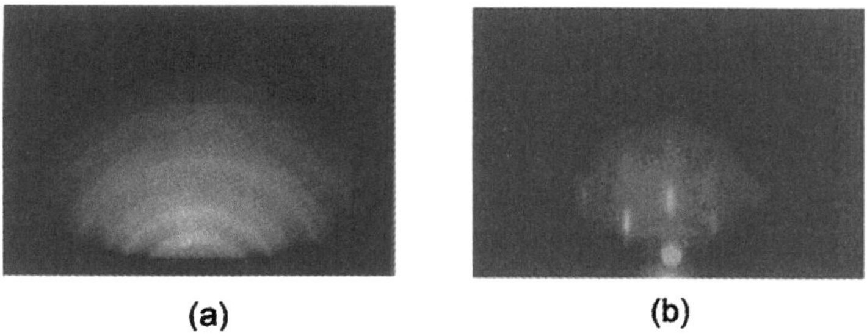

Fig. 11. RHEED patterns taken from NiO/(100)MgO structures. (**a**) Conventional eclipse PLD and (**b**) Eclipse-aurora PLD. PLD was carried out at room temperature in 0.01 Torr oxygen back pressure

remains a problem to be solved for perfect epitaxy of NiO/YBCO at room temperature.

A step-edge Josephson junction consisting of NiO/YBCO bi-layered structure has been long desired because of its suitability for the low noise SQUID sensor. In the preparation, however, we must pay attention to the fact that the junction part made of YBCO grain boundary is unavoidably degraded via additional heating or thermal strain introduced during overlying NiO film deposition at elevated temperatures. Thus, it is thought that room temperature epitaxy of the NiO layer is the best solution for the Josephson junction formation. In the experiment, prior to the film deposition the MgO substrate was patterned and partially ion etched with an ion beam angle of 30 degrees, resulting in the step-etched substrates. Owing to the mild ion etching by the Kaufmann-type etcher, damage to the etched surface of MgO substrates was less pronounced. NiO/YBCO films grown by the eclipse-aurora PLD were followed by the junction patterning process with the ion etching. The completed junction image is given in Fig. 12a.

One Josephson junction in the SQUID loop was wire bonded with four connections, allowing four-terminal measurements. As seen in Fig. 12b, the prepared junction worked well at 77 K.

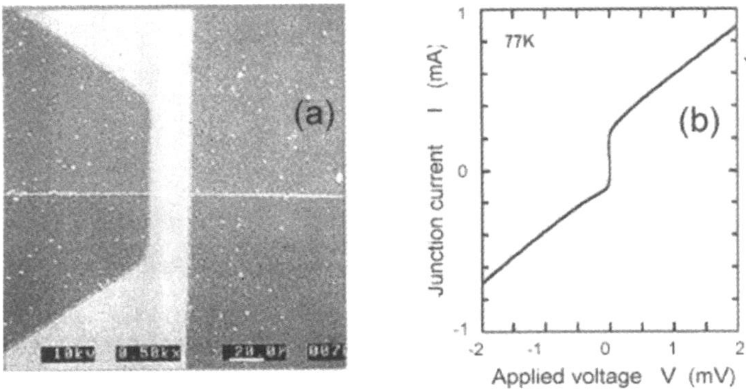

Fig. 12. Prepared SQUID Josephson junction of NiO/YBCO structure. (**a**) Junction image and (**b**) the current-voltage characteristics of one junction at 77 K

2.4 Summary

New PLD methods named "eclipse PLD", "aurora PLD" and "eclipse-aurora PLD" were proposed and developed. In all cases, new PLD methods utilize the characteristic dynamics of ablated particle motion. Eclipse PLD uses only a tiny shadow mask between the target and substrate, which completely hinders droplet formation on the grown films. One can thus solve the droplet problem inherent in the conventional PLD method. The aurora PLD works under application of a magnetic field from the substrate side. It is characterized by an enhanced ionization of the ablated particles during transport from the target to substrate. The main mechanism is thought to be electron recoil arising from the magnetic mirror effect which, in turn, gives rise to interaction of energetic electrons and atoms (or ions). It is a promising technique to obtain room temperature epitaxy of (100)NiO on (001)YBCO, which was applied to prepare the SQUID Josephson junction. Using a water cooled magnetic platelet located between the target and the substrate, the eclipse-aurora PLD was developed. It is capable of working well even at the heated substrate. A much better quality in $SrTiO_3$ film was obtained with this improved PLD. The deposition rate also increased in the aurora PLD.

3 MOCVD for Thin Film Growth

Various deposition methods have been applied to the preparation of thin films of oxide superconductors. For fabrication of integrated vortex devices, productivity and reliability of thin films are very important issues. Metalorganic chemical vapor deposition (MOCVD) is a manufacturing technology, particularly for the features of high throughput and large area uniformity. However, MOCVD of oxide superconductor thin films has several problems, e.g., poor reproducibility and outgrowth of a second phase. In addition, for device applications, poor reliability of YBCO, such as the deterioration of the superconductivity caused by oxygen deficiency in Cu–O chains, becomes a serious problem.

For improvement in reproducibility of MOCVD we have proposed and demonstrated precursor supply monitoring using an ultrasonic gas concentration analyzer and real-time characterization of superconductivity of YBCO thin films by spectroscopic ellipsometry. We also report advanced techniques for further improvement of reproducibility and reliability, i.e., real-time controlling of initial growth by spectroscopic ellipsometry and oxygen stabilization by substitution of Ga for Cu in the Cu–O chain in YBCO.

3.1 Atomic-Layer MOCVD System

Since vortex oxides such as $YBa_2Cu_3O_{7-\delta}$ can be regarded as having a layered crystalline structure, the atomic layer-by-layer growth method is effective to provide high crystalline quality. $YBa_2Cu_3O_{7-\delta}$ films of 14–24 nm thickness were prepared by the atomic layer MOCVD system shown in Fig. 13 [9,10]. Metalorganic sources used in this experiment were $Y(DPM)_3$, $Ba(DPM)_2 \cdot 2\,tetraene$ and $Cu(DPM)_2$ and the temperature of each vaporizer was maintained at 129, 158 and 120°C, respectively. The gas transfer tubes and the ultrasonic concentration analyzer, EPISON II made by Thomas Swan, were heated to 180°C so as to prevent condensation of precursors. Half of the vaporized precursors were transferred sequentially by Ar gas into the reaction chamber, a horizontal quartz tube, in which the deposition takes

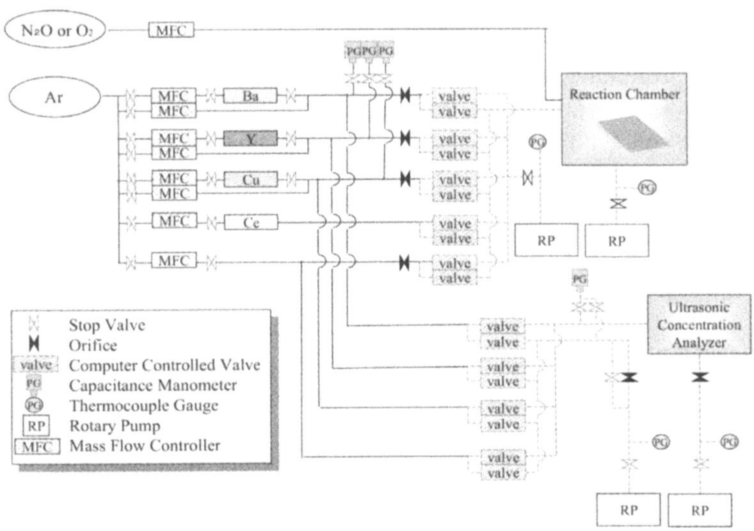

Fig. 13. Schematic diagram of atomic layer MOCVD system

place on $SrTiO_3$ substrates at 650°C. Also the rest was sequentially transferred into the analyzer. Pressures in the vaporizers and in the analyzer were about 150 Torr. The total reaction pressure was 2 Torr and the partial pressure of N_2O as the oxidizing agent was 1 Torr. In order to introduce sufficient oxygen into YBCO for obtaining superconductivity, annealing in O_2 at 400°C, 1 atm for 10 min was carried out after deposition of films. The zero resistive superconducting critical temperature T_c of the films was measured by means of the four-point probe method. The composition of the deposited films was determined by inductive coupling plasma spectroscopy (ICPS).

In-situ observations of the initial growth of YBCO were performed by means of a spectroscopic phase modulated ellipsometer, UVISEL made by Jobin-Yvon. Light was introduced at an angle of 70 degree from the normal to the substrate through a quartz window attached on the reaction chamber. We choose three photon energies, 2.7, 3.9, and 4.1 eV for analysis.

3.2 Real-Time Process Monitoring

3.2.1 Process Monitoring in MOCVD of Oxide Thin Films

In MOCVD the source gas supply rate is determined by the saturated vapor pressure of precursors (defined by the vaporizer temperature) and the flow rate of carrier gas, which are measured and controlled precisely by using a PID temperature controller and a mass flow controller, respectively, as long as the vapor pressure of the precursor and the efficiency of mass transport are constant as is the case for MOCVD of common semiconductors. However, this is not the case for MOCVD of vortex oxide films. Precursors of alkaline earth metals and rare earth metals are not as stable as those of group IV or group III elements. Therefore, in-situ monitoring of the precursor flow rate is desirable for reproducible growth of vortex thin films.

The ultrasonic sensors applied for concentration monitoring the binary gas mixture, are based on the principle that the velocity of sound is determined by the molecular mass and the ratio of the specific heats (C_p/C_v) of the species in the gas phase. Researchers in STI used the ultrasonic sensors to control the composition of metal precursors and successfully prepared high quality YBCO films for microwave applications [11]. We have investigated the method to stabilize the precursor supply rate of Ba by employing feedback systems [12].

Ultraviolet absorption of precursors was reported for in-line process monitoring of MOCVD of YBCO [13]. Optical reflectance measurements also provided a method for in-situ diagnostics of the crystal growth process in MOCVD of vortex thin films [14]. The thickness of films and the mode of crystal growth are deduced from the reflectance measurements. Recently, we investigated spectroscopic ellipsometry for real-time monitoring the crystal growth process of YBCO [15]. In the initial stage of the growth the crystal growth mechanism is distinguishable from the measurement of the

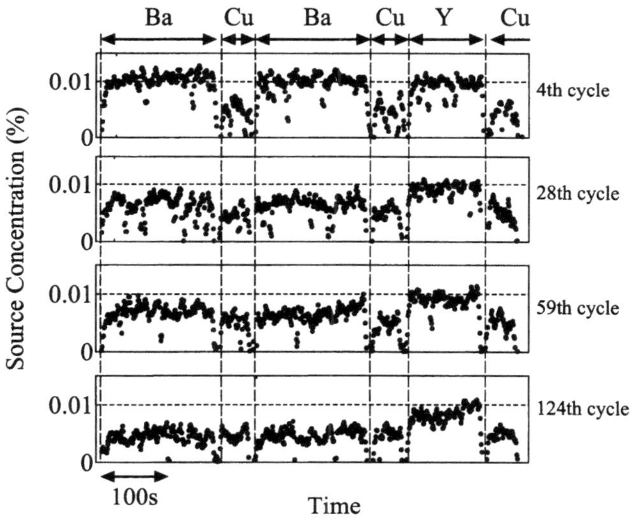

Fig. 14. Time dependence of source gas concentration during deposition

time-dependent ellipsometry signal. A phase transition due to oxygen doping, which is closely related to the critical temperature, is also observable.

3.2.2 Precursor Concentration Monitoring by Ultrasonic Cells

Figure 14 shows the time dependence of the source concentration, the ratio of precursor molecules with respect to Ar carrier gas atoms, during deposition of a YBCO film. Concentrations decrease gradually as deposition time proceeds. In particular, the Ba source concentration of the 124th cycle was about half of that of the 4th cycle. From ICPS analysis, we clarified the correlation between chemical composition in the film and gas flow rate. If we can control the concentrations by employing a feedback system, the reproducibility of MOCVD growth will be improved significantly.

A feedback system using the concentration analyzer was developed based on a basic understanding of precursor properties. In order to maintain deposition conditions including source flow rate, a system having two mass flow controllers for each source was designed. In this system, the sum of the flow rates of vaporizer and bypass mass controllers is kept constant. Figure 15 shows the time dependence of the averaged concentration and flow rate of the vaporizer mass flow controllers during YBCO deposition for 70 cycles. The set point for the temperature controller for the vaporizer of Ba precursor was elevated from 155°C to 157°C at the point indicated by an arrow in Fig. 15 because the flow rate of the vaporizer mass flow controller almost reached the upper limit due to a large decrease of Ba concentration in the vaporizer. During the whole deposition period, the precursor concentration was clearly stabilized by this feedback operation. When feedback operation

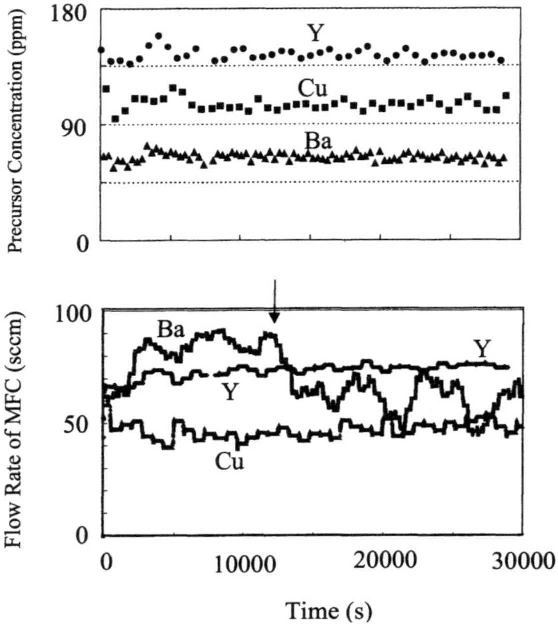

Fig. 15. Time dependence of averaged concentration and flow rate of vaporizer mass flow controllers during YBCO deposition for 70 cycles. The set point for the temperature controller for the vaporizer of Ba precursor was elevated from 155°C to 157°C at the point indicated by an *arrow*

was used, the long-term variation of source concentration was eliminated and reproducibility was improved significantly [12].

3.2.3 Initial Growth Monitoring by in-situ Ellipsometry

Although a report on in-situ measurement of pseudodielectric function in YBCO thin films by sputter deposition was published by *Bijlsma* et al., only a single photon energy of 3.5 eV was used in their experiment [16]. We measured the in-situ pseudodielectric function, $\epsilon = \epsilon_r + \epsilon_i$, of YBCO thin films on $SrTiO_3$ substrate during MOCVD deposition at five photon energies, where ϵ_r and ϵ_i represent the real and imaginary part of pseudodielectric function, respectively. A difference of trajectories on $\epsilon_r - \epsilon_i$ planes was observed due to different film compositions. An especially distinctive difference was obtained at 4.1 eV because in the tetragonal phase YBCO, which is thermodynamically stable in our deposition condition, interband electron transitions at Cu(I) sites cause a strong peak in ϵ_i at 4.1 eV [17,18]. If the pseudodielectric function around this peak is systematically observed, information about creation of YBCO in the initial growth stage is obtained. The difference between the imaginary part of the pseudodielectric function at 4.1 eV and that

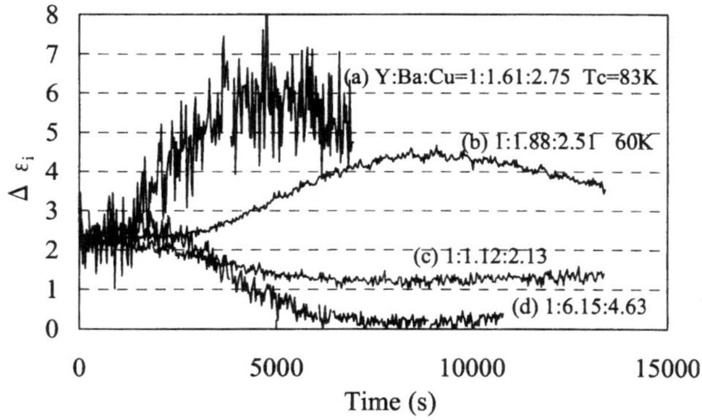

Fig. 16. Time dependence of the imaginary part of pseudodielectric function measured at 4.1 eV subtracted by the value at 3.9 eV. (**a**), (**b**) YBCO thin films showed superconductivity after oxygen doping, while (**c**), (**d**) showed non-superconductivity. Film compositions and zero resistive superconducting critical temperatures are indicated in the figure

at 3.9 eV, $\Delta\,\epsilon_i = \epsilon_i(4.1\text{eV}) - \epsilon_i(3.9\text{eV})$ as an indicator of the peak height at 4.1 eV, varies as time goes on as shown in Fig. 16. The $\Delta\,\epsilon_i$ have a peak point at a film thickness of about 20 nm. In this condition the YBCO thin film is superconducting after oxygen introduction. When the film thickness becomes greater than 20 nm, $\Delta\,\epsilon_i$ decreases because of reduction of the substrate contribution. As shown in Fig. 17, the peak height is strongly related to T_c, i.e., superconductivity, because both these parameters become high when the volume ratio of highly crystallized YBCO is also high. The superconductivity of films can be estimated from this signal during deposition even at a film thickness of 10 nm or less because the signal rises from a very early stage of deposition.

Figure 18a shows some typical trajectories of the pseudodielectric function at 2.7 eV of YBCO films in a Gaussian plane during deposition. The curvature of the trajectories corresponds to the amount of precipitates on the films, i.e., the film corresponding to the inner arc has large precipitates. The case of lower precipitate density corresponds to the outer arc. In order to clarify this correlation between the precipitates and the trajectories, we compared the total volumes of precipitates on YBCO films with the values of $(\epsilon_r + \epsilon_i)^{1/2}$ at the end of the 10th growth cycle, where ϵ_r and ϵ_i represent the real and imaginary parts of the pseudodielectric function, respectively. In Fig. 18b we can see clear correlation between these two parameters, and can decide that observation at 2.7 eV is useful to investigate the formation of precipitates. The varieties of trajectories originated from small differences in the very early stages of YBCO growth, so that we must control the initial growth stage precisely in order to avoid outgrowing of precipitates. Figure 19 shows two trajectories at the first cycle of the block-by-block growth. Although the

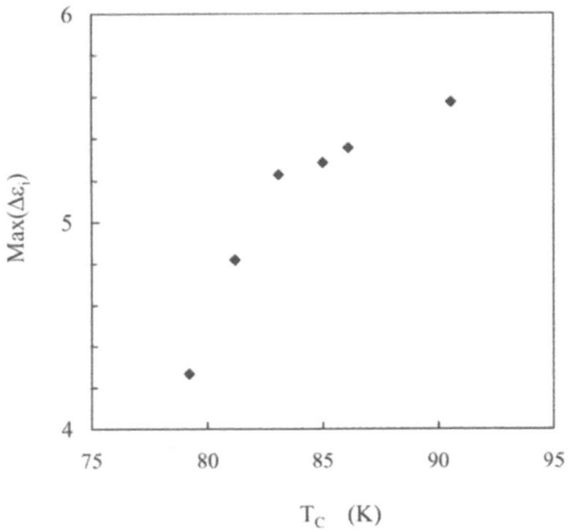

Fig. 17. Max($\Delta \epsilon_\mathrm{i}$), the maximum value of $\Delta \epsilon_\mathrm{i}$ in the first oscillation of the pseudodielectric function, versus T_c(on), the onset of the superconducting transition temperature

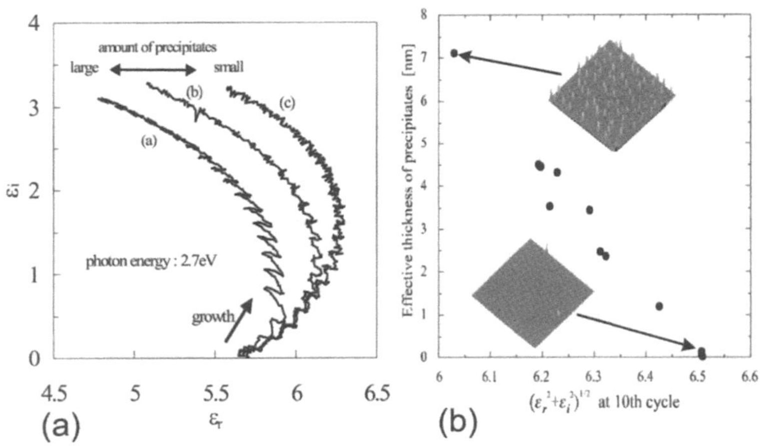

Fig. 18. Correlation between pseudodielectric function and the amount of precipitates on YBCO thin films. (**a**) Trajectories of the pseudodielectric function during deposition. The effective thickness of precipitates was 7.1 nm for (a), 2.5 nm (b) and 0.03 nm (c), respectively. (**b**) Relation between effective thickness (volume/surface area) of precipitates and $(\epsilon_\mathrm{r}^2 + \epsilon_\mathrm{i}^2)^{1/2}$ at the 10th cycle

two were deposited with the same source concentration and the same supply time, the route of sample (a) was quite different from sample (b) particularly for the period of Ba supply. These results indicate the necessity for real-time control based on in-situ growth monitoring.

We found that the initial growth model of BaO as shown in Fig. 20 could reproduce the experimental results of ellipsometry signal trajectories. Nuclei of BaO, formed on $SrTiO_3$ substrate at an average distance of d_{init}, grow as islands and then coalesce as a two-dimensional film structure. Growth pa-

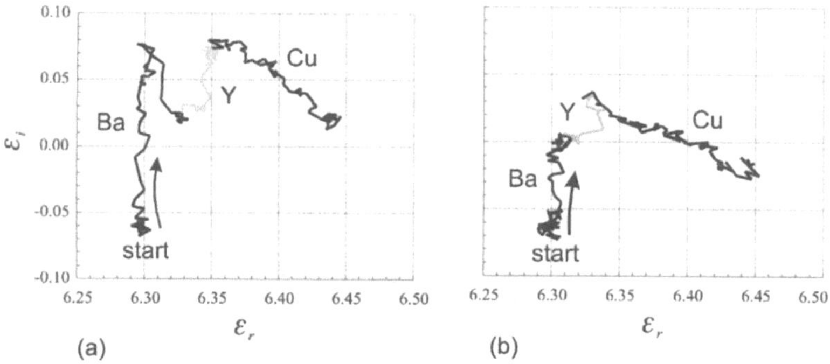

Fig. 19. Trajectories of the pseudodielectric function of YBCO thin films during the first block-by-block growth cycle. There are clear differences between (**a**) and (**b**) particularly in Ba

rameters deduced from experiments with $SrTiO_3$ substrates of various terrace width, prepared by vicinal cut surfaces, are plotted in Fig. 21. The distance between nuclei and the time to coalescence increase linearly with terrace width. However, the growth rate of cylindrical islands is independent of the terrace width. Nuclei may be formed at the edge of the surface step until about 450 nm, which may correspond to the migration distance of species on the substrate surface.

If an optimal supply time of each source were determined by means of iteration of measurements and computer simulations, it is expected to know the best timing to switch the precursor supply for the desired growth and to obtain good YBCO thin films with sufficient smoothness and superconductivity.

3.3 Ga Addition to YBCO Thin Films

Ga addition to YBCO thin films was investigated in order to confirm the hypothesis that thin films with Ga atoms substituted for Cu atoms in Cu–O chains became more stable than undoped samples and was effective in suppressing the diffusion of oxygen atoms. We first evaluated the solid solubility of Ga atoms in YBCO, i.e., how high a concentration of Ga atoms can be incorporated in YBCO films without causing serious deterioration. Figure 22 shows top-view AFM images of the surface of YBCGO ($YBa_2Cu_{3-x}Ga_xO_{7-\delta}$) films for $x = 0$ (undoped) to 0.15. A large number of outgrowths which may be due to the Ga-rich second phase, can be seen in the case of $x \geq 0.07$. The temperature dependence of resistivity also indicates degradation of crystallinity for samples of $x \geq 0.07$. We concluded that solid solubility of Ga into YBCO is around 0.06.

In order to confirm the effect of the suppression of oxygen diffusion, ellipsometric measurements were performed. As reported previously, the tetragonal phase YBCO has a distinctive optical absorption at 4.1 eV [18]. When Ga

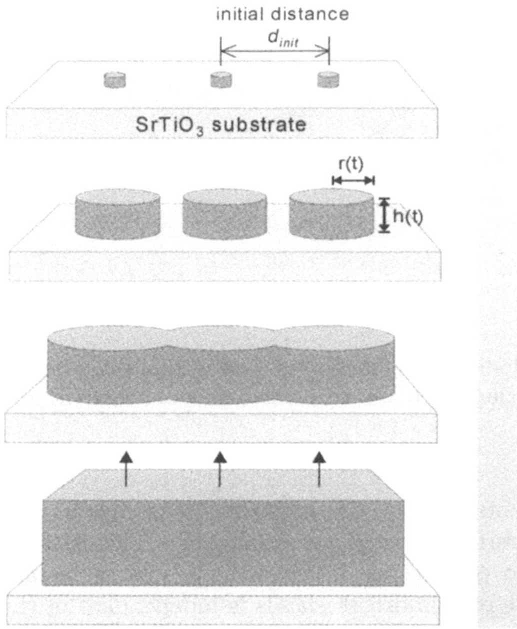

Fig. 20. Initial growth model of BaO

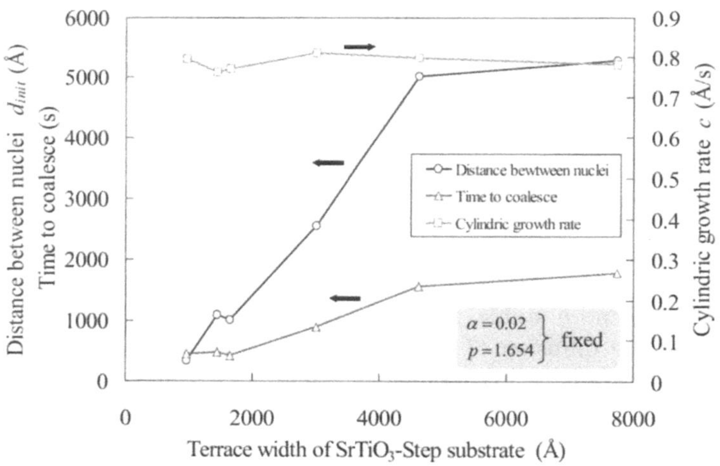

Fig. 21. Variation of growth parameter of BaO as a function of substrate terrace width

substitutes for Cu atoms in Cu–O chains, that Ga–O structure will not contribute to the absorption. In the temperature dependence of 4.1 eV absorption intensity, which was represented by the imaginary part of the pseudodielectric function ϵ_i, shown in Fig. 23, the onset of absorption for undoped and

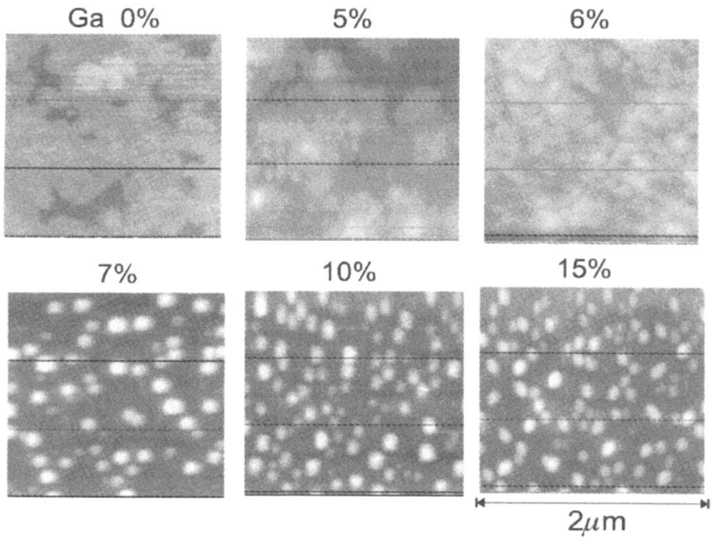

Fig. 22. Top-view AFM images of $YBa_2Cu_{3-x}Ga_xO_{7-\delta}$ thin films for several x from 0 (undoped) to 0.15. In case of $x \geq 0.07$ a large number of boulders were outgrown on the films

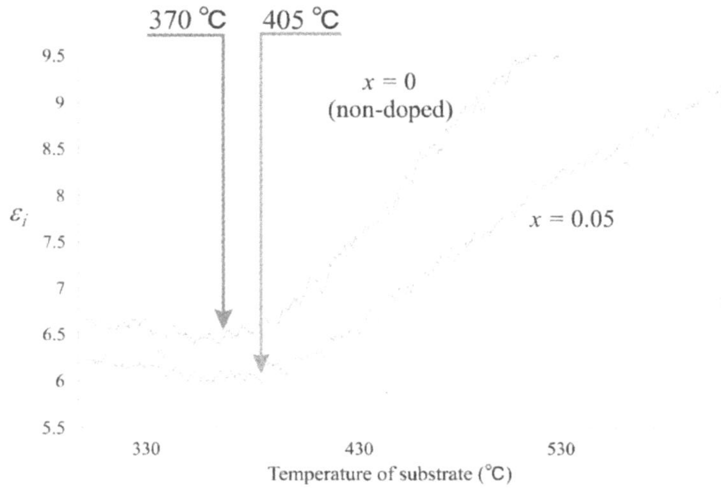

Fig. 23. Temperature dependence of the imaginary part of the pseudodielectric function at 4.1 eV, which corresponds to the optical absorption intensity of the films. The onset of absorption for the Ga-doped film is higher than that of the undoped film by 35°C

Ga-doped sample were 370 and 405°C, respectively. This result indicates that YBCGO requires much more heat than YBCO does to diffuse oxygen atoms. Therefore oxygen in YBCGO is more stable than in YBCO [19].

3.4 Summary

Two methods to improve the reproducibility of MOCVD growth of YBCO thin films are described. The precursor supply rate is stabilized markedly by employing ultrasonic concentration analyzers. In-situ spectroscopic ellipsometry is useful for real-time monitoring of the initial growth of YBCO thin films. The quality of thin films as well as the thickness is evaluated. Ga addition into YBCO is effective for improving the reliability of vortex thin films by suppressing the diffusion of oxygen atoms due to strong Ga–O bonds.

4 Sputter Growth of HTSC Thin Films

4.1 Fundamental Features of Sputtering

Sputtering involves high energy particles from a target surface. By repeating this process, a thin film is deposited on a substrate facing the target. In the synthesis of a high-T_c superconducting (HTSC) thin film, reactive sputtering is nearly always used in order to obtain a film with a stoichiometric composition containing oxygen atoms. In reactive sputtering, activated oxygen plasma that has been generated diffuses throughout the sputtering chamber, enabling a thin film with a thickness of over 1 μm to be deposited. Since sputtered particles from an HTSC target are not composed of large clusters, a thin film with a smooth surface is synthesized. However, the rate of deposition using a conventional sputtering method is low. Magnetically enhanced sputtering systems are used to increase the deposition rate. The magnetic field effects are useful for reduction of discharge voltage, electron bombardment of the substrate and sputtering gas pressure. Originally, the advantage of sputtering was to obtain a thin film with strong adhesion by bombardment of the substrate surface with high energy particles. However, the existence of high energy particles is a disadvantage when epitaxially growing thin films of HTSC materials with weak crystal lattices. Therefore, a magnetron cathode structure and configuration of the cathode and substrate that prevent bombardment of the substrate with energetic particles such as electrons, negative oxygen atoms and sputtered atoms has been investigated to obtain high quality thin films.

The form and quality of an HTSC target affect the quality of the thin film. Recently sintered targets have been used. The composition of a sputtered thin film is generally different from that of the target. In the initial stage of HTSC research, non-stoichiometric targets were used. However, since the clarification of sputtering conditions under which high quality thin films can be synthesized, targets that are almost similar to stoichiometry have been used. In both DC and RF sputtering, high quality HTSC thin films have been obtained.

4.2 Synthesis of High Quality Films and Large Sized Film Deposition

In order to obtain high-quality "123" thin films with T_c endpoints (T_{ce}) of about 90 K, the incidence of high-energy particles on the substrate surface must be suppressed. There are three types of sputtering method that enable this incidence to be supressed: single target planar magnetron sputtering, ring-target sputtering, and multi target sputtering. A disk target is currently used for single target planar magnetron sputtering in many research institutes. As shown in Fig. 24, there are three sputtering configurations: on-axis, off-center and off-axis [20]. An off-center configuration is when a substrate is moved parallel to a target surface from the on-axis position [21], and an off-axis configration is when a substrate is not in the on-axis position and the angle between the substrate axis and the target axis is nearly 90° [22]. In off-axis sputtering there is no ion bombardment of the surface and high quality thin films are obtained, but the deposition rate is small. On the other hand, in off-center sputtering using a target 9 cm in diameter, the area of the thin film without damage is about $1 \times 2\,\mathrm{cm}^2$, but the deposition rate is comparatively large. Energetic ion bombardment of the substrate surface at the on-axis position cannot be avoided and it is therfore difficult to obtain good HTSC thin films.

In ring cathode sputtering, inverted cylindrical magnetron sputtering [23] in which a cylindrical sintered HTSC target is used is representative. This is a kind of hollow cathode sputtering. The sputter deposition of HTSC films using a hollow cathode without a magnet has also been investigated [24].

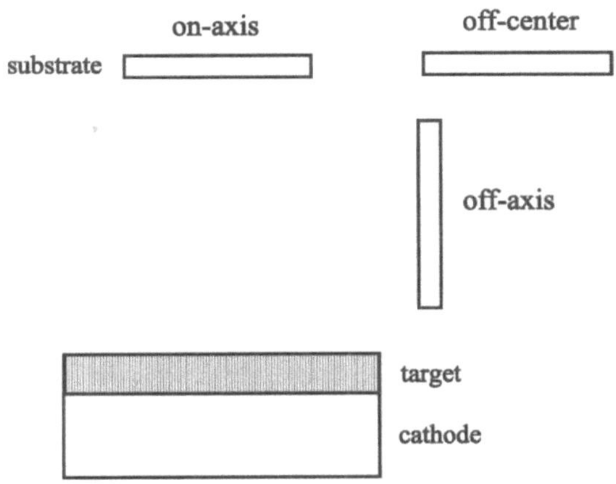

Fig. 24. Configurations of target and substrate

Since high energy particles are confined in the ring target in these sputtering processes, there is little ion bombardment of the substrate surface and high quality thin films are obtained. As the plasma is distributed in only one direction, the deposition rate is comparatively large. For deposition of a uniform thin film over large areas on the substrate, however, there is a problem in selection of the rotation axis of the substrate and the method of its reciprocating motion. Preparation of a large cylindrical HTSC target with a uniform sintered density is also very difficult.

The aim of multi-cathode sputtering is improvement in deposition rate and deposition of a thin film with a uniform thickness over a large area. The arrangement of cathode electrodes is classified into two groups. One is a sputtering method with aconfiguration in which the adjacent planar cathodes over two face a substrate holder. A substrate is placed at the off-center position between the cathodes and sputtering is carried out simultaneously. In this deposition, comparatively large sized thin films are obtained, but high energy particles are apt to irradiate the substrate surface and so there is a limit to the size of the deposited film. The other is a facing targets sputtering method [25]. A top view of this sputtering configuration is shown in Fig. 25. The ion bombardment effect on the substrate surface in this sputtering is fundamentally the same as that in ring cathode sputtering. The deposition rate is smaller than that in inverted cylindrical cathode sputtering. Deposition of a large sized thin film is easy because the distance between target surfaces (D_{t-t}) can be optionally chosen and a disk target is used. In the facing target configuration there are two sputtering methods, one in which the two targets

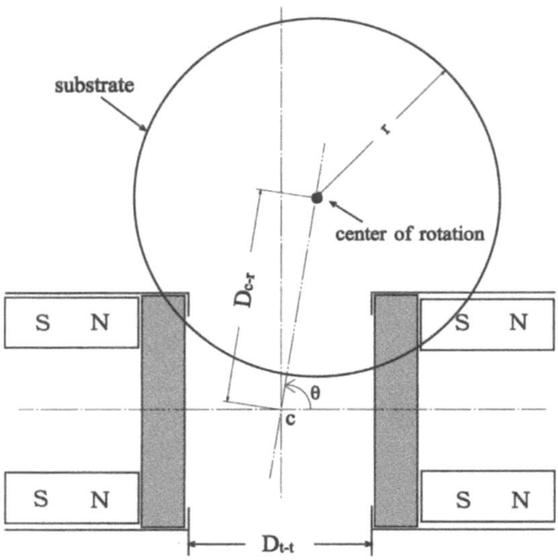

Fig. 25. Facing targets sputtering

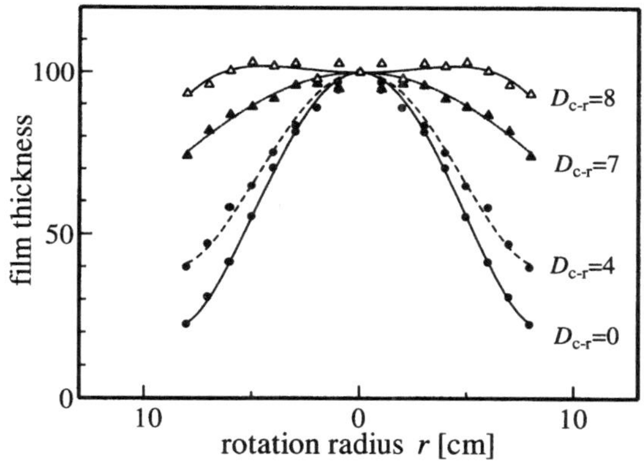

Fig. 26. Dependence of film thickness on eccentric distance ($D_{\mathrm{c-r}}$), rotation radius (r) and at $D_{\mathrm{t-t}} = 8\,\mathrm{cm}$

operate as a pair and one in which the two targets operate independently. The selection is dependent on the distance between targets. In order to obtain a uniform thickness of thin film over a large area, the optimum position of a substrate rotation axis must be determined. Using data on the thickness of a thin film deposited on a static substrate located horizontally and 4 cm above the top of the targets ($H_{\mathrm{t-s}}$) at $D_{\mathrm{t-t}} = 8$ cm, the distribution of thicknesses of $EuBa_2Cu_3O_7$ (EBCO) films on substrates rotated at various positions was simulated. The distribution of thicknesses was examined with substrates rotated at various eccentric distances ($D_{\mathrm{c-r}}$) away from the vertical line that passes through the center between targets at $H_{\mathrm{t-s}} = 4$ cm. The relationship between thickness fluctuation and rotation radius (r) for various eccentric distances is shown in Fig. 26. When $D_{\mathrm{c-r}} = 8\,\mathrm{cm}$ a uniform thin film with thickness fluctuation within $\pm 3\,\%$ was obtained over $\phi 12\,\mathrm{cm}$.

4.3 Epitaxial Growth of $EuBa_2Cu_3O_7$ Films on R-plane Sapphires

Sapphire is promising as a substrate of an HTSC thin film because of its excellent microwave properties and because a large sized substrate is available at moderate cost. However, a buffer layer is needed for reaction with HTSC materials at deposition temperatures (T_s). Using a single target planar magnetron, CeO_2 films were deposited on R-plane sapphires located at the off-center position and the sputtering conditions under which the film grew epitaxially were examined. A (001)-oriented CeO_2 film grown at 7 Pa ($Ar + 7\,\%\,O_2$) and $T_s = 650°C$ was used as a buffer layer [26]. CeO_2 has good wettability for a sapphire. With an increase in CeO_2 film thickness, bamboo-like grains grew and EBCO thin films exhibited a unique behavior of the epitaxial growth on the buffer layer. The sputtering gases used in the synthesis of c-axis oriented EBCO (c-EBCO) and a-axis oriented EBCO (a-EBCO)

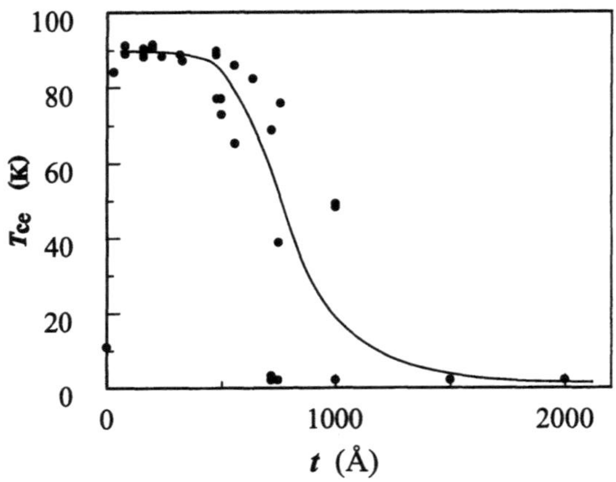

Fig. 27. Dependence of T_{ce} on thickness of a CeO_2 buffer layer (t) under the conditions in which a c-EBCO thin film is deposited

thin films were Ar + 8 % O_2 (7 Pa) and Ar + 20 % O_2 (7 Pa), respectively. EBCO thin films were deposited at off-center positions. Then 2000 Å EBCO thin films were deposited in Ar + 8 %O_2 at $T_s = 650°C$ using a stoichiometric sintered target. Figure 27 shows the dependence of T_{ce} on CeO_2 film thickness (t). c-EBCO thin films are obtained on CeO_2 buffer layers of 30–400 Å. The T_{ce} values greatly fluctuate for buffer layers of 500–1000 Å. Over the range of film thicknesses in which this large fluctuation occured, (103)-EBCO grains were observed. As-grown c-EBCO thin films on a 50 Å CeO_2 buffer layer exhibited T_{ce}'s of about 91.7 K.

An a-EBCO thin film grows at a lower T_s than a c-axis one. A single-phase a-EBCO thin film was deposited on a 50 Å CeO_2 buffer layer at $T_s = 600°C$ and exhibited $T_{ce} = 65$ K. The dependence of T_{ce} on CeO_2 thickness was also observed under the conditions of a-EBCO thin film deposition. The results are shown in Fig. 28. The values of T_{ce} are maximum at a CeO_2 thicknesses of 100–200 Å and 700–800 Å. a- and (110)-EBCO thin films grew on the former and the latter, respectively.

For the purpose of improving T_{ce}, an a-EBCO thin film was deposited using the self-template method [27]. After an 80 Å seed EBCO layer had been deposited at $T_s = 610°C$, sputtering was stopped for 15 min. Then T_s was increased to 670°C at 10°C/min. In this way, a single phase a-EBCO thin film 2000 Å in thickness exhibited a T_{ce} of 85.4 K [28].

Next, attempts to improve T_{ce} were made using a template layer of an a-axis oriented $PrBa_2Cu_3O_7$ (a-PBCO) layer [29,30] instead of the seed layer [31]. The PBCO template layer was deposited in the same atmosphere as the a-EBCO thin film.

When a PBCO film was deposited on a 50 Å CeO_2 buffer layer at T_s = 620°C, a mixture of a- and c-axis oriented grains grew. The content of a-axis oriented grains depended on the deposition rate (R_d). Figure 29 shows the

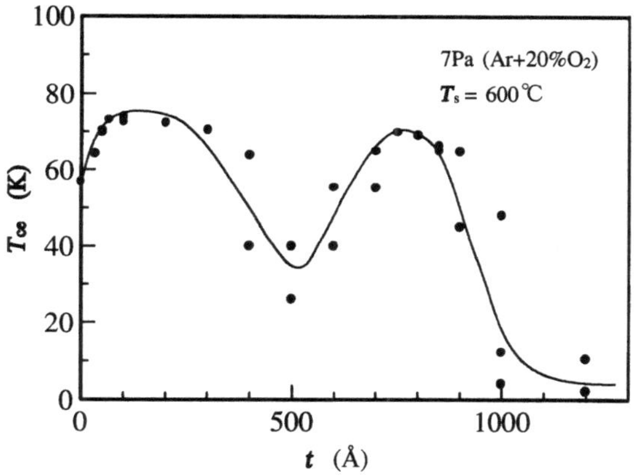

Fig. 28. Dependence of T_{ce} on thickness of a CeO_2 buffer layer (t) under the conditions in which an a-EBCO thin film is deposited

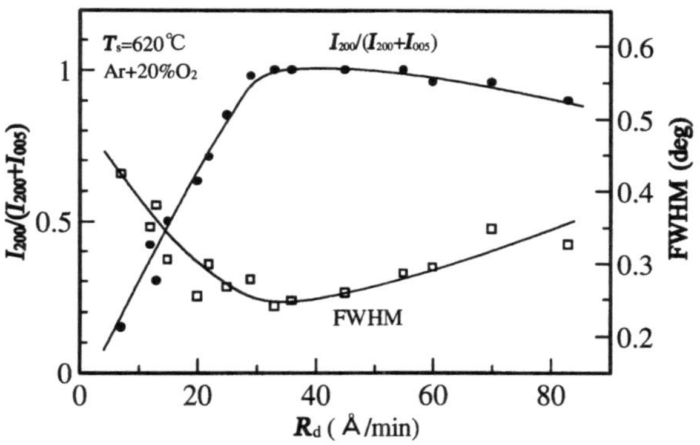

Fig. 29. Deposition rate dependence of diffraction intensity ratio $[I_{200}/(I_{200}+I_{005})]$ and FWHM using (200) reflection of a PBCO thin film

R_d dependence of the intensity ratio $I_{200}/(I_{200}+I_{005})$ and the full width at half maximum (FWHM) of a (200) peak. At R_d = 30–55 Å/min, a single-phase a-PBCO thin film grew. Then a-PBCO templates with various layer thicknesses were deposited at R_d = 35 Å/min, and 2000 Å EBCO thin films were deposited on the template layers at T_s = 650°C. It was found that the orientation of the EBCO film depended on the a-PBCO layer thickness. When the a-PBCO layer was thin, a c-EBCO thin film grew preferentially. A single-phase a-EBCO thin film was obtained on the a-PBCO template layer over 700 Å. The a-EBCO thin film on a 1000 Å a-PBCO template layer exhibited T_{ce} = 86.7 K and R_r/R_{100} = 1.5.

The in-plane orientation relationships of multi-layer epitaxial thin films were clarified using RHEED or the ϕ scan method. Figure 30 shows the epitaxial structure of an a-EBCO thin film with an a-PBCO template layer. The

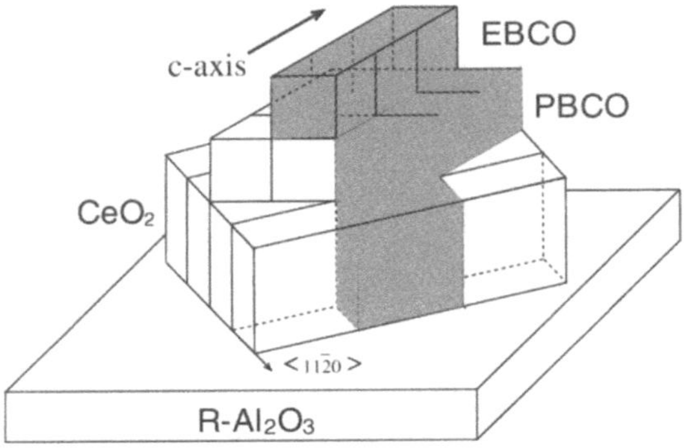

Fig. 30. Epitaxial structure of a multi-layer of an *a*-EBCO thin film

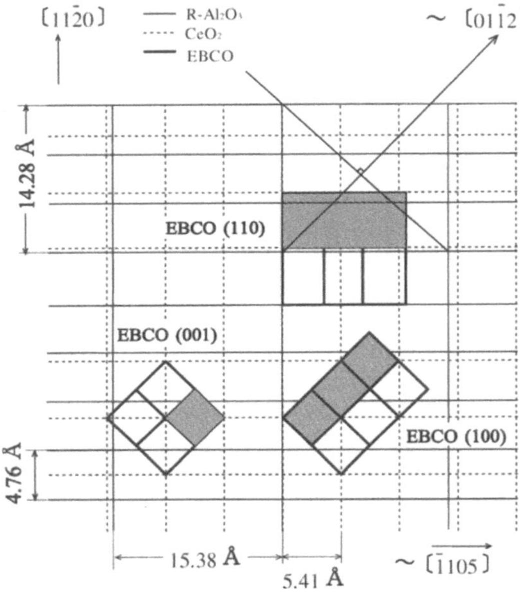

Fig. 31. Epitaxial relationships among the sapphire R-plane, $CeO_2(001)$ plane, PBCO(100) and EBCO(100), EBCO(001) and EBCO(110) planes

in-plane orientation relationships among the sapphire R-plane, $CeO_2(001)$ plane, *a*-PBCO plane and *c*-, *a*- and (110)-EBCO planes are shown in Fig. 31. Figure 32 shows cross-sectional views of multi-layers of *a*- and *c*-EBCO thin films and their *R*-*T* curves.

4.4 Application of Sputter Plasma to Recovery Treatment

In reactive sputtering using an oxide sintered target, activated oxygen atoms are generated. In the deposition process, the activated oxygen atoms enable

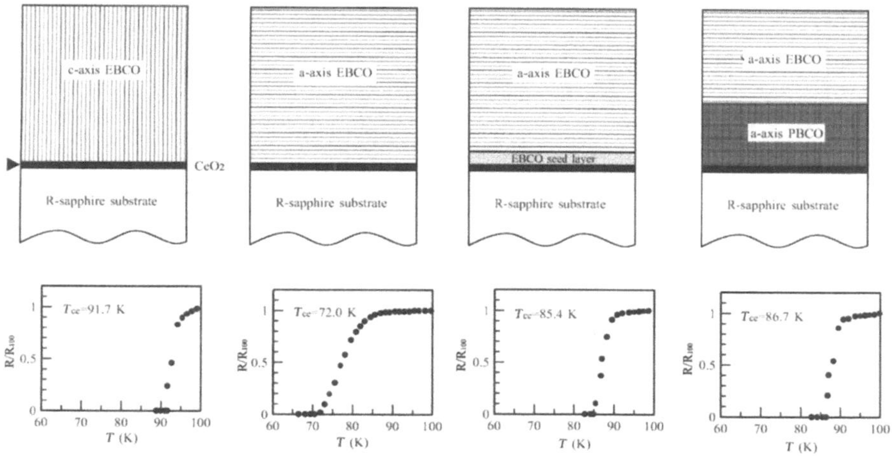

Fig. 32. Cross-sectional views of multi-layers of *a*- and *c*-EBCO thin films and their *R*-*T* curves

high quality as-grown thin films to be easily grown. The activated oxygen atoms are effective for restoration of HTSC thin films. Since leaving a "123" HTSC film as it is in the atmosphere or heating it at a high temperature causes oxygen deficiency, a reliable recovery treatment had been desired. Therefore activated oxygen plasma annealing (AOP annealing) was used for recovery treatment of the *c*-EBCO thin films that had deteriorated due to oxygen deficiency [32]. This treatment was carried out as follows. First, an as-grown thin film 3000 Å in thickness with $T_{ce} = 90\,K$ was heated at 520°C for 15 min in vacuum and became a non-superconducting thin film with a c_0 of about 11.86 Å. The degraded thin film was set in a sputtering chamber and a shutter which prevented sputtered particles from depositing on the thin film was set. The chamber was then set in $Ar + 8\,\%\,O_2$ (7 Pa) and sputtering and heating operations were started. After annealing of the film at a fixed temperature (T_{sa}) and for a fixed time (t_p), the sputtering and heating operations were stopped and pure oxygen was introduced into the chamber to a fixed pressure (P_{O_2}). Finally, the film was cooled naturally. Figure 33 shows the T_{sa} dependence of T_{ce} and c_0 during the AOP annealing of $t_p = 30\,min$ and $P_{O_2} = 0.2$ and 2 kPa. When T_{sa} exceeded 400°C, the superconducting properties of the film began to recover rapidly. The optimal recovery conditions were found to be $T_{sa} =$ 450–700°C, $t_p \geq 30\,min$, and $P_{O_2} \geq 2\,kPa$. The AOP annealing was also effective for recovery of the *a*-EBCO and *c*-YBCO thin films.

The wraparound effect of the AOP diffusing to the back side of the shutter was examined at a target-substrate distance of 5 cm and a shutter-film distance of 1 cm [21].

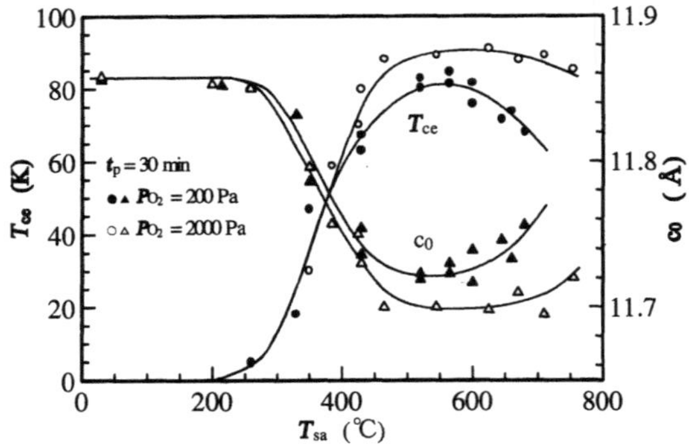

Fig. 33. T_{sa} dependence of T_{ce} and c_0 under the conditions of AOP annealing of $t_p = 30$ min and $P_{O_2} = 0.2$ and 2 kPa

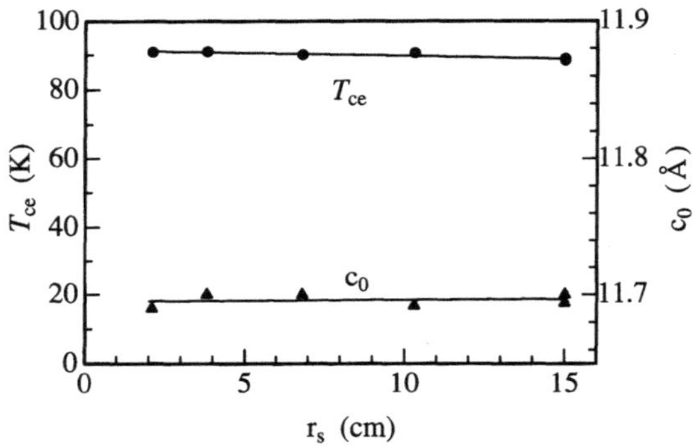

Fig. 34. Shutter radius (r_s) dependence of T_{ce} and c_0 of a 0.5 cm square EBCO thin film OAP annealed at $T_{sa} = 520°C$, $t_p = 30$ min and $P_{O_2} = 2$ kPa. ©1993/2001 IEEE

Figure 34 shows the shutter radius (r_s) dependence of T_{ce} and c_0 of a 0.5 cm square EBCO thin film AOP annealed under the conditions of $T_{sa} = 520°C$, $t_p = 30$ min and $P_{O_2} = 2$ kPa. When the value of r_s exceeded about 2 cm, EBCO particles were not observed on the film. Even when a large shutter with a radius of 15 cm was used, the superconducting properties of the EBCO thin film recovered completely. These results indicate that AOP has a strong wraparound effect and a long life.

4.5 Summary

A recent problem in the synthesis of HTSC thin films is how to reliably obtain a high quality thin film on a large sized substrate. The unique features of sputtering are the existence of high energy particles and high gas pressure. As HTSC materials contain oxygen atoms that easily escape from the crystal lattice and have structural sensitivity, it is essential to prevent energetic particles from irradiating the substrate surface during the synthesis of a thin film. Off-axis or off-center planar magnetron sputtering, ring target sputtering and facing targets sputtering all enable deposition of high quality HTSC films, and facing targets sputtering is a promising method for large-sized film deposition.

By using off-center planar magnetron sputtering, a CeO_2 buffer layer was deposited on an R-plane sapphire and epitaxial growth of the EBCO thin film on it was controlled. A *c*-EBCO thin film with $T_{ce} = 91.7\,K$ was synthesized on a 50 Å CeO_2 buffer layer. *a*-EBCO thin films deposited using the self-template and *a*-PBCO template methods exhibited T_{ce}'s of 85.4 K and 86.7 K, respectively. The in-plane orientation relationships of the multi-layers were clarified.

High gas pressure is effective for the wraparound effect of light elements. The activated oxygen plasma generated in reactive sputtering diffused even to the back side of a large shutter and enabled the degraded EBCO thin film to completely recover its high quality characteristics.

5 Preparation of Ultrathin Films and Superlattices of high-T_c Oxides by MBE

The discovery of the high-T_c superconducting oxides has renewed an interest in the preparation of epitaxial films of oxides for the study of basic physics as well as for wide applications. Most preparation methods including molecular beam epitaxy (MBE), reactive sputtering and pulsed laser deposition are essentially co-deposition of the elements. The major point of interest concerning the growth process is to find what the growth unit is. All the element atoms coexist on a substrate Surface. Reflection high energy electron diffraction (RHEED) specular intensity oscillations are well known in the MBE growth of semiconductors and metals [33]. RHEED oscillations have played a significant role in understanding the microscopic processes involved in the epitaxial growth [34]. It has aided in the development of extremely precise control over the crystal growth of these materials on the atomic scale. The RHEED oscillations are interpreted as a consequence of the nucleation of 2D islands and their growth into a flat terrace in a cyclic process. The growth unit for single-element metals and semiconductors has been confirmed to be one atomic layer from observation of RHEED oscillations. However, in ionic oxide systems, the growth unit may be subject to charge neutrality and therefore may not be an atomic single layer. We have observed for the first time RHEED oscillations during the epitaxial growth of oxides caused by two-dimensional growth and have found that one period of the RHEED oscillations corresponds to the height of the minimum unit which satisfies the chemical composition and electrical neutrality of the oxide [35]. The appearance of the RHEED oscillation during the epitaxial growth of the oxides means that a definite stacking sequence of atomic layers should exist in the growth unit and therefore the top atomic layer in the growth unit should be always the same. In this paper we describe the stacking sequence of the growth unit in $YBa_2Cu_3O_{7-\delta}$ (YBCO) and discuss the superconductivity in ultrathin YBCO films, especially for one unit cell thick (1-UCT) film, in relation to the microstructures [36]. Since high-T_c superconductors have layered structures based on the two-dimensional CuO_2 planes, it is significant to investigate the intrinsic nature of the individual CuO_2 plane and the role of interlayer coupling between the CuO_2 planes. Superlattices composed of YBCO and nonsuperconducting $Pr_2Cu_3O_{7-\delta}$ (PrBCO) and ultrathin YBCO films have been intensively studied focusing on these subjects. It has been reported for the [YBCO 1 unit cell thick (1-UCT)/PrBCO] superlattices that the zero-resistance temperature decreased rapidly with increasing PrBCO thickness but did not go down to zero. In the superlattice, the degree of interlayer coupling can be changed by varying the PrBCO thickness. In this section we describe the fabrication of YBCO/PrBCO superlattices and their resistive transitions in magnetic fields.

5.1 MBE Growth of Oxide Thin Films

Figure 35 shows a schematic diagram of the deposition system. Oxide thin films were grown by co-evaporation of metal elements in an oxygen atmosphere [37]. The oxygen gas was introduced to the substrate surface through a commercially available ozonizer system (O_3 content $< 5\%$). The local gas pressure at the substrate surface was 10^{-2}–10^{-1} Torr and the background pressure was kept at 10^{-4} Torr during the deposition. The low background pressure enables us to use in-situ RHEED observation during the growth of the films. The acceleration voltage of the incident electron beam for the RHEED observation was 20 keV. The intensity of the RHEED pattern was measured via an optical fiber using a photomultiplier. Metal elements with high melting temperatures such as Y, Pr and Ba metals were evaporated from e-beam heated sources and Cu metal was evaporated from thermal sources. The substrate was $SrTiO_3$ (100). The substrate temperature was 680°C and the deposition rate was 0.05 nm/s.

5.2 RHEED Oscillations

Figure 36a shows RHEED oscillations observed during the epitaxial growth of $BaTiO_3$ (001) on $SrTiO_3$ (100), where the direction of the incident beam is [100]. The intensity I decreased rapidly as soon as the growth was initiated and then continuous oscillations were observed whose amplitude decreased gradually. If the growth was interrupted, the intensity increased back to the initial levels and showed definite oscillations again. We have determined the period of the RHEED oscillations by comparing the number of the oscillations with the film thickness on an atomic scale. One way to realize this is by the measurement of X-ray diffraction peaks due to the finite size effect.

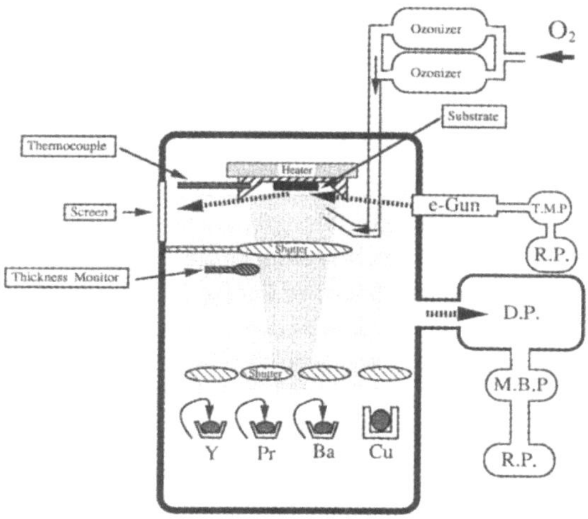

Fig. 35. Schematic illustration of the deposition system

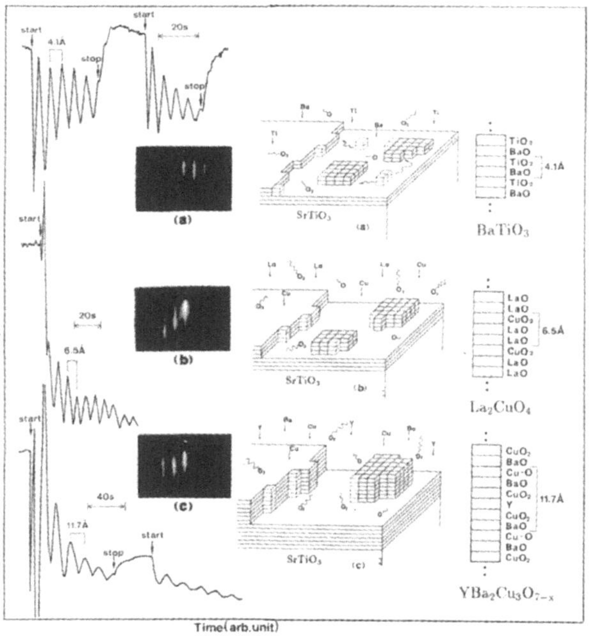

Fig. 36. RHEED patterns and RHEED intensity oscillations of the specular beam during epitaxial growth for (**a**) $BaTiO_3$ (001); (**b**) La_2CuO_4 (001); (**c**) YBCO (001) on $SrTiO_3$ (100). *Insets* are schematical illustrations for the deposition and growth process of each oxide

The RHEED oscillation period thus precisely corresponds to the height of one unit cell of $BaTiO_3$, i.e. the thickness of two atomic layers, one each of (BaO) and (TiO_2). As mentioned already, the intensity recovered and the oscillation amplitude increased after the growth was interrupted. This recovery of the intensity is commonly observed in the growth of GaAs by MBE. It is interpreted to be due to an increase in the mean terrace width of the surface and, hence, a reduction in the surface step density by the migration of surface adatoms to the flat terrace edge. From the remarkable recovery of the RHEED intensity, it is assumed that the migration of surface adatoms easily occurs in the $BaTiO_3$ system. The inset in Fig. 36a shows a schematical illustration of a deposition-and-growth process model for $BaTiO_3$. Figure 36b and c shows RHEED oscillations observed during the growth of La_2CuO_4(001) on $SrTiO_3$(100) and YBCO (001) on $SrTiO_3$(100). For this film, the thickness was determined from the Laue function of the X-ray diffraction to be 6.60 nm. From this result, one period of the RHEED oscillations was determined to be 0.66 nm. For La_2CuO_4, the chemical repetition unit consists of three atomic layers (LaO), (LaO) and (CuO_2), as shown in the inset of Fig. 36b, and this unit is stacked on itself with a relation $(a/2 + b/2)$ shifted in the (001) plane. The lattice spacing of this unit along the [001] direction is determined to be 0.65 nm. This value is in good agreement with one period of the RHEED oscillations. It has been also revealed that a period of the RHEED oscillations observed during the growth of YBCO (001) corresponded to the height of the unit cell (Fig. 36c). The present results have significance for understanding the epitaxial growth of ionic oxides. The observed period in the RHEED oscillations indicates that the 2D nuclei of YBCO, La_2CuO_4 and $BaTiO_3$ all have the height of one minimum unit needed to satisfy the chemical composition and electrical neutrality. That is, the chemical and neutral unit is the unit of crystal growth of ionic materials. The deposition and growth process models for La_2CuO_4 (001) and YBCO (001) are schematically shown in the insets of Fig. 36b and Fig. 36c. An atomic force microscope (AFM) image of a 12-UCT ($\sim$ 14 nm) YBCO film is shown in Fig. 37a. The surface appears to be multi-terraces, implying 2D nucleation and growth (Fig. 37b). A line scan across the film surface (Fig. 37c) reveals that the height of the terraces is one unit cell (1-UCT) ($\sim$ 12 nm) and the maximum roughness of the surface is less than two unit cells ($\sim$ 2.4 nm). AFM images indicate that a 10 nm thick ultrathin YBCO film has no 3D island or no spiral grain.

5.3 Superconductivity of Ultrathin YBCO Films

In the following we describe the superconductivity of ultrathin YBCO films, especially focusing on the superconductivity of a 1-UCT YBCO layer grown on nonsuperconducting PrBCO. Since a 1-UCT YBCO layer grown on a 11-UCT PrBCO layer is assumed to have a surface morphology like that of the 12-UCT YBCO films, the structure of a 1-UCT YBCO layer can be depicted as illustrated in Fig. 37d. A continuous conduction path can be established

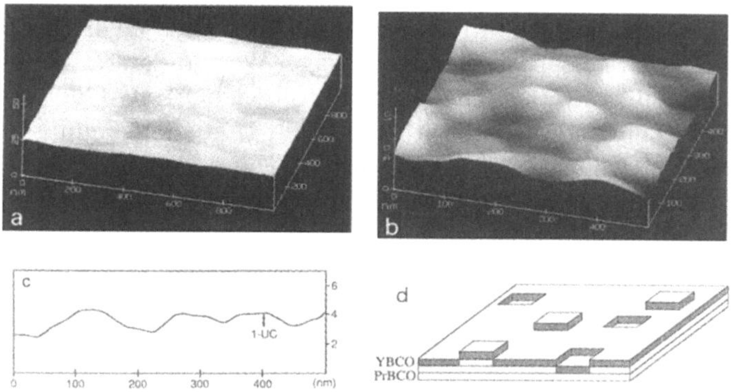

Fig. 37. (**a**) AFM image for a 12-UCT (~ 14 nm) YBCO film. (**b**) Higher magnification of the film shown in (**a**). (**c**) Line scan profile along the surface of the film shown in (**a**). The height of one unit cell of YBCO is shown by a marker. (**d**) Sketch of the structure of 1-UCT YBCO (*hatched*) on PrBCO layer

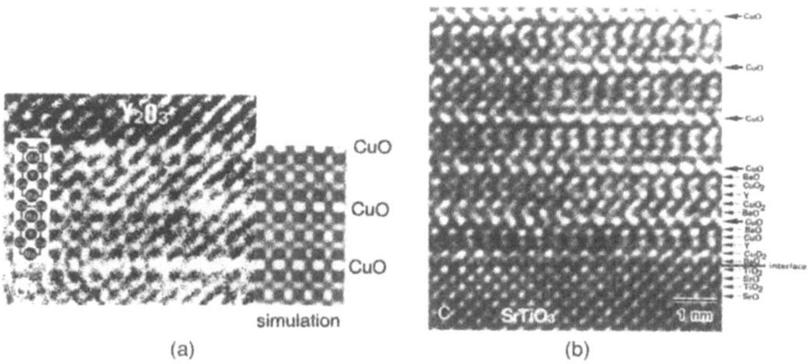

Fig. 38. (**a**) High resolution TEM image for the surface of YBCO covered with Y_2O_3. Calculated image is shown on the *right* side of the observed image. (**b**) High resolution TEM image for the YBCO/$SrTiO_3$ interface. Atomic species of the layers at the interface are indicated on the right side of the image

in the 1-UCT YBCO. Figure 38 shows high resolution TEM images for a 10-UCT YBCO film grown on $SrTiO_3$. Figure 38a shows a cross-sectional TEM image near the surface of the ultrathin YBCO film, which has been covered in situ with Y_2O_3. In the TEM image of YBCO, bright contrast is attributed to the CuO chain layer through simulation of the image. The bright contrast of the CuO chain layer would be caused by a lower concentration of oxygen atoms in the layer than that of the CuO_2 layer and a relatively large distance between two BaO layers interposing the CuO chain layer. The terminating layer of the ultrathin YBCO film is revealed to be a CuO layer through simulation of the image. The termination by the CuO layer implies that the stacking sequence of atomic layers of the unit for the 2D nucleation and growth is CuO-BaO-CuO_2-Y-CuO_2-BaO. Figure 38b shows a high resolution TEM image at the YBCO/$SrTiO_3$ interface. The CuO chain layers with a spacing of 1.2 nm are clearly seen in the image. We can determine the stacking sequence of atomic layers at the interface to be CuO-BaO-CuO_2-Y-CuO_2-BaO-TiO_2-SrO, being consistent with the result that the terminating layer of the ultrathin YBCO film is a CuO layer. The same stacking sequence has been reported by *Wen* et al. [38]. An X-ray standing wave (XSW) study of an ultrathin film of $GdBa_2Cu_3O_7$ (GdBCO) grown on $SrTiO_3$ has revealed that an interface layer of GdBCO is BaO when the interface layer of $SrTiO_3$ is TiO_2 [39]. In [40], we have shown that 1-UCT YBCO grown on a nonsuperconducting PrBCO buffer layer never shows superconductivity until 1-UCT PrBCO is grown on it. We can give an explanation for this result as follows: the growth of 1-UCT PrBCO guarantees formation of a BaO-CuO-BaO charge reservoir block above the CuO_2 bilayer interposing an Y layer (Fig. 39a). When the terminating layer of YBCO is a CuO layer, the growth of 1-UCT PrBCO provides a BaO layer for completion of the BaO-CuO-BaO charge reservoir block (Fig. 39b). If that is the case, a capping with BaO would make a superconductor of 1-UCT YBCO. Figure 40 shows the temperature dependence of resistance for 1-UCT YBCO with cap oxides of (a) BaO, (b) $BaTiO_3$, and (c) $BaBiO_3$. We have fixed the thickness of the cap oxide layer at 1.5 nm in order to sufficiently cover the YBCO surface with them. The surface of the cap oxide was covered with Pr_4O_7 (5 nm) for protection against surface degradation. Hereafter, we present

a normalized resistance, $R/R_{100\,K}$, for all films: above 100 K the resistance of the trilayer film, cap oxide/YBCO(1-UCT)/PrBCO(11-UCT), is dominated by the contribution from the 11-UCT PrBCO layer. The sheet resistance of every film at 100 K was lower than 3 kΩ. Clearly, capping with BaO makes 1-UCT YBCO superconductive. After the film is grown at 680°C, the CuO chain oxygen order is formed on cooling in the oxygen atmosphere. The growth of a BaO layer on the CuO terminating layer has an important role in the formation of the framework needed for ordering the CuO chain. It is seen that the film capped with $BaTiO_3$ exhibits a drop of resistance due to the superconducting transition, although zero resistance is not obtained. On the other hand, the film capped with $BaBiO_3$ shows semiconducting behavior. The results in Fig. 40 can be understood by considering the lattice matching between cap oxides and YBCO. When a perovskite type (ABO_3) oxide is grown on YBCO, it is reasonable to consider that an interface layer of the cap oxide becomes an AO "rocksalt" type layer; many studies of heterostructures of perovskite related oxides, $A'B'O_3/ABO_3$, through TEM [38], RHEED [41], and X-ray observations [39] have revealed that the interfacial structure prefers to be $A'O/BO_2$ rather than $B'O_2/BO_2$ when the surface layer of ABO_3 is BO_2. When the lattice mismatch is sufficiently small ($\leq 2\,\%$) the cap oxide would be strained elastically to match the lattice of YBCO, which guarantees complete matching of the BaO lattice on the CuO lattice (Fig. 41a). On the other hand, when the lattice mismatch increases more than $\sim 2\,\%$, misfit

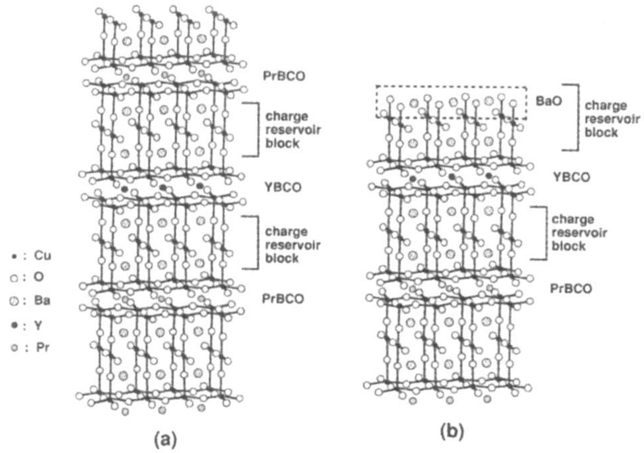

Fig. 39. Schematic illustrations of the atomic layers in 1-UCT YBCO capped with (**a**) 1-UCT PrBCO, a complete BaO-CuO-BaO charge reservoir block exists between 1-UCT YBCO and 1-UCT PrBCO cap layer; (**b**) BaO, capping with BaO layer also forming a BaO-CuO-BaO charge reservoir block above the CuO_2 bilayer interposing an Y layer

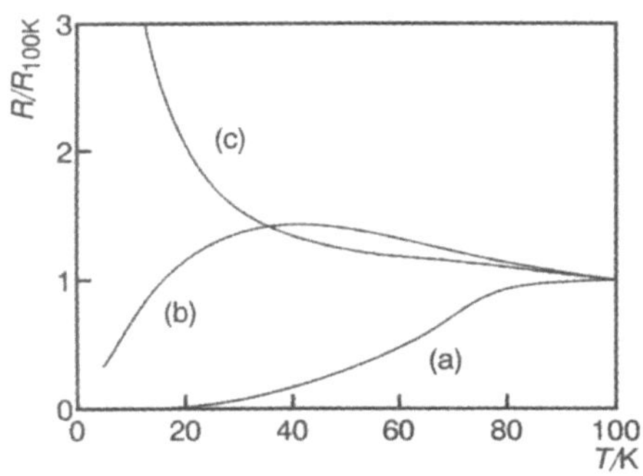

Fig. 40. Normalized resistance [R(T)/R(100 K)] vs temperature for 1-UCT YBCO layers capped with (**a**) BaO, (**b**) $BaTiO_3$, and (**c**) $BaBiO_3$

dislocations should be generated at the interface between the cap oxide and YBCO (Fig. 41b). Since a matching of the BaO lattice to the CuO lattice may not be so good around the dislocations, small domains with ordered structure which are superconductive would be isolated and no superconducting path would be established over the sample. The lattice mismatch increases from +1.20 % for BaO to +4.46 % for $BaTiO_3$; a non-zero resistance observed for the cap oxide of $BaTiO_3$ may occur in this situation. When the lattice mismatch becomes much larger than 6 % (Fig. 41c), BaO lattice would never match a CuO lattice all over the sample. The lattice mismatch of +12.83% for $BaBiO_3$ may be too large to play a role for the charge reservoir block. We have examined various cap oxides having lattice mismatch values in a wide

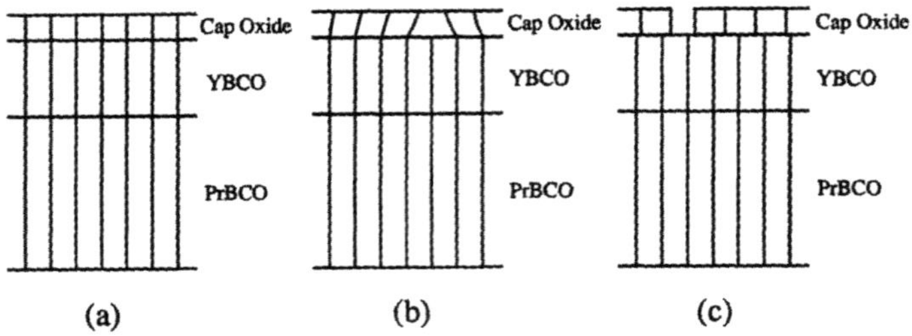

Fig. 41. Schematic illustrations for the types of lattice matching of the cap oxides on 1-UCT YBCO: (**a**) mismatch $\leq 2\,\%$, the lattice of cap oxide matches that of 1-UCT YBCO; (**b**) $2 \leq$ mismatch $\leq 6\,\%$, the misfit dislocation is generated at the interface of cap oxide/YBCO; (**c**) mimatch $\geq 6\,\%$, the lattice of cap oxide does not match that of YBCO and the large mismatch may also enhance 3D nucleation of the cap oxide

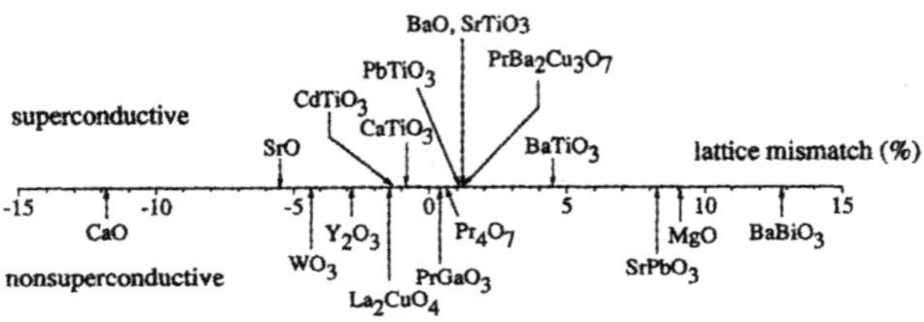

Fig. 42. Plot of the lattice mismatch between the cap oxide and YBCO vs the conducting property of 1-UCT YBCO

range. The results are summarized in Fig. 42. We have found that the cap oxide should contain an MO "rocksalt" type layer to produce the superconductivity in 1-UCT YBCO, where the lattice mismatch between the cap oxide and YBCO is sufficiently small ($\leq 6\,\%$) and the valence of M is 2+. BaO and SrO with NaCl type structure and $BaTiO_3$, $SrTiO_3$, $CaTiO_3$, $PbTiO_3$, and $CdTiO_3$ with perovskite type structure have the ability to produce superconductivity in 1-UCT YBCO. The result presented here also indicates that two CuO chain layers are needed above and below the CuO_2 bilayer interposing a Y layer for the occurrence of superconductivity in 1-UCT YBCO.

5.4 Superconducting Transition of Ultrathin Films and Superlattices in Magnetic Fields

Figure 43 shows RHEED intensity oscillations during the growth of a [YBCO (1-UCT)/PrBCO(1-UCT)] $\times$ 10 superlattice. The oscillations in the 5th to 7th periods are shown. Each oscillation corresponds to the growth of a 1-UCT layer. An interruption time of 20 s has been taken after each unit cell growth in order to improve the surface flatness. Figure 44 shows the X-ray diffraction profile scanned along the direction normal to the film surface for a [YBCO(1-UCT)/PrBCO(1-UCT)] $\times$ 10 superlattice. Strong sharp satellite

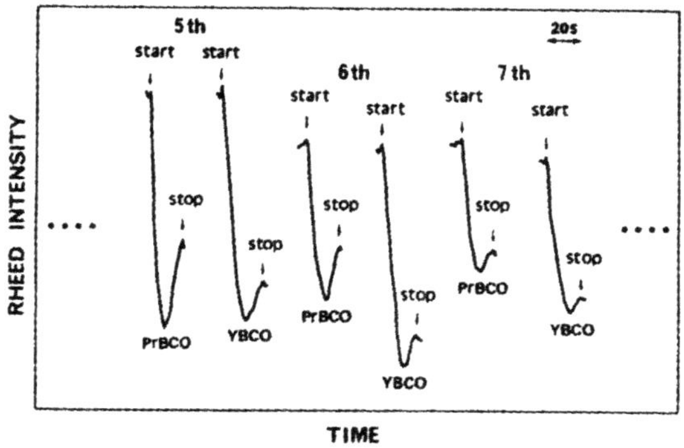

Fig. 43. RHEED intensity oscillations during the growth of [YBCO(1-UCT)/PrBCO[1-UCT)] superlattice

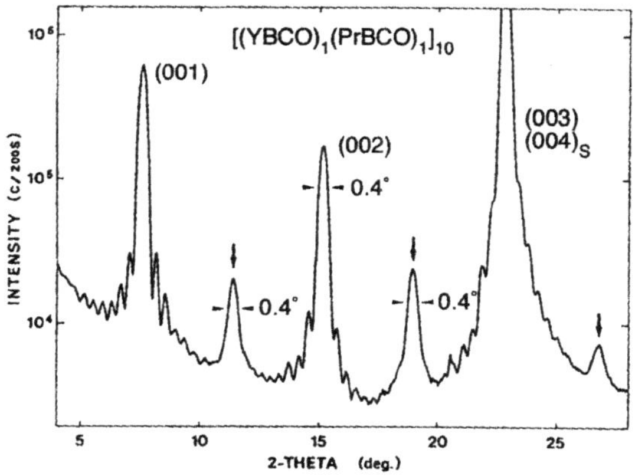

Fig. 44. X-ray scattering spectrum scanned along the [001] direction for the [YBCO (1-UCT)/PrBCO[1-UCT)] $\times$ 10 superlattice

peaks indicated by arrows due to compositional modulation of the superlattice and the oscillatory peaks between the fundamental peaks are seen. The modulation wavelength calculated from the satellite position is 2.34 nm. The widths of the satellite peaks are the same as those of the fundamental peaks. This result indicates that there is no fluctuation of periodicity and no intermixing of YBCO and PrBCO due to the interfacial roughness. The oscillatory peaks are attributed to the finite layer thickness of the film and called Laue function peaks. When the film consists of N unit cells, $(N-2)$ submaximum peaks are seen between the fundamental peaks of (002) and $(002+\ell)$. The number of unit cells in the superlattice (ten) derived from the number of Laue peaks (eight) agrees with that of RHEED oscillations. The sharp Laue peaks indicate that the superlattice has good coherence in crystallinity and compositional modulation. Before discussing the resistive transitions of the superlattices in the magnetic fields, we consider the properties of the single

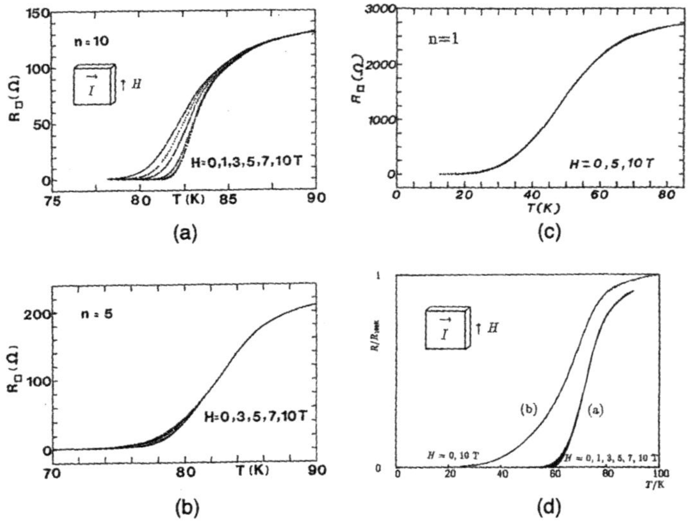

Fig. 45. Sheet resistance versus temperature curves in magnetic fields parallel to the *a–b* plane for (**a**) 10-UCT, (**b**) 5-UCT, and (**c**) 1-UCT YBCO films. (**d**) Resistive transitions in magnetic fields parallel to the *a–b* plane for (I) [YBCO (1-UCT)/PrBCO(1-UCT)] $\times$ 10 superlattice and (II) [YBCO(1-UCT)/ PrBCO(2-UCT)] $\times$ 10 superlattice

layer films consisting of a few unit cells thick YBCO. Figure 45a–c shows the resistive transitions of 1, 5, 10-UCT YBCO layers sandwiched between PrBCO layers in magnetic fields parallel to the *a–b* plane. The current is perpendicular to the fields. The field induced broadening of the resistive transition is clearly observed in the 10-UCT YBCO film. This broadening is supposed to be caused by vortex motion under a Lorentz force. The field induced broadening is markedly suppressed in the 5-UCT YBCO film. In a 1-UCT YBCO film no broadening is observed even at 10 T. When the film thickness is much smaller than the magnetic penetration depth, the field penetrates almost uniformly in a small magnetic field range below the lower critical field H_{c1}. The energy gain due to the vortex creation is thus significantly reduced in a thin film to remarkably increase H_{c1}. *Sonin* calculated H_{c1} for ultrathin layered superconductors using the Lowrence–Doniach model [42]. H_{c1} was determined to be $\sim 0\,\mathrm{T}$ and $\sim 2\,\mathrm{T}$ for 1-UCT ($D = 1.2\,\mathrm{nm}$) and 5-UCT ($D = 6\,\mathrm{nm}$) YBCO films, respectively. These results provide quantitative support for the present observation. The resistive transitions of 1-UCT YBCO film are broad. The origin was suggested to be a Kosterlitz–Thouless (KT) transition of vortex–antivortex pairs excited in the CuO_2 bilayer interposed with a Y layer [43]. Above the KT transition temperature the pairs start to dissociate and give rise to finite dissipation (resistance) in the presence of transport current. Figure 45d shows the resistive transitions for [YBCO(1-

UCT)/PrBCO(1-UCT)] × 10 and [YBCO(l-UCT)/PrBCO(2-UCT)] × 10 superlattices in magnetic fields parallel to the a–b plane. A slight broadening can be seen in the YBCO(1-UCT)/PrBCO(1-UCT) superlattice. The slight broadening may be caused by the dissipation associated with the flux motion, where the vortex cores would be located in the PrBCO layers and the shielding current along the c-axis should be a Josephson current through the PrBCO layer. In contrast to YBCO(1-UCT)/PrBCO(1-UCT) superlattice, YBCO(1-UCT)/PrBCO(2-UCT) superlattice does not show field induced broadening and its zero resistance temperature is markedly reduced. These results mean that the interlayer Josephson coupling between the 1-UCT YBCO layers is almost suppressed by the 2-UCT PrBCO layer.

5.5 Summary

We have observed RHEED oscillations during the epitaxial growth of oxides caused by layer-by-layer (unit-by-unit) growth and have found that one period of the RHEED oscillations corresponds to the height of the minimum unit which satisfies the chemical composition and the electrical neutrality of the oxide. AFM observation has shown multi-terraces with a height of a unit cell of YBCO, indicating 2D nucleation and growth. TEM observation has revealed that the terminating layer of YBCO film is a CuO layer and the growth unit of YBCO has an atomic stacking sequence of CuO-BaO-CuO_2-Y-CuO_2-BaO-(substrate). We have found that capping with an oxide which contains an MO "rocksalt" type layer is needed for the occurrence of superconductivity in 1-UCT YBCO grown on PrBCO buffer layer, where the lattice mismatch between the cap oxide and YBCO is sufficiently small($< 6\%$) and the valence of M is 2+. The cap oxide provides the CuO terminating layer of YBCO with the MO layer to form an MO-CuO-BaO charge reservoir block. BaO and SrO with NaCl type structure and $BaTiO_3$, $SrTiO_3$, $CaTiO_3$, $PbTiO_3$, and $CdTiO_3$ with perovskite type structure have an ability to produce superconductivity in 1-UCT YBCO. We have grown YBCO/PrBCO superlattices while monitoring the RHEED intensity oscillations. It was found that monitoring the RHEED oscillations makes it possible to control the film thicknesses on a unit cell scale. We have found striking enhancement of H_{c1} in ultrathin YBCO films. A resistive transition study of superlattices in magnetic fields has suggested that the 2-UCT PrBCO layer is thick enough to suppress Josephson coupling between the YBCO layers.

References

1. M. Tachiki, M. Noda, K. Yamada, T. Kobayashi: J. Appl. Phys. **83**, 5351 (1998)
2. M. Sugiura, Urago, M. Tachiki, T. Kobayashi: J. Appl. Phys. **90**, 187 (2001)
3. K. Kinoshita, H. Ishibashi, T. Kobayashi: Jpn. J. Appl. Phys. **33**, L417 (1994)
4. M. Tachiki, T. Kobayashi: Jpn. J. Appl. Phys. **38**, 3642 (1999)
5. M. Tachiki, T. Hosomi, T. Kobayashi: Jpn. J. Appl. Phys. **39**, 1817 (2000)

6. T. Kobayashi, M. Tachiki, H. Akiyoshi: Appl. Surf. Sci. **197–198**, 294 (2002)
7. H. Ishibashi, S. Arisaka, K. Kinoshita, T. Kobayashi: Jpn. J. Appl. Phys. **33**, 4971 (1994)
8. T. Kobayashi (unpublished)
9. S. Oda, S. Yamamoto, A. Kawaguchi: J. de Physique IV **C5**, 379 (1995)
10. S. Yamamoto, S. Oda: Chem. Vapor Depos. **7**, 7 (2001)
11. J. Musolf: J. Alloys Compounds **251**, 292 (1997)
12. S. Yamamoto, K. Nagata, S. Sugai, A. Sengoku, Y. Matsukawa, T. Hattori, S. Oda: Jpn. J. Appl. Phys. **38**, 4727 (1999)
13. B. J. Rappoli, W. J. DeSisto: Appl. Phys. Lett. **68**, 2726 (1996)
14. S. Oda, K. Sakai, H. Zama: Appl. Surf. Sci. **75**, 256 (1994)
15. S. Yamamoto, S. Sugai, Y. Matsukawa, A. Sengoku, H. Tobisaka, T. Hattori, S. Oda: Jpn. J. Appl. Phys. **38**, L632 (1999)
16. M. E. Bijlsma, D. H. A. Blank, H. Wormeester, A. van Silfhout, H. Rogalla: J. Alloys Compounds **251**, 15 (1997)
17. J. Kircher, M. K. Kelly, S. Rashkeev, M. Alouani, D. Fuchs, M. Cardona: Phys. Rev. B **44**, 217 (1991)
18. K. Maruyama, K. Shinagawa, T. Saito, T. Tsushima: Jpn. J. Appl. Phys. **34**, L734 (1995)
19. S. Sugai, Y. Matsukawa, K. Shimosato, S. Oda: Jpn. J. Appl. Phys. **39**, L1032 (2000)
20. S. Miyazawa, Y. Tazoh, H. Asano, Y. Nagai, O. Michikami, M. Suzuki: Adv. Mater. **5**, 179 (1994)
21. O. Michikami, T. Hashimoto: IEEE Trans. Appl. Supercond. **11**, 3197 (2001)
22. R. A. Rao, C. B. Eom, M. Santer, M. Anlage: IEEE Trans. Appl. Supercond. **7**, 1278 (1997)
23. X. X. Xi, G. Linker, O. Meyer, E. Nold, B. Obst, F. Ratzel, R. Smithey, B. Strehlau, F. Weschenfelder, J. Geerk: Z. Phys. B **74**, 13 (1989)
24. Th. Schurig, S. Menkel, Z. Quan, J. Beyer, B. Guttler, S. Knappe, H. Koch: Physica C **262**, 89 (1996)
25. O. Michikami, M. Asahi: Jpn. J. Appl. Phys. **30**, 939 (1991)
26. O. Michikami, A. Yokosawa, H. Wakana, Y. Kashiwaba: Jpn. J. Appl. Phys. **36**, 2646 (1997)
27. H. Ozawa, N. Terada, S. Kashiwaya, H. Takashima, M. Koganai, H. Ihara: IEEE Trans. Appl. Supercond. **7**, 2161 (2001)
28. H. Wakana, A. Yokosawa, O. Michikami: Jpn. J. Appl. Phys. **38**, 939 (1999)
29. A. Inam, C. T. Rogers, R. Ramesh, K. Remschnig, L. Farrow, D. Hart, T. Venkatesan, B. Wilkens: Appl. Phys. Lett. **57**, 2484 (1990)
30. G. Y. Sung, T. D. Suh: Appl. Phys. Lett. **67**, 1145 (1995)
31. H. Wakana, O. Michikami: Physica C **350**, 29 (2001)
32. O. Michikami, H. Wakana, K. Atsumi: Jpn. J. Appl. Phys. **38**, 6674 (1999)
33. J. J. Harris, J. Joyce, P. Dobson: Surf. Sci. Lett. **103**, L90 (1981)
34. T. Sakamoto, T. Kawamura, S. Nago, G. Hashiguchi, K. Sakamoto, K. Kuniyoshi: J. Cryst. Growth. **81**, 59 (1987)
35. T. Terashima, Y. Bando, K. Iijima, K. Yamamoto, K. Hirata, K. Hayashi, K. Kamigaki, H. Terauchi: Phys. Rev. Lett. **65**, 2684 (1990)
36. K. Shimura, Y. Daitoh, Y. Yano, T. Terashima, Y. Bando, Y. Matsuda, S. Komiyama: Physica C **228**, 91 (1994)
37. T. Terashima, K. Iijima, K. Yamamoto, Y. Bando, H. Mazaki: Jpn. J. Appl. Phys. **27**, L91 (1988)

38. J. G. Wen, C. Traeholt, H. W. Zandbergen: Physica C **205**, 354 (1993)
39. M. Nakanishi, H. Hashizume, T. Terashima, Y. Bando, O. Michikami, S. Maeyama, M. Oshima: Phys. Rev. B **48**, 10524 (1993)
40. T. Terashima, K. Shimura, Y. Bando, Y. Matsuda, A. Fujiyama, S. Komiyama: Phys. Rev. Lett. **67**, 1362 (1991)
41. K. Iijima, T. Terashima, Y. Bando, K. Kamigaki, H. Terauchi: J. Appl. Phys. **72**, 2840 (1992)
42. H. J. Jensen, P. Minnhagen, E. Sonin, H. Weber: Europhys. Lett. **20**, 463 (1992)
43. Y. Matsuda, S. Komiyama, T. Terashima, K. Shimura, Y. Bando: Phys. Rev. Lett. **69**, 3228 (1992)

Index

Topics in Applied Physics

73 **Hydrogen in Metals III**
Properties and Applications
By H. Wipf (Ed.) 1997. 117 figs. XV, 348 pages

74 **Millimeter and Submillimeter Wave Spectroscopy of Solids**
By G. Grüner (Ed.) 1998. 173 figs. XI, 286 pages

75 **Light Scattering in Solids VII**
Christal-Field and Magnetic Excitations
By M. Cardona and G. Güntherodt (Eds.) 1999. 96 figs. X, 310 pages

76 **Light Scattering in Solids VIII**
C60, Semiconductor Surfaces, Coherent Phonons
By M. Cardona and G. Güntherodt (Eds.) 1999. 86 figs. XII, 228 pages

77 **Photomechanics**
By P. K. Rastogi (Ed.) 2000, 314 Figs. XVI, 472 pages

78 **High-Power Diode Lasers**
By R. Diehl (Ed.) 2000, 260 Figs. XIV, 416 pages

79 **Frequency Measurement and Control**
Advanced Techniques and Future Trends
By A. N. Luiten (Ed.) 2001, 169 Figs. XIV, 394 pages

80 **Carbon Nanotubes**
Synthesis, Structure, Properties, and Applications
By M. S. Dresselhaus, G. Dresselhaus, Ph. Avouris (Eds.) 2001, 235 Figs. XVI, 448 pages

81 **Near-Field Optics and Surface Plasmon Polaritons**
By S. Kawata (Ed.) 2001, 136 Figs. X, 210 pages

82 **Optical Properties of Nanostructured Random Media**
By Vladimir M. Shalaev (Ed.) 2002, 185 Figs. XIV, 450 pages

83 **Spin Dynamics in Confined Magnetic Structures I**
By B. Hillebrands and K. Ounadjela (Eds.) 2002, 166 Figs. XVI, 336 pages

84 **Imaing of Complex Media with Acoustic and Seismic Waves**
By M. Fink, W. A. Kuperman, J.-P. Montagner, A. Tourin (Eds.) 2002, 162 Figs. XII, 336 pages

85 **Solid-Liquid Interfaces**
Macroscopic Phenomena – Microscopic Understanding
By K. Wandelt and S. Thurgate (Eds.) 2003, 228 Figs. XVIII, 444 pages

86 **Infrared Holography for Optical Communications**
Techniques, Materials, and Devices
By P. Boffi, D. Piccinin, M. C. Ubaldi (Eds.) 2003, 90 Figs. XII, 182 pages

87 **Spin Dynamics in Confined Magnetic Structures II**
By B. Hillebrands and K. Ounadjela (Eds.) 2003, 179 Figs. XVI, 321 pages

88 **Optical Nanotechnologies**
The Manipulation of Surface and Local Plasmons
By J. Tominaga and D. P. Tsai (Eds.) 2003, 168 Figs. XII, 212 pages

89 **Solid-State Mid-Infrared Laser Sources**
By I. T. Sorokina and K. L. Vodopyanov (Eds.) 2003, 263 Figs. XVI, 557 pages

90 **Single Quantum Dots**
Fundamentals, Applications, and New Concepts
By P. Michler (Ed.) 2003, 181 Figs. XII, 352 pages

MIX
Papier aus verantwortungsvollen Quellen
Paper from responsible sources
FSC® C105338

If you have any concerns about our products,
you can contact us on
ProductSafety@springernature.com

In case Publisher is established outside the EU,
the EU authorized representative is:
Springer Nature Customer Service Center GmbH
Europaplatz 3, 69115 Heidelberg, Germany

Printed by Libri Plureos GmbH
in Hamburg, Germany